THE CALL OF THE Furies

SUNA FLORES

CITIOFBOOKS, INC.
3736 Eubank NE Suite A1
Albuquerque, NM 87111-3579
www.citiofbooks.com
Hotline: 1 (877) 389-2759
Fax: 1 (505) 930-7244

Ordering Information:
Quantity sales. Special discounts are available on quantity purchases by corporations, associations, and others. For details, contact the publisher at the address above.

Printed in the United States of America.

ISBN-13:   Softcover   979-8-89391-161-9
                        eBook   979-8-89391-162-6

Library of Congress Control Number: 2024912708

TABLE OF CONTENTS

THE CALL OF THE *Auras*

SUNA FLORES

The Call of the Auras

THE CALL OF THE AURAS

THE CALL OF THE AURAS

BEFORE WE BEGIN:

What is an Aura?

Auras are the essence of evil. Who's evil? Yours and mine. We can't see them or hear them, but they are always nearby. You can hear them calling and recognize them as *your* selfishness and greed... *your* jealousy and dishonesty... and then there's always *your* prejudice or just plain power. Auras are *your* evil thoughts that tug on your arm and keep asking you, "why not?"

Although most humans have never thought about Auras, their evil influence on earth has been obvious since the beginning of time and particularly now. Look around you.

Auras are frighteningly real. Ordinary rules are being broken that never would have been broken before. Prejudice and meanness grow, even among people that you think would know better. Why?...........It's all due to The Call of the Auras.

Fortunately, every human is equipped with a way to cope with Auras. That coping mechanism is called "a Conscience". Consciences are, of course, inside all humans and available for good decision making. Good people know this.

It is amazing however, how few people take advantage of their Conscience. They just blunder on, doing stupid things without asking their Conscience, *"Is this the right thing to do?"*

Some unenlightened people do not even know that Consciences exist. For those unfortunates, we should explain what a 'Conscience' is. The dictionary says, "A Conscience is an internal voice which assists a person to determine right from wrong." That definition is true but insufficient.

I must add my belief that Consciences are also tiny invisible beings who live in human hearts and work with humans as sort of built-in guides to good decision making...like an always-present 'best friend'. A Conscience is always inside every person, available to offer advice or just discuss things, but whether you take advantage of yours is up to you.

The first recorded Consciences existed in medieval times. Consciences of that time were 'high verbal'. In other words, there was much healthy discussion between humans and their Consciences. For that reason, this is the time frame for our story.

Suna Flores

Note: Consciences here are identified with the letter 'C' in front of the name of their host human. Queen Willa's Conscience is Cwilla. Drum's Conscience is Cdrum. The Auras are nameless. Auras and Consciences are both invisible to humans, but humans can hear both. (Be careful who you listen to.)

*****

THE CALL OF THE AURAS

AN EMERGENCY CONSCIENCE ALERT:

A rare Conscience Alert Meeting was taking place near River Kingdom Castle. This was not going to be a long meeting for the assembled Consciences, but a 'requested notification of concern'. Local Consciences occasionally needed to be made aware of unusual dangers growing in their area.

"Thank you for gathering on such short notice," a young female Conscience began. "My name is Cwilhelmina. I asked for this Emergency Alert Meeting because I am an inexperienced Conscience, but I have heard some things that sound frightening, and I thought everyone should know. I believe that my human is surrounded by Auras, and I don't know what to do about it.

"There is no real emergency today, but I feel like something bad is going to happen. River Kingdom is not actually being invaded by foreign kingdoms, but..."

"So, you are saying that we have no reason for an emergency Conscience Alert meeting but we're having one anyway, right?" growled Chorace. He was an old curmudgeon Conscience whose human, Horace, worked in The Castle of River Kingdom, itself. "So why *exactly* are we here?" Like his human, he was not noted for patience.

"I am here because what I see and hear around me are talks of violence and greed... prejudice and power...all the time! That makes me worried that my human, and perhaps the whole Kingdom, is being invaded by Auras, but my human doesn't understand the danger of any of it," Cwilhelmina responded.

"I thought that helping our humans was what we Consciences were supposed to do. There is an evil man called Drum who is always at Thorndike Manor. I can feel that he brings extra Auras with him whenever he's there, but my human doesn't understand any of that. I think her home is infested with Auras and they are getting stronger. It feels like an Aura invasion! That's why I'm here."

Chorace shrugged. "No offense, but there are always Auras and plots of evil everywhere. They come and go every day...no big deal. We Consciences just handle them as they come up. That's our job. What's so special about this?

"If you would be still and listen, maybe you could find out," snorted Cagnes, Conscience for the Queen's lady-in-waiting. She had been working with Horace in The Castle for years. "This young Conscience is sensitive and concerned. She is trying to warn us," Cagnes continued. "In addition, I know where she lives. She lives in Thorndike Manor, the home of a noble. That means that Auras of evil are not growing so much among townspeople. They are growing among wealthy noblemen. That could spell trouble for The King."

"Noblemen have everything they need, but they never have everything they want," snickered Chorace. "Their greed Auras must feel right at home at that manor. Greed is their middle name."

"You are so right," contributed Chorace's brother, Chector. "Greed and *power*. As you know, I farm near the property of Wilhelmina's nobleman father, Lord Thorndike. He's rich and has a lot of land but, from where I sit, he has spent his life being dissatisfied. He was *born* dissatisfied. He and his noblemen friends are always afraid that someone who doesn't deserve anything will swoop in and take away what <u>they </u>have...someone like Lowlanders, for instance," he added, rolling his eyes.

 "That indicates Auras of power and prejudice together," commented Cagnes. "That combination of Auras is the very worst kind. Those evils make humans want to control other humans and take away their rights for no reason. Hmm...I may have seen some of those Auras myself recently when a few nobles visited The Castle. They smiled and acted friendly, but I heard them grumble that The King isn't strong enough to s*tand up for* 'their rights' against Lowlanders. I didn't think of it as prejudice against Lowlanders at the time, but now that you mention it..." Agnes sighed thoughtfully.

"Being a Lowlander by birth, you know how I feel about prejudice against Lowlanders, but now I am worrying about Lowlander Consciences, too," injected Carmando. Carmando was Prince Armando's Conscience. (Armando was known as 'Great Leader' among the Lowlanders.) "Their Consciences don't counsel them to stick up for their own rights, enough sometimes. Lowlanders should be here at this meeting. Everyone needs to be aware of Auras. Lowlander Consciences never show up at Conscience Alerts."

"They don't seem to be joiners," commented Cagnes. "Your people are very independent."

"Independence is admirable," growled Carmando, "but someone needs to tell them that they have the right to say no."

"Back to my question," Cwilhelmina reminded them softly, "Should I worry about the rich noblemen meeting all the time at Thorndike Manor? They think they are better than Lowlanders and even Urhonordorians. I think they'd like to take charge of *all* those folks, and take over the whole Kingdom themselves. I think some of them even think Lowlanders should be slaves. They're the ones that worry me the most. "

"I hear you and I think yer right, Chector continued. "I hear, myself, that Auras are increasing with nobles, especially up north and down south. My human is a Central farmer-landholder. Him and me feel it in our bones. Them nobles are getting' ready for somethin' big and it ain't good."

Young Cwilhelmina looked around at the motley crew of Consciences before her. Those who lived in The Castle didn't follow the gossip as much as the common folk outside The Castle walls. She shrugged and sighed, but at least the old Conscience from the country knew. *Somebody* would be watching.

When the Alert concluded, all the Consciences agreed that they needed to keep their eyes open. Something might be brewing. It was time to pay attention.

AN AURA SUMMIT:

Even as the Conscience Alert was taking place in the northern part of River Kingdom, a gaggle of Auras was growing at Drumsworth Lake, down river. Their nameless Aura leadership was present. The presence of Leadership was evidence that a new drive for evil was under way. Evil control and turmoil were the primary goals of Auras and word had it that an excellent opportunity was available.

Although this Aura Leader had no knowledge of Consciences, he was aware that some humans showed an ability to resist evil. He was determined to marshal his forces against what he called 'resistant goodness'... (Consciences).

"Hail, Auras! It is time for all Auras of River Kingdom to gear up for a great opportunity to use our skills to create mayhem. Prejudice, jealousy, power, and revenge! These are the essential weapons that we have already planted in those most deserving influencers in River Kingdom. These elements have grown strong enough now to support our final drive. Our takeover for evil is at hand!"

"You've said that before," a swirling Aura hissed in return. "King Solem is still on the throne. His queen is faithful to him. Prince Solem Junior is content to wait in the wings for his turn to rule.

Urhonordo is a female stronghold that kisses up to the foreigners of the Lowlands. They all get along... You're dreaming."

Leader scowled down on his troops. "You doubt me? You dare to question my knowledge after all the tools for conquest that I have provided for you?

"Has jealousy and lying suddenly gone out of style with humans?

"Have greed and violence disappeared?

"And what about prejudice? Have you forgotten prejudice? Huh? Huh?

"Well, I haven't. Look around you. The number of unwanted Lowlanders amid Birther whites grows. Their impact on the kingdoms increases. Their dark-skinned presence is evident everywhere. Fear of 'others' will always be a source of Aura power! That is the horse we will ride to glory now! Fear of 'others' will never go out of style!"

The wily Aura Leader had sent out an advance team of Auras to encourage prejudice and anger in the nobility. Quotes from their early report were read aloud to increase Aura enthusiasm:

"The Lowlanders think they're as good as everybody along The Great River."

"Boo-oo!" shouted the Auras.

"The nobility believes that Lowlanders are already overrunning the white population. They complain, 'It's time to put them in their place."

"Yea-ah!" cheered the Auras.

"Lowlanders have already married up with some Urhonordorians. Next, they will want to marry Birthers from River Kingdom."

"Boo-oo!"

"The nobles didn't have these problems back in the time of King Drum. There was strong human leadership then. We need Drum!"

"Yea-ah!"

The reaction to the report was conclusive. The advance team had done its job. Areas to exploit that would maneuver people into insurrection were prejudice, envy, control, and just plain power! Most of all, River Kingdom must understand that its coveted white culture and superiority were at grave risk!

"Yea-ah!"

"As with all combat preparation, we Auras must understand the strengths and weaknesses of our position", the Leader continued. "Here are your official recon-reports. Pay attention. Do you understand me?"

The Auras erupted into chants of…"'Fear! Selfishness! Control! Prejudice!" Their targets were set.

We are…***The Auras!***

The Call of the Auras

The Auras settled in for vital recon information. Aura reports and observations are as follows: (See map of area.)

*****

<u>AURA REPORTS-</u>

WHO WE MAY BE INVADING, THEIR LOCATIONS AND NOTABLE CHARACTERISTICS:

<u>REPORT 1-THE ROYALS OF RIVER KINGDOM:</u>

King Solem was a thoughtful and serious monarch as his father and grandfather had been before him, but River Kingdom was more difficult today (and more complicated) than it had been in the past. The landed nobility was very opinionated and constantly challenging his rulings. Much diplomacy was required.

King Solem's kingdom included a few villages, but much of the population lived along the western shore of The Great River (locally called The River). Control of The River's flow was an essential part of the King's duties. River Kingdom was an agricultural Kingdom, and its economy was based on farming which got its water from The River.

In many ways, Solem was looked upon as a good king as he was neither mean nor violent. Auras had shown little interest in him.

As the sole authority and ruler, Solem had never taken his obligation lightly. He believed in the rule of law, and, as King, he believed that his word *was* law...if he was fair. Solem was a King, after all. *He obeyed the laws of the realm and assured those* laws were obeyed by his family, and his kingdom.

And there was always protocol, a right and a wrong way to do things, of course. Protocol was the way things got done and protocol had always been honored during his reign.

King Solem was the father of two young adults, Prince Solem, II (Solem Junior, of course), and Milinda, named for her mother, Queen Linda. Prince Solem had asked to be called 'June', short for 'Junior', but King Solem had been named for *his* father, so 'Junior' it was...protocol.

<u>Prince June- an extra observation</u>- Although Prince Solem Jr. loved his father and mother, he had already proven to be a very challenging young royal. The story goes that he was changed as a child by slipping over the Royal Wall that surrounded The Castle, looking for playmates. (No protocol there, thank goodness.)

June found many friends in town, and, without his father knowing, the 'townies' had become a regular part of his world. These hidden adventures, of course, would have been against King Solem's rules and laws, but good deeds were of no interest to Auras.

Playmates were much more interesting than stuffy royals, so the Prince continued his regular escapes over the wall, but during this bit of rebellion, the young Prince discovered a different world on the other side. There were rich people and poor people, brown people and white people...and, among them all, there were hungry people.

As he continued to play with his new, secret, 'townie' friends, June discovered that their hunger was something that he couldn't ignore, primarily because of his young Conscience.

At an earlier age than most, his Conscience started to nag June, asking him why it was fair that a Prince had so much more food than he needed, and the poor children had so little. His Conscience told him it wasn't fair. Right then, June decided to do something about it and talked to his father.

"If they haven't worked for it, they shouldn't get free food," grumbled King Solem. "That's my rule... also my law." (The Aura thought that maybe *this* attitude had promise for selfishness.)

"There's no law that says it's good to leave hungry people hungry, Father. If people are too hungry, how can they work?" countered June.

Cjune and Csolem, their young and middle-aged Consciences, huddled on the subject. Then Csolem talked to King Solem. "The boy has a point," he advised.

Little by little, the Royal family began handing out their leftovers to the poor. That progressed to farmers turning over some extra harvest. Then, someone suggested gleaners and soon, edibles which usually rotted in the fields went into hungry mouths, instead. Auras didn't notice. Poor people were boring.

King Solem looked around. *There is less waste and there are fewer gaunt faces among the people,* he noted to himself. *The healthy work more and more work gets done.* King Solem was not above change...good change.

King Solem and his Conscience not only accepted these practices but became proud of them and spoke about them to others. "Sharing is good for the economy. That's my new discovery...my new law..." The Consciences were proud. The Auras shuffled off...too much niceness.

The Call of the Auras

REPORT 2---THE ROYALS OF URHONORDO:

Queen Luna was the ruler of Urhonordo, a Kingdom (actually a Queendom) on the east side of The River and directly across The River from River Kingdom. The two kingdoms had a peaceful, friendly relationship.

Queen Luna and her kingdom were very different from King Solem and his. Queen Luna was an active monarch. Although a beautiful woman of means, she spent little time thinking about laws and had no patience with protocol.

Queen Luna was frequently found on horseback, traveling from village to village along the riverbanks of Urhonordo, speaking with her subjects, noting their difficulties, helping them solve problems and visiting the little schools that dotted the commonwealth.

Urhonordo was also an agricultural Kingdom. Working with King Solem, Queen Luna carefully monitored the flow of The River twice a year. This was one of her most important duties.

In the same way that River Kingdom Consciences were proud of their humans for sharing their food, the Consciences of Urhonordo also had a reason to be proud of Queen Luna's insistence on the new idea of 'universal education'.

"Education is for everyone. The more our people know, the more they can take care of themselves and the more they can survive happily in the world," Luna told her advisors when they had questioned her. "Education is like armor," she continued. "Education helps people make better decisions."

"I thought that was what we Consciences were for," commented Cluna.

"Of course," Luna laughed. "You Consciences are our first line of defense. But you must admit that education probably makes smarter Consciences, too," Luna offered.

"Good point," replied Cluna.

Consciences did notice that thoughtfulness and honesty seemed to improve when the population understood more about the world around them...could read ...could think ...could discuss. For this reason, almost everyone in the kingdom attended some sort of school.

There was something else that was unique about Urhonordo. Unlike the fair-haired, white- skinned citizens of River Kingdom, some Urhonordorians now possessed darker skin and hair. This was because, over the years, Lowlanders had migrated for work there and eventually married white friends and fellow Urhonordorian workers.

The result among the Urhonordo population was a delightful mixture of skin shades and hair textures, even new manners and speech patterns. No Urhonordorians seemed to notice or care, but River Kingdom and particularly the Birthers (white nobles), were not in favor of all this mixing of cultures and races.

Queen Luna admired King Solem for both his generosity and his lack of prejudice but found him a bit stuffy and overly affected by tradition. There were no schools in River Kingdom, and, as far as she knew, there were no females among his advisors. Luna found these facts strange and limiting.

Many of the official members of Luna's staff were educated women and very involved in decision-making. They were essential. King Solem was 'missing out' on female assistance, in Luna's opinion.

Urhonordo's rise in female influence was not common outside of Urhonordo and a bit disconcerting to King Solem, but it was not a problem for the men of Urhonordo Kingdom. The strength of female input was never going to be an issue for them. The Queen's husband, Crown Prince Consort Armando, a man born in the Lowlands, demonstrated the behavior of choice. It was called respect. Urhonordorians were not at all interesting to the Auras. Too much education.

Manda- another extra of interest- Queen Luna and Prince Consort Armando had only one child. Her name was Amanda...'Manda', for short. As a toddler, she had been their 'wild child'. The people of the kingdom shook their heads but laughed. They knew why she was the way she was. She was the personification of both of her very active parents.

Manda was a handful. The story goes that she had ridden off by herself on her mother's horse when she was three. She bounced off into a ditch but did the same thing the next day. Her mother had finally strapped the toddler to her back as Lowland mothers frequently did when they worked. In this way, Queen Luna took Manda along with her through the countryside. This was the only way that she knew to keep an eye on the girl.

By the time she was four, however, Manda had become such a confident rider that she had her own pony, first following her mother and soon traveling all over the countryside alone. She was an explorer. She was fascinated by everything and everyone around her. By the time she was twelve, she knew most of the citizens of Urhonordo by name and some of River Kingdom as well. Although she was very active, the Auras had no interest in an under-age female.

SPECIAL SUB-REPORT---FEMALES AND CHILDREN---A NOTED AURA-LESS, TWO-KINGDOM RELATIONSHIP:

What were the ladies of the Kingdoms like as people? Luna's daughter's wild, tomboy behavior had frequently been a topic of conversation between Queen Luna and King Solem's Queen, Queen Linda. Luna and Linda were good friends. They discussed personal matters that concerned them both. One of those matters was their daughters.

Luna's daughter was the tomboy.

Queen Linda's daughter was not.

Manda was an independent explorer of everything.

Mimi was a happy-go-lucky social butterfly.

The conversations between the two mothers frequently ended in a combination of laughter and concern. The girls were so different and yet they got along famously.

Mimi loved to dress up and fix her hair. She loved to laugh and play with friends and she had many of them.

Riding out on wide pastures or tromping the marshlands of Urhonordo, Manda wanted to know about everything. On her horse, she was often mistaken for a boy. She wore britches. She never sat gracefully like a lady. She was frequently off by herself, but she *did* have a few good friends. Unpredictably, Mimi, the social butterfly, was one of them. Their friendship started when they were twelve.

Manda had met 'Princess Milinda' (Mimi) several times over the years when she accompanied her mother on formal, Queenly visits. Mimi had seemed pretty and nice, but Manda had wondered what she did all day, dressed up fancy like she always was. Manda had been forced to dress up when visiting, too in those days. Ugh.

If Manda wasn't being forced into school by her mother, Manda would have wanted to spend more time with her father, Prince Armando. "Father is a person of action like Mother, but mother is usually in Urhonordo. I've seen all there is to see of Urhonordo. I need to see and do more" she had argued.

Finally, at age twelve, Manda made her break. Secretly she followed her father on an official mission to The Castle and a visit with King Solem, camouflaged in the garb of one of his aids. Her father caught her immediately, but she had refused to go home when discovered, of course. Armando sent her home twice, but she did not go.

The Call of the Auras

Her Conscience had chattered incessantly against her disobedience, but Manda had persevered. *I am twelve and almost a woman now,* she thought stubbornly. *I need to explore things I don't know yet*. Auras had no interest in her. She was just a child.

"You are stubborn and willful, and destined to get in trouble because of it," growled Cmanda, maintaining a good grip on Manda's hair as the girl now blended quietly into her father's group of aids again.

Manda's disobedience led to the first opportunity that she had ever had to visit Princess Mimi *without either mother.* Manda had raced through the halls of The Castle looking for Mimi.

Agnes, the Queen's lady in waiting, spotted her and clucked nervously. *Who is this young boy who is brashly dashing from room to room?* thought Agnes. *Should I call The Castle Guard?*

Manda was finally happy. She found Mimi, another female her own age. She was only a year younger, alone in her sitting room, stabbing at her handwork with aggravation. Hearing a small ruckus in the hall, Mimi looked up just in time to find Manda sneaking in, looking furtively from side to side.

"Manda! Is that you? I thought you were a boy. Why are you dressed like that?"

Manda peeled off her hat and shook loose her dark curls, continuing to absorb what she saw around her. "I had to dress like this or my father wouldn't have let me come. He has such old-fashioned ideas about me. He always wants to protect me."

"I know what you mean. My father is the same way" growled Mimi. "My mother has me doing needle point. Can you imagine anything more boring?" The girls gave each other a hug. Then they laughed and jumped on the bed.

That day had been the beginning of a friendship that had no boundaries. These young girls, although completely different, had learned, discovered and shared ideas endlessly together. Auras didn't see anything interesting in friendships.

Together the two shared thoughts and schemes. They laughed and giggled at their opposite ways, promising to be friends forever. The girls felt that they had a secret connection, and it would have continued to be secret forever if June had not discovered them.

June had his own secret. He continued to climb over the wall and meet with *his* friends from town. They had a different way of looking at things. From them, he could discover what the people in the town and the countryside were thinking.

June had lots of friends that his father didn't know, and he loved this. He also loved his independence. His Conscience thought this was valuable, even though his father didn't know about it.

Then came the day when his town friends told him of seeing another young man climb over the wall of The Castle, sneaking *into* The Castle grounds! They had noticed that the boy was daring and obviously not afraid of being caught. Someone inside The Castle was allowing this and keeping it a secret. Who could it be? Whoever it was, was obviously up to no good.

June was worried. He was independent but he was also his father's son. Someone was sneaking into The Castle without permission? Why? What was he doing there? This was not proper protocol and surely against the rules...even the law.

June felt he must get to the bottom of this dangerous breach, if not invasion. He was a young man of spirit. *He* would investigate and make his father proud.

June hid by The Castle wall daily, waiting to catch the intruder in the act. He finally succeeded. The culprit was quiet and cautious. Slowly, June saw the sneak's head pop up over the edge of the wall to look around. There were bushes up against the wall so that the invader thought he could enter undetected.

Smart, thought June. *That's what I would do. But I have found him. He doesn't look too big.* At thirteen, June believed he could capture this intruder easily.

The culprit made his stealthy move and dropped down behind the bushes. June was on him like a cat. Grabbing him with all his youthful energy, June slung his opponent to the ground and sat on him.

"Now I have you, you thief! Did you think you could come in and out of the Royal Grounds of River Kingdom without permission any time you wished? Well, think again. I have you now." June glared down at the startled face in front of him.

"Get off me, June, or I'll beat you up," snarled Manda.

June couldn't believe it. This girl was his size and strong. She punched him, flipped him off her body with a twist, punched him and stood over him, ready to fight some more.

June was humiliated and disgusted. Girls shouldn't do things like this. It wasn't proper. He was going to tell...but he couldn't. *She* was a *girl* and she had bested *him.*

Manda pulled June to his feet and explained. "Mimi has *her* friend...me, and you have yours in town. I've seen you. It's the same thing. "

June kept this encounter secret, of course. From his 'adjusted' point of view, his pretty sister was 'secret pals with an ugly (strong) girl from Urhonordo'. She probably had no friends on her side of The River. Mimi was just being kind. He could be kind, too. He decided to forget the whole thing. It wasn't out of friendship. It was more like respect.

June's and Manda's trips over the wall had stopped as they got older, but the girls' friendship had continued. June paid no attention to his sister's activities. Were Mimi and that Urhonordo girl still friends? Who knew? Youthful situations had no effect on anything, Auras concluded...or did they?

REPORT 3---THE LOWLANDS:

The Lowlands were a territory at the Southern end of The River just below River Kingdom and Urhonordo. Because The River's flow was small by the time it arrived in The Lowlands, the territory was very dry and growing anything was difficult.

The Lowlands had no royal leadership but preferred a nomadic life to that of the organized kingdoms to their north. Lowlanders were good people and hard workers, however, and many of the farmers and nobility of the kingdoms were dependent on their labor. This was a beneficial arrangement for all concerned but it was also a source of extreme interest to the Auras.

The Lowlander population was dark skinned. Not every white skinned person in the kingdoms above them was happy about that. Many of the nobles of River Kingdom were very *un*happy about that. They didn't trust people who didn't look or act like them, and they made that very clear. This dislike and 'fear of others' was more than just interesting to Auras. It was an open invitation to the prejudice and control that was their goal.

REPORT 4- THE DRUMSWORTH LEGACY

There was another area on The River that was very important. It was lovely to look at, but it was one of the areas that was most interesting to the Auras. This area was called 'Drumsworth'. It was located at the southern base of The River and on River Kingdom land, but it touched Urhonordo on the east. It was in a very strategic position.

Doran Drumsworth lived here in Drumsworth Castle on Drumsworth Lake. But the truth was that Drumsworth Lake wasn't a real lake. A century before, a deep hole had been hand-carved into the southern base of The River to form a pool. Its sole purpose at that time had been to please the wife of a wealthy River Kingdom nobleman.

Its function was to catch what it needed of The River's water as it flowed toward the Lowlands to form a private pool. This practice tended to flood Urhonordo in the spring and leave less water for The Lowlands. This presented a major problem for both...an Aura delight.

Urhonordo farms were frequently flooded. The Lowlanders always needed more water to stay alive. But Auras had been strong one hundred years ago when that pool was created. " The Lowlands are filled with shiftless people of color" reasoned the nobles of River Kingdom back then.

"Who cares what happens to them." No one worried about blocking the sparce and very necessary Lowland water when the pool was first built. "Tell them to go home and eat dirt," Lord Drum had declared. "The Lowlands and Urhonordo are not part of River Kingdom so who cares?" Lord Doran Drum (who had been 'acting King' of River Kingdom at the time) was a Birther, so Lowlanders and Urhonordorians did not present meaningful problems to him.

That was then but times changed. Rules involving the rights of Lowlanders and Urhonordo and River Kingdom had fixed many of these problems. But that fix brought up another problem...the interference of 'others' in decisions that should only belong to River Kingdom, according to Birthers. Birthers were white skinned persons whose ancestors were born in River Kingdom, of course, and considered (by Birthers) to be superior to all others.

This feeling of superiority was, of course, the mindset that Auras loved and cultivated. This feeling of superiority was key to bad decisions and misery...fuel for all kinds of evil.

Drumsworth Lake was now owned by Lord Doran Drumsworth, a descendant of the Drum ancestor who had been 'acting King' only one hundred years ago. He was also the owner of Drumsworth Castle. As far as he was concerned, Drumsworth Castle, Drumsworth Lake and the water near it all belonged to him.

For one hundred years, the lake was the private playground of River Kingdom's noble families. In Lord Drumsworth's mind, it was a 'local treasure' and needed to be maintained and protected from 'those that were not worthy of it'. "No Lowlanders Allowed" had always been on a sign.

Drumsworth Castle was not really a real castle, either. It only looked like one. Like the lake, it was built originally as a country home for another noble's wife. That noble mysteriously died and Doran Drum eventually parlayed the influence and connections he had garnered as acting king into his sole ownership.

Over the years, the sprawling manor was improved and enlarged and embellished until it became an imposing structure and a haven for vacationing Birther noblemen. It now belonged solely to Lord Doran Drum's descendants.

One hundred years ago, Lord Doran Drum, the grandfather of the current Drum, had been an interesting gentleman. As a wealthy landowner owner, this Drum had maneuvered himself into the position of 'Royal Caretaker of River Kingdom' for more than a year while the real King was called away to a Dragon Quest. (See "Castle, Crown and Conscience".)

No one really cared to evaluate or remember how the position of Royal Caretaker had suspiciously morphed this Drum ancestor into a position that very much resembled that of a king, but it did. In his own mind and that of many of his followers, Drum became *'King'.*

The Call of the Auras

'King Drum's 'reign' was fabled to have been an exciting year for the noble citizens and Auras of River Kingdom. There had been lavish rallies and investments with which they had been made richer at the expense of the absent King's financial holdings.

Doran Drum had been the popular center of all activities during which he and his fellow noblemen had prospered by morphing the existing rules of The Kingdom to benefit themselves and their own enrichment. During this time, the Auras had flourished. The nobility's memories of these times still inspired them.

'King' Doran Drum's manner of governance had been flamboyant and completely Drum-centered. There had been avoidable flooding in Urhonordo that wasn't avoided, and Lowlanders had been jailed for simply being Lowlanders. There had even been a near takeover of Urhonordo by 'King' Drum's version of the Royal Castle Guard, The Thorndike Militia.

Why was this legacy important? Because some Birthers still thought that Drum's permanent takeover *should* have occurred then. "Make River Kingdom Great" had been the ambitious motto for the time. 'King Drum' had been greatly admired and appreciated by his fellow Birthers then and his story was still admired and longed for to this day.

Back in the day, when the actual King, King Sole, King Solem's grandfather, had surprisingly survived and returned from his Dragon Quest, River Kingdom's finances were depleted and the relationship with Urhonordo needed repair badly. Today no one talked about that.

'King' Doran Drum had reluctantly retired to what became Drumsworth Castle, richer and completely unrepentant for his lawless extravagance. To Drum, his royal experience was a success, at least financially. It could have been even more successful, he thought, if he could have remained in his 'king-like' position. But he couldn't. King Sole was back and so was his armed Royal Castle Guard. One can now understand the motivation of the current Auras

AURA REPORTS CONCLUSION: *Aura intervention in human life along The Great River advised. Mayhem definitely possible. Permission to proceed granted.*

LIFE ALONG THE GREAT RIVER

THE WORLD OF DRUM

Today's Drum was accustomed to the fine life of a nobleman and felt entitled to that life and the power attached to it. He had never worked but had inherited huge sums, *accumulating* even more through shifty financial dealings and land grabs. He now occupied the *expanded* Drumsworth Castle and Lake and an extensive swath of the questionable border land between the Lowlands and the rest of River Kingdom. His methods had made Drumsworth into 'Aura Territory'.

The only negative feature of Drum's holdings in his mind was the fact that the land that bordered his on the south was The Lowlands. That place was too full of 'Lowlander lowlifes' in his opinion.

Promoting himself and increasing all that he had accumulated financially was Drum's most outstanding skill set. He knew how to manipulate situations to his benefit. He knew how to market himself and his ideas, legal or not. He also knew how to intimidate anyone that might turn against him. He was much more skillful at manipulation than he was at honesty, and he had become an Aura favorite for that reason.

Lord Drumsworth now called himself 'Drum' like his grandfather whom he admired. His father had lengthened the family name to Drumsworth because it sounded more dignified, but the new name didn't stick to anything consistently other than Drumsworth Lake and Drumsworth Castle.

Today's Drum had also accumulated an odd human following. This group was a collection of influential Birther noblemen who admired Drum's wealth and his grandfather's regal fame and hoped some of the power of both would rub off on them.

These individuals met with Drum on a regular basis to discuss how that might happen. They were not fans of King Solem, whom they considered to be bogged down by protocol and attention to others, some of whom were not even born in River Kingdom. The King is 'a Royal without River Kingdom Vision', they said.

Drum agreed. "The Royals have no interest in our ideas because, as Royals, *they* have the power to do what they want without asking permission and *we* do not," Drum would explain confidently to his only daughter, Balynn. "Their type have control. They were *born* having it just because they are *Royal*. They think that their Royalty is key to power and the way things must be. (Drum's jealousy made Auras swoon happily.)

"Sadly, their belief is true," Drum announced to his fellow nobility. "Being Royalty *is* the key to power and control, but it does not make them better than us. That belief is *not* true. They don't understand how to manipulate people the way I do," Drum continued confidently.

And the Auras agreed. Unnoticed, Auras of pride and envy swirled lazily around his feet, feeling right at home but urging Drum to action. He was the leader of their quest for mayhem.

"Indeed", nodded his daughter with an innocent smile. Balynn was the only one of Drum's many offspring that he acknowledged and kept nearby. She was pretty and ready to adopt her father's positions on anything, no matter how they changed ...or lacked reason.

Balynn had always lived in Drumsworth and had never seen much of the outside world, but she was adaptable and, as a pretty girl, she had marketability with other noblemen...maybe even Royalty. She was a Drum political asset.

"The Royals of this world think that *their* way is the only way, don't they, Daddy?" offered Balynn. "They have rules and laws in place just to keep us nobility down, right?" Balynn motioned a Lowlander servant forward with a fresh pitcher of peach ale for her father.

"All we need to do is find a way around The King's silly laws. And he doesn't even know about The Plan, does he, Daddy?" she chirped, smiling her father's favorite smile.

Drum jumped to his feet in fury, knocking the servant aside and sloshing Balynn with ale. "I told you not to mention 'The Plan', ignorant girl...particularly not in the presence of others!

"Whatever you think you understand, you must *not* talk about 'The Plan' in front of anyone. You are only a pretty girl. Talking about important things is not your function.

"And for heaven's sake don't mention such a thing *in public. The Plan* is 'evolving *on its own with the noblemen'.* This Plan is '*of the people'.* Therefore, it is *never my idea. It is theirs.* Can you, as a mere *female,* understand that?"

"Oh, goodness, Father dear... My mistake...slip of the tongue...it won't happen again." Balynn sidestepped to a corner of the room and safely out of reach.

The Lowlander housemaids stood back fearfully with her. They had been severely punished for such small errors as spilling ale before. *Should we run now? Are we in danger again?* they thought.

Lord Drum sneered at the servants around him. He had been doing that a lot more, lately. Their heritage was the obvious cause. Their skin was so dark. Their hair was even darker and curled in an unruly fashion.

Disgusting. Thank goodness they are at least functional as slaves, he mused thoughtfully, *and soon their kind will be under Birther control. My Plan will see to that.*

"The Plan" that Balynn had mentioned was not maturing as quickly as Drum had anticipated. King Solem should already be out of the way by now. His current reign was a disaster, making him ripe for replacement and Auras' condemnation.

Drum growled at his trembling Lowlander servants and glanced at a message he had just received from Lord Thorndike up north. "The King now advocates *food sharing, for goodness' sake,*" he snarled. "Lowlanders like you twits are all over the place in the villages up north with no direction or control and now Royalty is giving away good food to some of them. If someone smart doesn't take control soon, River Kingdom will be empty of hard-working white Birthers. It will simply be taken over by Solem Junior, lazy Lowlanders and the sniveling poor."

"Yes, indeed, Daddy," agreed Balynn, hoping to get back in her father's good graces. "But that won't happen when Solem Junior becomes my 'Dream-Prince', right Daddy?"

Yes, but this needs to be understood about his Conscience. This Drum was born into a wealthy Birther family with little use for Consciences. Purposefully not acknowledging Consciences was a Birther trait. They didn't care if a decision was good or bad...only if it served their purposes. The Drums were born narcissists.

Drum had a theory about 'the poor and unfortunate' that he had inherited from his father and grandfather: "Poor people are poor and unfortunate because they *deserve* to be poor and unfortunate. The wealthy have wealth and power because they know how to get it, know how to keep it and, therefore, deserve it.

"More than all others, *I* deserve wealth and power because my family taught me how to *take it* and *control what would keep it.* That is why Grandfather Drum was King!" The Auras loved this saying.

Cdrum was unhappy about his lack of use as a Conscience, but he had known this would be true since Drum was born. Humans can not choose their Consciences. They are born with them. Cdrum still lived in hope in Drum's heart but had very little access to Drum himself. The Auras around Drum were too strong.

Note: Consciences are always present but can only do their job when acknowledged by their humans and not blocked by Auras.

A VERY BAD DAY

Princess Mimi was a few years younger than her brother June and was also blessed with some of her brother's gentle kindness. She was sweet but not creative and thoughtful, like June. She was much like her mother, more of a listener and interested in the social side of life.

 The practice of food sharing was the first new idea that King Solem had had in years. The nobles of the Kingdom thought it was foolhardy and wrongheaded. But Solem told his wife about the wonderful changes sharing had made in the lives of the poor in the kingdom. She listened and agreed with him, of course, and shared the news with her friends, the Thorndikes, at the manor down the road.

The word began spreading among the ladies of the kingdom and Queen Linda reported back. "Your popularity is actually rising among the ladies," she told The King happily.

King Solem laughed. It would probably take more than ladies chatter to get approval from their noble husbands, but he was pleased. And he was right. The men folk were not as thrilled about sharing food as the ladies were. In fact, they took this new action as a stunt that Solem had done just to gain popularity and waste their tax money.

Now the Auras paid attention. The stupid King was coddling the worthless poor. Even some of the Lowlanders were getting some 'good, River Kingdom produce' that they hadn't earned. The Auras spread the word. "Somebody ought to do something about this!" The nobles agreed.

But Queen Linda had laughed and continued to enjoy the good that was coming from the new practice... right up to the day she died.

Wait... Queen Linda died? *Yes...Queen Linda died.*

On a lovely, sun-drenched day, Queen Linda requested a carriage to go on a visit with Queen Luna of Urhonordo. She would tell her of the Kingdom's new practice of food sharing that she was so proud of relating. It was a wonderful day for such an outing, and the Queens hadn't chatted for a long time. It had all been arranged.

King Solem was proud, as well. He even purchased new horses, especially elegant ones that were trained and recommended by Lord Thorndike himself.

Lord Thorndike was an Aura favorite and a Birther friend of Lord Drum. He trained horses. His special interests were war horses for his militia, but he was well known for training horses of all kinds. He was also well known as having the largest private militia in the kingdom.

Although Lord Thorndike and King Solem were not friends, everyone knew that Thorndike horses were the best in the Kingdom. King Solem wanted the best for his wife. Kings always needed 'the best'. It was protocol. It was almost a law.

The Call of the Auras

As the Queen's carriage drew away, King Solem smiled proudly at the sight and waved. The horses appeared elegant and frisky, but The King's experienced driver had them well under control.

This was the last time he saw his wife alive.

Tragically, the horses had slowly returned hours later, calm and in control, but with loose reigns and broken pieces of the royal carriage dragging in the dust behind them.

The Castle Guard and King Solem, in person, had ridden out immediately, only to find Queen Linda and her driver lying at the edge of The River, mangled, and floating at its shore as though thrown there by horses gone wild. Both were dead, leaving no one to tell what had happened.

The Auras slunk off with contentment into the underbrush. This could be helpful.

The cause of the tragedy was never found. The King did not expect foul play. The Queen was loved by The Kingdom. The King had no major enemies that he knew of. And the returning horses had seemed docile and tame, not at all wild or uncontrollable. What had gone wrong?

There was no logical cause for this horrible incident. Lord Thorndike had been quick to assure the King that the horses were well trained and had not caused the accident even though it had occurred on the road near the gate of Thorndike's Estate. "It was the terrible condition of the road," Thorndike had declared over and over. The Auras heard his complaint and decided to remember it for future use.

The road along The River *was* rough, Salem had noted. It needed to be repaired. The carriage had probably hit a big rough spot or a drainage ditch and toppled. In his misery, Solem did not discuss it further.

Although the gossipers of the Kingdom seemed to relish the idea that an evil conspiracy somewhere had caused the Queen's death, no suspicious plot appeared during the next month of mourning and gossip subsided. But the Auras smiled and knew what to do.

WHAT TO DO WHEN A QUEEN DIES

King Solem suffered from the loss. He missed his gentle wife. He missed her listening and her support. And he felt guilty. Should he have let her travel with new, frisky horses? Should he have known that that part of the road needed fixing? He became increasingly withdrawn and no longer appeared to have interest in the affairs of the state. His Conscience could not console him.

A month went by...then another and another. King Solem began to lose interest in his work noticeably, even though he had seemed focused and engaged before. The nobles watched with heartless glee. So did the Auras. They rushed in immediately and encouraged the Kingdom wags to gossip with a negative, evil focus. A miserable king was ripe for replacement.

"He's lost it."

"He's 'over the hill'."

"If our kingdom is ever attacked, he will not be fit to command."

"We need a new king."

"The Prince is next in line."

"June is young and virile. He should just do away with his useless father and become King. We can take it from there." (This was the Auras' first choice.)

June was a young man now and he could see that The Kingdom's dreary gossip was not good for his father's strong Royal leadership.

"It's not good for *you*, either" June's Conscience reminded him. "A strong King is essential to every kingdom, but I hope you know that *you* are not ready yet to do the job. On the other hand, you *should* do *something* to help your father."

Although he was young, June *did* believe that he needed to do *something* to improve this sad situation, and one the Auras was more than ready to offer him a solution.

"Taking over as king could be great," the Aura began smoothly. "*You* could be in charge. *You* could make the rules. You're a big hero with the poor in town because of food sharing..."

Taking over as King is not the answer, June thought stubbornly...even though the thought *had actually* occurred to him more than once. *Maybe...*

The Auras' support of his potential for power continued to waft invitingly around June's ears and whisper...

"I'll bet you always wanted to be King, didn't you June?"

"Lots of princes do that. No one will blame you."

"The King is too boring to have real enemies."

"The well-fed towns' people will love you."

"You wouldn't have any rivals."

"You would be a good King, too, June."

The Call of the Auras

June's Conscience finally noticed the Auras and took action. "What are *you* thinking, June?" Cjune asked quickly. "You aren't *really* thinking of taking your father's place while he is in this weakened condition, are you?"

"Of course not," June snorted. "What kind of a son do you think I am?"

"Ease up, June. I was just checking."

"Well," June grumbled, "As a Conscience, you would be more useful if you came up with a good idea instead of asking stupid questions." (Unlike today, speaking aloud with one's Conscience was common in these times.)

"Right. You're absolutely right," answered Cjune. "And I am here for you, June. I actually *do* have a pretty good idea."

"Fine. That's better. What is it? I have no idea what to do. My father depended on my mother to listen and support him. Since he lost my mother, he's sad and he doesn't even enjoy our practice of food sharing any more. What can possibly make my father feel alright again?"

"My idea won't make him feel 'alright'," Cjune countered slowly, "but maybe 'alright...*er*'."

June's Conscience offered his idea carefully. "Your father is a man, right? Now he is a *lonely* man. What might make life more tolerable for someone who misses a good woman could be...*another* good woman."

June gulped. What a shocking thought... *Replace my mother?*

The Auras had not left June, so they paid attention. He seemed to be talking to himself. They circled. They needed a new position to go for a different kind of win. This king needed a new queen, and these local Auras thought they knew exactly where to find one. The Auras hurried off to work on their idea.

Cjune counseled June. "This wouldn't exactly be replacement, June. Your mother would want your father to be happy and do a good job as king. He just needs a female companion like your mother. Surely you can understand that."

June frowned and kicked the dirt. He now believed that his Conscience was right. He nodded slowly. His father needed a new wife. The Auras snickered. This was also *their* idea.

Slowly the odd suggestion of finding another queen became less odd. First, June asked his sister. Mimi, as her brother called her, was still in mourning and not ready to 'replace mother.'

Mimi's Conscience was ready, however. "Your father is lonely, Mimi. A new partner would not be *replacing* your mother," she suggested gently. "She would only be helping your father to be less lonely...helping him be a good King again. That's important, right?"

The royal siblings talked to each other as did their Consciences and finally all agreed that a new partner might be a good idea for their lonely father. But finding the right person would not be a simple task. A good, 'new Queen' for a King who wasn't looking for one was not an easy thing to accomplish.

*****

Wilhelmina Thorndike arrived at The Castle unexpectedly a few days later. King Solem and his children were surprised. Mimi was sure that her father, Lord Thorndike, did not particularly like The King or even know his queen so why was she here?

This young woman was only a few years older than Mimi. She seemed shy, out of place and unsure. *Is it out of guilt?* King Solem wondered. *Thorndike-trained horses were involved in my Queen's terrible accident. Maybe Thorndike sent his daughter with condolences so that he would not be blamed.*

"Did your father send you?" King Solem asked, although not unkindly.

The girl hung her head and answered sadly. "Yes, Your Majesty. That is my father's way. I think he is afraid that you fault his horses for the terrible accident. He has too much pride. He insists over and over again that the accident was not his fault but probably the fault of a rough road."

Without anyone noticing, the Auras who had heard June's thoughts about a new partner for his father, quietly nudged their newest protégé, Wilhelmina. "Say what we told you to say," the Auras hissed. (Were Auras being kind? No. They had set this up.)

The young woman raised her eyes and stated an unexpected truth. "Queen Linda was kind, and everyone knew that. And she was everyone's Queen, including mine. I am really sorry that she died so tragically. I lost my mother too when I was very young, and I know about loss. I was so sad to hear that The Kingdom's Royal Family had lost their mother that I was glad to come here to say that to you when my father suggested I come."

Solem and his children were charmed.

Dang, thought the Auras. *She sounds like she really means it. We didn't[expect that kind of sweetness coming from a Thorndike. Old Thorndike may have wangled a family member into The Castle like we Auras planned, but she may not be as useful as we wanted. She's not at all like her father.*

Surprisingly, Wilhelmina stayed a while longer than anticipated that afternoon and engaged in an interesting dialogue with both siblings and The King. Slowly she began to voice opinions on things. And she really had opinions. She was articulate. She had read many books and wasn't afraid to share ideas. June and Mimi were fascinated.

The young royals had heard about Wilhelmina in their earlier years, but they had never been close. They had heard that *her* mother died when she was young, and she had become shy and bookish. 'The Thorndike girl' was seldom seen.

"Everyone knows that her mother died," Mimi murmured. "She's probably a little bit boring and kind of an 'old sole' so she's more like our father than most young people. But she does seem kind and nice. Maybe she will like father. I wonder how she feels about sharing food with the poor. I hear that her father and the other nobles hate the idea."

"Let's talk to her and find out," June suggested.

And they did. Soon they felt that they had possibly discovered a solution to their search even though she was a young Thorndike. She was intelligent. She could talk. She was not afraid to do a kind thing. As the child of a wealthy noble, she was unique.

 With a bit of research, they also discovered that Wilhelmina had had a more difficult life than appeared to be so. Wags had reported that she had never been valued by her father, Lord Thorndike, a pompous sort who raised horses and liked to be known as 'Commander of The Militia'.

Although his militia was merely an independent troop that he had pulled together to help 'fight for the rights of The Kingdom' when needed, it was particularly large and appreciated by many nobles as a sign of Birther independence. That reputation meant *everything* to Lord Thorndike...and the Auras.

Wilhelmina was not close to her father, gossips had told them. The primary problem was the fact that she was not a boy. As a girl, Wilhelmina Thorndike could never become part of her father's vaulted militia. Her mother had not been able to produce the desired male heir for that purpose and so she had been cruelly shunned by Thorndike and had finally died. It was a sad story.

That had happened when Wilhelmina was only a toddler. It was said that the little girl had been left to loneliness and the care and companionship of servants while her father ignored her completely.

Yes, this was a quiet young woman, but she was not boring. She was aware. The royal children of The King approved of all they saw and heard.

After much discussion with his children, King Solem decided to consider this young woman as a possible partner. It was true that he was lonely. It was also true that he missed his wife. He said he would need to think about it...and he did.

Solem's Conscience was hesitant and very direct. "What would a pretty young woman like her want with an older man like you? Will she continue to care for you as you age? Will her father feel he can influence you?"

King Solem talked to his Conscience for several days. "I guess there's really only one way to find out", they both concluded.

At last, a quiet courtship began. King Solem paid Lady Wilhelmina a visit. To the gratitude and joy of the young Royals, the shocked young woman listened to the idea with an open mind...and a lonely, open heart.

"He is too old for you, Wilhelmina," her Conscience warned. "I always thought that June might be a good match for you, not his father. Whose nit-wit idea was this?"

""It was June's and Milinda's idea in the first place," countered Csolem.

Slowly Wilhelmina's Conscience forwarded this amazing information to her human. "I hear that the King's children suggested your name, themselves", she whispered to Wilhelmina, "but they are calling you 'Willa' for short. I find that fitting. Wilhelmina was your mother's name."

'Willa' smiled in a shy but thoughtful manner. "Your children must love you and be very caring to approve of your attention to someone after their lovely mother, my King," she had said quietly to King Solem. He took her hand and smiled in return.

The hovering Auras grinned snidely. They approved of this. The usually negative Lord Thorndike readily approved as well...almost as though he had planned it. There would be a Thorndike in The Castle. This meant influence.

 Within a year, a quiet royal wedding occurred, and Willa was crowned Queen of River Kingdom.

River Kingdom citizens grumbled when they heard it at first. They had not seen this coming. For a while, the gossips were buzzing about the age difference and the Auras stirred up rumors among the nobles of the possible negative influences of such an unlikely pairing.

Drum looked upon the union particularly carefully. "Lord Thorndike is a shrewd businessman as well as Commander of the Militia", he mused. "He is not a friend of the King. Did he allow this to happen? Or... did he *connive to make this happen?* Is there a secret plot? Did he think he could get richer...or get access...or exert influence with his daughter as Queen? The Auras just smiled and said "Of course."

Gradually, after a few weeks, however, the *new* idea of the *new,* young Queen lost its *newness.* The Auras watched for an opportunity to push Thorndike's possible advantage or increase the rift that had always existed between Thorndike and The King, but they finally lost interest. If Thorndike wasn't ready to make a move yet, they would just wait.

Communication between Urhonordo and River Kingdom stopped when Queen Linda died. Their habits had changed. Their comfort level with The Castle had changed. The King had been so sad. Then, before they could decide how to talk to a King with no Queen, a new Queen had arrived. They heard that the new Queen was the young offspring of the Thorndike family, and no one really knew her.

Manda and Queen Luna hadn't talked about it much. Both of their Consciences nagged that they should make contact. "Mimi has been your friend for years," Cmanda chided Manda. "You are twenty now. You are a self-directed woman. Get on a horse and go tell Mimi in person how sorry you feel about her mother."

"But she has a new mother now, and that Thorndike girl is the same age as me. That is so weird."

"How do you know? It's already been a few weeks since the wedding and you haven't even met her. How do you know this is weird?"

"I will see her at the Water Rights Gathering," Manda countered stubbornly. "Everybody will get to meet her then. Mimi will still be my friend."

THE GATHERING

It was that time again...time for the semi-annual Water Rights Gathering of representatives from the entire region. There were at least two Water Rights Gatherings each year, one in the spring and one now, in the fall. These meetings were essential to the agricultural health of both kingdoms and the Lowlands.

The object was for everyone with land along The River to present their planting, and levee needs for the upcoming season to the leadership of both kingdoms. Queen Luna, Prince Armando and King Solem would read over the submissions, combining ideas and issuing directives, usually within a week or two. It was quite an efficient system.

The Auras also gathered with enthusiasm. This was an opportune time to sow discontent, jealousy, and selfishness. And everyone would be here, even Urhonordorians and Lowlanders. Aura leadership was primed and ready.

Consciences were always in attendance at these gatherings, of course, because their humans were. Everyone was talking about their needs and their livelihoods. There were inevitable tensions and disagreements. Consciences would remind folks of honesty and patience, throughout the event, trying to keep things civil. It was an unrequested but necessary duty.

Auras knew nothing of the Consciences, and this Gathering showed promise. They had done some additional recon since their meeting at Drumsworth Lake. There were new problems and conflicts just waiting to stir up anger and resentment… terrific for the general goal of mayhem. Conflict was definitely possible. And there would be a new Queen on the River Kingdom throne…a Thorndike! Despite her apparent sweetness, this could present a new opportunity. They were looking forward to this.

Equally important to agricultural planning, the Water Rights Gatherings were social affairs. They always drew a great crowd. The ladies dressed in their finest and the men fastened their medals to their chests with pride. It was a time to see and be seen. This was the main social event of the fall.

Upstairs in her bedroom, new 'Queen Willa' looked at herself in the mirror and sighed. "I don't look like a queen. I look like somebody's awkward daughter whose playing 'dress-up'." She cinched in her bodice while Agnes, now *her* lady in waiting, looked on disapprovingly.

"If you would stop trying to do everything for yourself and let me help you, I could make you look queenly in a flash," Agnes grumbled.

"She's right, you know," whispered Cwilla. "You are a queen now. The staff of this Castle expects you to be cared for like royalty."

"My Conscience says you are right, Agnes," Willa laughed. "I'm just not used to it. I've always dressed by myself and being dressed up in corsets where you can hardly breathe seems like a stupid thing to do. How can royal women think when they can't breathe?"

"I'm probably speaking out of turn, but I don't think *you* are attending this meeting as 'a thinker'," Agnes offered with a sniff. "The King will do all the thinking *and* the talking for the Kingdom. *You* will be a lovely and cordial decoration. That is your role today."

Willa closed her eyes as a handsome burgundy gown billowed effortlessly over her head and was fastened snugly to her waist. "Hogwash. Solem didn't take me as a wife because I am mindless or speechless. We have an understanding, he and I. I can say what I wish."

Agnes smiled knowingly. She had attended King Solem's first queen, and she liked this new one just as much. She was much more verbal, however. Agnes considered herself an old hand at King Solem's ways and she knew what she knew. King Solem was a fan of tradition and protocol.

"You had better say something more persuasive and say it quick," cautioned Agnes's Conscience. "This is Willa's first real meeting of consequence. The King is not used to women 'thinking out loud', especially in public. This girl will need your assistance all day today."

"I know the King respects your opinion," Agnes began, fastening Willa with a stern look. "But with affairs of state, *he* is accustomed to taking the lead. You might want to consider holding your opinions back until you and the King can be alone. That's just my thought on the subject." Agnes stepped back but looked at Willa without blinking.

"Well done," commented Cagnes.

Willa nodded. She and her new husband had had wonderful conversations these first few weeks of married life, but they were in a more public situation today. She must use restraint. She changed the subject.

"Queen Luna will be there, today, right? Maybe she and I can finally meet and talk, woman to woman," she said with a sensible tone.

Agnes sighed. "Don't count on that. I haven't seen Queen Luna or her daughter since Queen Linda died. Besides...the Queen will be hearing Water Rights submissions all afternoon. She won't have time for chit-chat."

Willa looked at Agnes in dismay. "Isn't there anywhere that I can just get to know people?"

Agnes felt a pang of sympathy for the lonely young queen.

"Agnes is not suggesting that you change," advised Cwilla. "And you know me. I'm never one to tell you to keep your mouth shut. She's just suggesting that today, a quiet, thoughtful pose might be appropriate for a new queen in a big important meeting."

"Maybe I should just stay in my room, like I used to back home."

"You don't need to do that anymore." whispered Cwilla.

Willa did want to make Solem proud, even though she was inexperienced in royal social affairs. Stoically she stood still and allowed the determined fussing of Agnes to continue unhampered. Now she turned to her mirror.

'Queen Willa' was shocked at her final appearance. "Thank you so much, Agnes," she whispered gratefully. "I look like a queen. This dress is stunning. Who knew I could look like this? Do you think Solem will like me this way?"

"He will love you," Agnes answered reassuringly. "You are beautiful."

Willa blushed, hugged the surprised Agnes, and swept toward the grand staircase of The Castle's upper hall. Taking advantage of the confidence given her by her new attire, she descended the wide stairway with head held high and no hesitant steps.

"Queens should not appear shaky", Cwilla counseled firmly.

The crowd of dignitaries at the bottom of the stairs looked up and inhaled in unison. Regal and confident, Willa floated down the massive stairway in burgundy splendor, smiling broadly.

Few, other than family and Castle dignitaries, had really seen Queen Willa yet. Her wedding had been a quiet, private affair so all assembled here today were ready to observe the new arrival and take measure.

The ladies gathered below in the reception hall made note of Willa's burgundy satin skirts and the voluminous white lace that peeked from beneath it. The men folk focused on the tight-fitting bodice and the mounds of curving breast that rose above it. All were impressed and fascinated.

"Queen Linda would never have worn such a thing," snorted one wag.

"Queen Linda was fifty years old and a bit skinny."

"True. How true. How old is this beauty, do you think?"

"Old enough to be your Queen. She is your King's choice."

Prince June was also waiting at the bottom of the stairs. He swallowed hard and grinned. "Willa, my friend, you look marvelous!" he whispered.

It was not as though June had never seen Willa before. He had just never seen her looking like *this.* As a Prince, he was in his required position now, ready to escort her into the throne room as he would have done for his mother. There his father would be, waiting on his throne, his sister obediently flanking him on his left. King Solem had always insisted on reception protocol.

The Call of the Auras

Gathering his dignity quickly, June held out an arm to this strangely beautiful person that he knew he had not ever seen before... not *really* seen. *I wonder if father knows that this stunning creature is the shy girl he married,* he wondered to himself.

"This is *your father's* new wife," Cjune reminded June curtly.

From the back of the entry came an impatient voice. In a harsh, booming tone, the voice called out, "Huzzah! Make way for the new Queen. Huzzah Queen Wilhelmina! Huzzah!"

Prince June solemnly extended his arm again. Willa blinked and smiled, placing her hand on June's elbow. She knew who belonged to the voice they heard. It was her father, the crude and demanding, Lord Thorndike. She had forgotten that he might be here. She shuddered.

Lord Thorndike had not immediately approved of Willa's nuptials when the King had sent emissaries to declare his interest. He definitely *did not like* King Solem and The Kingdom knew it...but there might be benefits, the Auras whispered.

He and the nobles knew that having a daughter who was Queen had made his approval a useful decision. The Auras slid in among the guests today, snugly ready for action.

June smiled at his new 'Queen Mother' as they proceeded together into the throne room. They were followed by the nobility, the landholders, and finally assorted towns people, all of whom lived and worked along both sides of The River.

Mimi's social behavior was well known in the Kingdom. She had always attended all these Gatherings just to visit with everyone. And everyone knew her frivolous nature, now, including Willa. But Mimi was a lighthearted treasure. No one was surprised when she ran to Willa and hugged her.

"You look smashing, Willa! That dress is perfect! Agnes is a genius!"

"You're supposed to be acting like an adult, Mimi," snorted brother June with an arched brow. He looked straight ahead toward his father and the Queen's throne that was waiting to the right of it. "She's the Queen now, not a girlfriend. Back off."

Willa grinned and made use of her growing familiarity with the pair. "Ease up, June. Mimi's just being herself," she whispered.

June and Willa proceeded to the elegant throne where Willa now took her place to the right of her husband. Stoically she tried to remember who she was supposed to be and how she was supposed to act. June nodded to his father with solemn respect and took his seat between his father and sister on the left. Protocol.

A gaggle of the wealthy and privileged noblemen of River Kingdom and citizens of Urhonordo followed the Queen into the throne room, jostling and shifting for position along polished tables that reached from front to back along the sides of the enormous throne room.

The Royal Guard took their places at the back of the hall. The center was left open for additional arrivals and presentations. All were intent on seeing and being seen.

The next contingent entered grandly. There were only three of them, but their elegant presence filled the spacious floor. This was the beautiful Queen Luna of Urhonordo, resplendent in velvet and accompanied by her impressive escort, the tall, dark Prince Consort Armando. Whispered appreciation of the handsome couple buzzed through the gallery until they focused on the immediately following, Princess Amanda.

At least they thought it was Princess Amanda...the way she looked now, many were not sure. The usual tomboy they had watched and clucked about for years had undergone a swan-like transformation, dressed as she was today in a white gown that set off her amber skin and halo of black, silky curls.

"Who would have ever guessed she could look like that," snarked the old wags.

The Call of the Auras

Prince June was stunned again. Who was this woman? He had never seen her before. She was a vision, rivalled only by Willa.

He was just recovering from *really seeing* Willa when this new goddess moved forward and stood before him. Then he noticed something. There was a slight grin curving the lips of this goddess as her eyes met his.

This had to be Manda, his sister's friend, the 'ugly girl' from over the wall long ago. June smiled, hoping that she wasn't noticing his sweat.

"Grow up, June," his Conscience murmured.

The three Urhonordo Royals now glided to the foot of the platform on which King Solem's throne stood. The elegantly attired Prince Armando bowed low with a gracious wave of his arm. The two women curtsied gracefully but did not lower their heads. Instead, they looked straight ahead in pride as was Urhonordo custom. Female equality and importance should always be honored and displayed.

"Thank you for accepting us to this auspicious gathering, Oh Gracious King and Queen of River Kingdom," intoned Armando. His eyes were warm with friendship beneath the black curls that escaped from his crown. "My family and I wish to extend our congratulations and best wishes to you and your new Queen, the honorable Queen Wilhelmina.

"Today our contingent represents both Urhonordo *and* the Lowlands, as usual," he continued, gesturing toward a small troop of Lowlanders still framed by the doorway at the far end of the hall.

Amanda now beckoned the waiting Lowlanders forward with a flourish saying, "We also bring with us some of the fruits of the fall harvest from our combined orchards. We offer these for your enjoyment and in thankfulness for the water of The Great River that produced them."

She swirled back to the Lowlanders and proudly marched with them, carrying baskets heaped with gleaming apples and pears. Ceremoniously, they laid their bounty with pride at the foot of the throne's platform.

King Solem smiled and nodded. "Thank you, my friends," he responded with a bow.

Queen Luna and Prince Armando bowed again and stepped toward the elegant seating arrangements that they knew would be set for them on the left, indicating their Royal status. Protocol.

Overcome by the loveliness of the presentation, Willa rose from her seat without thinking and curtsied as they passed before her. "Thank you so much for the lovely fruit! I am so thrilled to finally meet all of you," she sang out spontaneously. "I have always wanted to meet you in person."

Quickly, almost invisibly, Agnes, still in attendance to her new Queen, leaned deftly from behind to 'adjust the Queen's gown' and whispered sternly, "Sit down, Your Majesty..."

Willa nodded obediently and sat down.

It only took a second or two, and King Solem appeared not to notice, but Willa had given the crowd something to discuss among themselves as they took their seats.

Urhonordorian Royalty chatted as well, but with definite smiles of understanding. Now they thought they knew what they needed to know. They would accept and like this new Queen.

THE ENTRANCE

Suddenly there was a commotion at the entrance to the hall. A rotund gentleman dressed in dignified, regal attire strode into the hall. He came to a stop alone at its center, smiling and looking expectantly from side to side. Applause sprang up immediately from a corner of the room on the right and echoed into its high ceilings.

The gentleman grinned broadly, responding to the apparent enthusiasm and clapped his hands in acknowledgement. It looked as though he had anticipated this reception and was pleased. And this was true. The attending Auras had helped him orchestrate it. This was Lord Drum.

How strange, thought Willa, straining to see who the new arrival was. *I am sure this type of reception isn't the usual protocol here.* Solem was staring ahead solemnly. Did he know this person?

Those who were not clapping looked around in astonishment. This was rude behavior and contrary to the King's dignified protocol, but the nobility from the north and south regions seemed to be celebrating this man for some reason.

The reason was Drum's private contingent of Auras who had also entered unseen and spread out around the hall for support. He was now the center of attention. He was working his plan. This bravado would get the Auras what they wanted... turmoil in the castle of The King and fame for Drum.

The Call of the Auras

Drum swaggered to the foot of the throne, kicking the fruit baskets aside and nodding his head in an imitation of a bow. "Good afternoon, Your Highnesses," he said casually as he stopped clapping and waved at others to do the same. He smiled at his control.

From behind Willa's throne, Cagnes noticed Drum. She now realized that she had seen him in The Castle before and this time she could see that he had brought Auras with him. The Conscience Alert had warned of this. This was not good.

Willa stared at Drum in disgust. Now, she also recognized this irritating person. He had been a frequent visitor at her father's meetings. She shivered and backed into her chair. Memory of him was not pleasant. What was *he* doing here?

Drum walked directly to her and took brazen inventory of what he saw. "You must be the lovely new *young* Queen that I have heard so much about," he said with a lude wink toward King Solem. "Nice going, Solem," he continued. "She's a beauty."

Solem rose and stepped forward in one movement. "Please take your seat at a proper location and show some respect for your Queen, Lord Drumsworth." He glared at Drum. The Castle Guard at the rear of the hall came to attention.

Drum stopped his performance momentarily and looked around. There was an uncomfortable silence. This was not the usual, polite gathering that Agnes had described to Willa earlier. She did not know what a 'usual' Water Rights Gathering was like, but she was already aware that dealing with this rudeness was probably not her husband's experience in the past.

What does Solem think of this man? Willa wondered. *I know him. Crude remarks and this strutting self-importance is the way I have seen him behave in my father's manor. He is crude. I have felt that in his presence for years. My father likes him but does Solem like him, too?* Carefully she straightened her skirt and sat up straight. Since her first outburst had been tamed by Agnes, she counseled herself now to be quiet and exhibit queenly behavior.

She knew that dinner would be served in the Royal Dining Hall after everyone's presentations of their fall water rights issues. Willa decided she would wait until dinner to find out what Solem felt about Drum...and tell him how *she* felt about him.

But seeing Drum now and his ogling manner of addressing her was disconcerting. Obviously, he didn't recognize her. Drum had been a frequent visitor with her father in recent years, but she had managed to stay out of his way. He had been regular and welcomed by her father although he always seemed to act superior and in charge. He was definitely disliked among her father's female servants with whom she was friends. They had deemed him 'disgusting'.

I had no idea that Solem had anything to do with this man, Willa thought uncomfortably to herself. *I was hoping to never see him again. He is so rude and full of himself.*

"Your husband is the King of everyone in the Kingdom," Cwilla reminded her. "He has to deal with everyone, no matter who they are."

But Drum is conniving and dishonest, Willa moaned inwardly. *I've always known it. I don't know how, but I do. He's disgusting. He does not value anyone but himself. And did you see how he looked at me? Ugh!*

"I know. I know. I remember him, too", whispered Cwilla. "But you have a husband now. I'm sure Solem will keep him in line. Don't worry."

Willa glanced around the room. Hopefully, Drum's noisy arrival had run its course, and he would disappear into the crowd.

But that was not going to happen. To her chagrin, the clapping and cheering began again as Drum joined his welcoming group. They all appeared to be noblemen and what was worse, *her* father was leading their cheers. How humiliating. King Solem, her wonderful new husband, was being disrespected in his own Castle by this Drum person *and* her own father.

Drum proceeded to amble in front of additional seated guests, clapping and greeting them as though they had come here just to appreciate him. "Everyone's glad to see you, Drum," shouted the familiar voice from the crowd. "Looks like we might actually get some approval on our project tonight. Tell The King what we want, Drum. Show him what we came here for." It was her father, again.

Drum's daughter now entered the throne room, as well, joining her father and acknowledging additional clapping with a broad smile and a wave to the crowd. She was very pretty and dressed beautifully like her father. Like Drum, she continued to wave as her father had done while King Solem scowled at the renewed disruption.

Protocol? Where was protocol? It was more like a rally.

The usually sedate King Solem and other members of the court simply stared at the disrespectful and unprecedented behavior. The din grew. The Auras ran among the crowd of nobles encouraging even more upheaval.

What should a loving, respectful wife do in a situation like this?

The Call of the Auras

Willa lowered her head but peeked up at her husband. She wanted to help but how?

"Sit up straight, Willa. You are the Queen now," Cwilla hissed. "You know Drum. He and his daughter are being rude to your husband and so is your father. Drum and his family love the attention. That's what they came here for."

They are obviously being encouraged by Auras, Cwilla now realized. *There's a big Aura presence here that wants to disrupt this meeting. They have an agenda. I must warn Willa.*

"Willa, beware! There are evil forces at work here," she whispered hoarsely. "They are trying to intimidate your husband."

The former Wilhelmina Thorndike would have remained quiet. She would have gone to her room, unnoticed. But this boor was disrespecting her beloved husband. But Wilhelmina Thorndike was now Queen Willa. She thought about that for a moment.

I should be helping my husband, she decided. *I am his queen now. If Drum and my father think this meeting is a joke, let's see if a joke from me will quiet them down,* she thought. On impulse, she raised a queenly hand. "Excuse me, Lord Drum... Excuse me, please."

The clapping noblemen stopped in amazement. What did the new, *very young* Queen think she was doing?

Slowly Queen Willa rose unsteadily from her chair and smiled. "Good evening, Lord Drumsworth and family. How nice it is to see you here all the way from Drumsworth Castle. But I see that you are clapping loudly. That can only mean that you are very hungry or you would not be interrupting this Gathering as rudely as you are now doing, right?" She smiled as warmly as she could muster.

A few chuckles escaped from the town folk in the back of the room. Cwilla gasped and Agnes tried to pull Willa back, but Willa stepped forward and continued.

"Now that you are quiet for a moment, I feel that I must explain something to you. Clapping for *yourself* is very rude. It appears that they don't practice good manners down at Drumsworth Castle, so you don't understand such things. I forgive you." She smiled again sweetly, and stared at Drum. Her knees were shaking but she went on.

Drum and his daughter froze in place.

"I am terribly sorry, Lord Drum, but here *at River Kingdom Castle* we must listen to everyone's presentations regarding Water Rights, *before* we have dinner. That is what we call 'manners' here at *'The Real Castle'.*" More of the crowd picked up on the jibe and the snickers increased.

"We will be feeding you and your 'hungry clappers' soon, of course, but if you are too hungry now, perhaps our Royal Castle Guard can *escort you* all into the banquet room for an early start on dinner. *Then* those of us here who know how to behave in a Water Rights Gathering can begin their presentations.

More chuckles.

"I know that my father is a friend of yours, Lord Drum. Perhaps my father would like to escort you and your fellow-clappers into the dining room, himself."

The throne room erupted in hearty laughter. Men guffawed and slapped each other on the back while the ladies giggled behind their fans. Urhonordorians nodded and smiled and nudged each other. All but the nobles applauded. This young woman was alright.

Drum and his daughter now looked around in awkward confusion. Was the joke now on them? Yes.

What had just happened? Drum and Balynn stood still together in the splendor of the glorious hall, looking momentarily lost. Their grand entrance performance was stalled, and a young Queen had come of age.

Willa's heart was beating fast. *What have I just done?* She asked herself.

"I appreciate the return to quiet," Solem whispered without looking her way. The throne room crowd continued chuckling, and Drum and his daughter hurriedly folded themselves into the group of noblemen that had been clapping for their entrance.

The nobles and their Auras were not chuckling. They had been caught in their own, pre-arranged show. Slowly they seated themselves and rolled their eyes in Willa's direction, mumbling with disdain. The foiled Auras retreated as the nobles muttered.

"This Thorndike brat doesn't know how to behave."

"The King let her get away with it."

"He showed a sure sign of weakness, right there."

Lord Thorndike scowled at his daughter. She was in the right position to be of assistance to him, but not acting helpful the Auras had said she would. Did she not understand her role?

The Call of the Auras

Willa heard the remarks of the nobles. She sat very still. Her father was embarrassed in front of his friends. As a daughter, she was sad.

Cwilla noticed the Auras that had accompanied Lord Drum as he entered. She now watched as they circulated again among the grumbling noblemen. Why were Auras so interested in this group? She crawled up next to Willa's ear. "I'm sorry, Willa. I think my suggestion may have caused your father some trouble. But, on the other hand, you were brave...and funny. I'm proud of you."

Agnes patted Willa with appreciation. Cagnes agreed with Cwilla, but a shaking Willa was not sure she would survive this day as queen.

Amazingly, Solem now turned to Willa and took her face in his hands. "Thank you. My Queen," he said softly. Willa sighed in relief.

Solem re-focused on the activity of the day. A business-like atmosphere replaced the rowdy bustle of the Gathering that had existed only a few minutes before. Several more participants arrived at the hall, paid their respects and found seating. King Solem greeted them cordially. He did not acknowledge the change in behaviors caused by his wife's contribution. Protocol resumed.

Willa folded her hands again and stared at them. *Why did I speak up like that?* she asked herself. *I have never been so outspoken in the past. I have always been a quiet person. I have stayed out of the way. What has happened to me?*

"You have become a woman, Willa," responded Cwilla. "You were just a lonely young girl who endured such things rather than trying to change them before. Your change is a good thing. Now you and I are participants in life."

Cwilla added another thought without hesitation. "You said 'yes' to Solem because you knew you could love him, but you also knew that he needed to be a strong king and you could help."

Did I help him or hurt him with what I did?

"Time will tell," Cwilla responded softly in her ear. "But your husband was being disrespected. I think you did what was right. And you did it with love.

New arrivals were no longer streaming into the throne room. It was time for the people from along The River to sort themselves into groups with similar concerns as was their custom.

Solem stood in front of his throne to make the customary announcements. Willa sat quietly and listened. Solem appeared to be in control, his voice calm and strong. *Maybe I did help,* Willa thought cautiously.

"Welcome to our Water Rights Gathering, one and all," said The King with a smile. "Today we gather to discuss The Great River as we have for years. Although we are all happy to see each other, our gatherings are held with purpose. Decisions made because of our discussions here will be important to all of us in the months ahead.

"This fall's decisions will help us work well with The Great River through the winter. Queen Luna, Prince Armando and I shall hear from all of you.

"Please gather yourselves for discussion. I will hear anyone who wishes to speak from any location on The River so that the best decisions can be made for all of us. Queen Luna, Prince Armando, Queen Wilhelmina and I will confer and contact you all when fair decisions have been determined. Enjoy your presentations and discussions."

The mass of humans from different regions began at once to shuffle into their gathering spots. There were two sides to The River so there were two basic groups and many sub-groups. This was serious and important business. Sometimes the Royal Decisions would take days and extra Royal Meetings to complete, but most would leave after dinner tonight, content that they had been heard and the eventual outcome would be fair.

With respect, the Urhonordo-Lowlander contingent nodded to King Solem and assembled on his left near Queen Luna and Armando. This was an eclectic group of many colors and manners of dress. All appeared to relax into the friendships that they were accustomed to renewing and enjoying twice a year. There was much hugging and laughing and shaking of hands.

To the right, a larger group bustled together with noticeably greater formality. This was the River Kingdom group. Some of these participants demonstrated the confidence of being on their 'home turf'. This was *their* kingdom's Castle. They were gathering before *their* King. They were the biggest group. They were proud... and one batch of them appeared to have an agenda.

What is their *agenda,* Willa wondered as she watched a small group within the River Kingdom group organizing separately. They were the noblemen of The Kingdom, obviously ready to support what their representatives were going to present to The King today. This group appeared to be very intense...and it included her father and Drum.

King Solem returned to his throne on the platform beside her. The River Kingdom group that had not been clapping arranged itself politely to speak directly to their King about their individual concerns. With a business-like air, Solem nodded to the first in line while Willa watched. *I must go back to being quieter as I was in my father's house. Speaking of that, what is my father's role in this?*

"You might take time to notice that *your* father is over there with Lord Drumsworth" Cwilla whispered. "He and your father are up to something. You can tell by their whispers. That whole little group of noblemen is up to something."

After a brief bit of discussion and jostling, a long line of landholders and serfs formed in front of the raised platform of the King. Presentations began. Many had come with requests to raise or lower the levees that protected their crops. Others were petitioning for larger allotments of seed. Most could not read or write, but with gestures and hand-drawn maps, they told of difficulties they had and described their concerns.

King Solem listened respectfully and asked questions about each presentation, careful to consider all points of view. Willa watched with admiration, but the afternoon was long and began to drag. Willa could not see a function for a Queen here.

Cwilla chided her mildly. "You are getting spoiled now that you are a Queen. Do you think *you* should be the center of attention? Get over it."

I suppose you're right, Willa thought. *I'm probably bored because I don't understand this whole Water Rights issue, yet. I think I've overheard Father talk about it, especially when Lord Drum came to visit, but, unfortunately, I didn't pay much attention.*

"It looks pretty important," Cwilla observed. "If you are going to be of help to your new husband, you should probably just listen and learn."

FRIENDS AND MAPS

Her Conscience was right...boring but right. But now Willa felt a nudge from behind. Turning with a start, she found Mimi and Princess Amanda (Manda) with sly grins on their faces.

Suddenly shy, however, Manda curtsied. "Pardon us, Queen Wilhelmina," Manda whispered.

"You don't have to call her 'Queen Wilhelmina', snickered Mimi. "She's just Willa now."

"I saw you looking my way a few moments ago," confessed Manda. "I could tell that you were bored out of your mind, like Mimi and I were. We always used to sneak out of these things when we got a chance since no one wants *our* opinion about anything. I hadn't talked to Mimi *or you*, since her mother died but Mimi said she understood. She came to get me so we could cry together...We just did that". The girls hugged each other again.

"But then we just figured we should get all three of us together while *your husband* and *my father and mother* work things out with The River. We women all need to be friends, after all."

Willa smiled gratefully and slid away from the throne. Solem was busy listening to the next landholder and didn't notice. With relief, she accepted hugs from the two young women but then stood back to look at them face-to-face.

They did the same with her. They all quickly took measure.

For a moment the three young women inventoried each other from their respective situations. They were all young but sensed that they weren't just carefree girls anymore. They were women with adult responsibilities. For an awkward moment, they simply stared at each other. Then young laughter took over.

"We're taking ourselves too seriously. Let's step outside," suggested Mimi, pointing to a side door in the throne room. Manda nodded and slipped away, pulling Willa by the hand.

Willa sighed. "Thank you for rescuing me," she began. "I love Solem. He is a good King, but I am not sure I am going to be a good queen. He is busy working with the landholders and their workers about things I know nothing about. And I might have made him mad when I spoke up to Drum. Do you think he's mad, Mimi?"

"I don't think so...surprised, maybe. My mother was very 'old fashioned. She didn't speak unless spoken to in public when my dad was around. When they were alone, I don't know. I think she may have been more of just a listening post than you are."

"I really *am* sorry about your mother, Mimi," said Manda, looking at the floor. "That's why I didn't come before. I didn't know what to say or how to say it. Your mother was always so good to me. I will miss her."

"I know." sighed Mimi. "I understand."

"And I know I will never replace her with your father," Willa added.

Mimi smiled sadly. "Of course not," you're very different...but in a good way," she added quickly.

The Call of the Auras

A snicker popped forward from Manda. *"I'll say* you are *different.* I laughed out loud when you spoke up to Lord Drum. And I couldn't believe how easy it was to put him in his place when he was strutting around, all full of himself. Your comments were hysterical...I really dislike that man."

"Me, too." There's just something about him," added Mimi.

"That something is called 'evil', growled Willa. "My Conscience says that he has Auras."

"What are Auras?" Manda asked.

Suddenly loud voices rose behind them from the throne room. The girls looked at each other with concern and ran back to the door of the hall. Peeking in, they saw a scramble that was not at all like the peaceful scene they had just left.

The Urhonordo contingent had retreated to the left wall of the hall as two River Kingdom groups shouted at each other and threatened with more than just words.

The Castle Guard moved to positions protectively in front of King Solem who now stood above them on his platform. His Kingly pride rose to the occasion. Firmly, he took a stance and issued commands in a tone that allowed for no dispute. He would not allow things to get out of hand again.

"Stand back unruly citizens of River Kingdom! You shame yourselves. You are not behaving as River Kingdom citizens but uncouth riffraff! Stand back, I say, or be escorted to my dungeon! You are in The Castle of River Kingdom now. You *will* behave with respect before your King and Queen, your Kingdom and each other. You will conduct yourselves with fitting protocol immediately or leave!"

The Auras scoffed. This lame king couldn't tell the nobles what to do.

Then the Castle Guard, swords drawn and ready, advanced to the edge of the River Kingdom crowd. The Auras changed their minds. Reluctantly, the nobles thought better of their current activities and backed away, but not as a single group.

They now became two tangled masses buttressed separately against the right wall. The Castle Guard had corralled both and stood between them forming a human fence. The Auras accepted the new positions happily. Their people were still angry.

The combatant groups appeared to be landholders vs. noblemen, an odd standoff since the landholders were, themselves the main support of the noblemen and the noblemen owned their land. Noble wealth came primarily from the landholders' taxes as did their food. Why were such interdependent groups now against each other? The Auras snickered. They loved this disruption.

Glancing at the opposing factions, the attending Auras distributed themselves through the crowd and became active. They were also ecstatic. Auras didn't take sides here. What they were fighting for was simply mayhem in general. Their evil encouragement was affecting tempers on both sides.

With no understanding of the situation that had arisen in such a few short minutes, each of the girls returned inconspicuously to their separate areas. When they had left, there had been boring discussions, but nothing was boring now. The room was filled with a tension that no one but Auras had apparently expected.

Willa tried to make sense of it. What had happened to cause so much anger in such a short time? Willa needed Solem to explain it to her, but he was standing in front of her demanding calm. His back was to her. She couldn't even see his face, but she could feel his determination.

Agnes slipped up behind Willa and whispered, "It was all because of that man you talked to before... that *Lord Drum*. This wild behavior is all *his* doing. You were right to call him out before, but don't say anything now. The King has got this. Just sit still and watch."

Cwilla added, "Agnes is right. Don't say a thing. There are warring Auras at work here."

Auras? Willa folded her hands in her lap again. Glancing to her left, she could see Mimi doing the same thing. She couldn't even see Manda at all.

Solem raised his hand and waited. Slowly, the room began to calm. Little by little, one man after another turned and looked at him. He smiled at them all, turned back toward his throne and Willa and took his place on his throne.

What had just happened? Solem looked calm, almost as though nothing violent had been occurring at all. Willa was confused.

"Protocol," Agnes whispered. "Royalty has used it for years. The King just raised his hand. It always works."

"Now that we have calmed ourselves, let us resume our discussions in a more civilized manner," Solem said firmly to the throng. "This is the way that, as my Queen says, we do things here at "The *Real* Castle." I will hear from you first, Hector. I believe that is where we left off. You noblemen can wait your turn."

Hector was a landholder and a farmer. He was a working man. Reluctantly he shuffled forward, glancing back at his fellow farmers for reassurance. They all nodded and urged him toward the King but made no moves to join him. They were obviously afraid of some of the noblemen.

The Call of the Auras

Hector was armed only with a map that he had rolled into a scruffy cylinder and carried in front of him like a sword. Trembling slightly, he placed his precious column on the floor before Solem's platform and stood back.

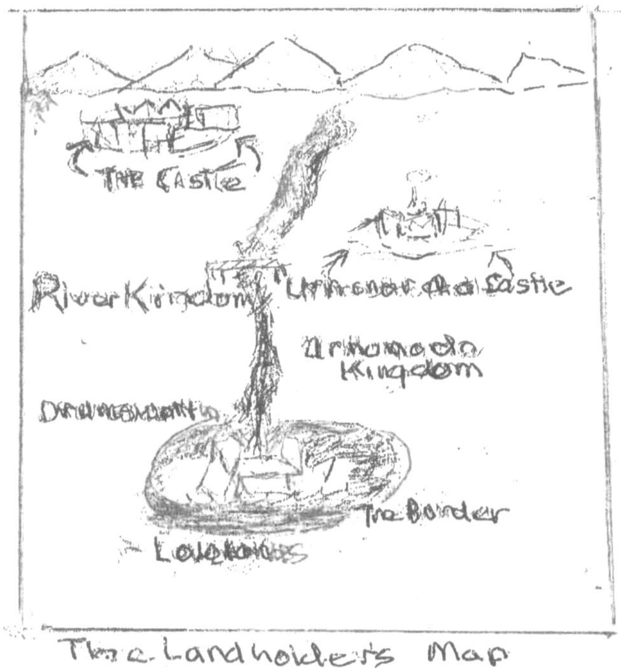

"I brung this from a meetin' that me and the rest of us Central landholders had last week," he declared. "This here is the map that shows what the problem is, sure enough. Do you want I should roll it out fer ya, yer Majesty?" Hector bowed respectfully.

"This ought to be good," snarked a well-dressed gentleman from the noblemen's group.

"Be still while Hector is speaking or find yourself in the dungeon," said Solem calmly. "Proceed, Hector. Let us see your map." The nobleman grumbled as did his surrounding cohort but said nothing more.

Hector carefully unrolled his map in front of the platform until it stretched to both sides of the room. It was rough and obviously scratched out with pieces of charcoal and splotches of dye, but it was very clearly a representation of The Great River and its west bank.

The peaks and spires of The Castle were strangely recognizable at one end and squarely in front of the noble group while the other end showed only a large blob of dye and a sign which read simply. 'Drumsworth.'

"This is it," announced Hector proudly. "This here shows the problem clear as day."

Solem rose again from his throne and looked down carefully upon the map below him. He scratched his chin and now stepped down from the platform for a closer look. Hector took a knee and humbly waited as Solem slowly made his way from one end to the map to the other.

"This is very well done," Hector," Solem said solemnly. "I see The Bridge that goes to Urhonordo. I see our Great River. And on the west side here, I can make out your farms, lined up one by one along the River's shore and back to the west. You and your men did a good job with this map. It is all very clear.

"There is only one thing that I don't understand. What is this strip that runs along The River between The Castle and Drumsworth, alongside your farms...and what are these two lines that run back from The River. Are these other rivers? Where do they go? How does that work?"

The group of farmer/landholders began to rumble loudly. This was their issue.

"That's the problem!"

"Now you see what they want to do!"

"The lords don't farm. They don't understand nothin' about farmin'!"

"They want to block the water from The River that makes 'em rich!"

"And keeps all of us alive."

"Silence, friends...Be silent or my dungeon will be filled by all of you," Solem said calmly. The farmers quieted immediately.

Willa strained forward on her throne and peered over the platform. She could easily see the farmers' map on the floor below her but knew nothing about what the farmers seemed to think was obvious.

She could see that their picture showed a flat strip along the River's edge that seemed to cover the levees, but what was the problem? She had traveled that road before. *Smoothing out the levees will be a good thing, won't it?* she wondered, acknowledging how little she knew.

Solem stepped further to the left and in front of the Lowlander and Urhonordo area. The south end of the farmers' map was still slightly curled where it had been rolled. The farmers hustled there to straighten it out as a few Lowlanders peered at it over their shoulders and nodded.

"And what is this area?" Solem questioned with eyebrows raised. "Is this Drumsworth Lake?"

"Indeed...That 'blob' is a shamefully poor drawing of an elegant idea, but it is, indeed, the area of the glorious Drumsworth Lake," a voice shouted from the other end of the map and the noblemen.

Drum stepped forward with a smile, brushing past the Castle Guard that was still planted firmly between his group and the King. "Would you like to see a better representation of our grand, new idea, Your Majesty?" he interrupted. "We have another map, and it shows the whole concept much more clearly."

Solem strode across the hall to Drum in seconds. "I will talk to you when I have spoken with Hector and his men, Drum. Stand back. Guard, you may help him if he does not obey."

Drum shrugged and obeyed.

King Solem stepped back up onto the royal platform and returned to his throne. Willa was awed by her husband's calm control. *He is strong and forceful. He is a man among men. He is nice to the farmer and puts Drum in his place,* Willa thought at once. Her thoughts were no longer on The River but on Solem. *This is why I love him*, she thought smugly.

Solem turned again to Hector who had returned to his fellow farmers for support. The King smiled and beckoned the quivering man forward again. "Tell me about your map, Hector. What is your concern?"

Hector swallowed hard but began. "We all heard about this plan or parts of it all summer when the noblemen came by from time to time. We know you can see The River that we drawed so that's not nothin' new. What's new is the *big road* that runs alongside it from end to end. Do ya see that, Yer Majesty? Do ya see that?

"And the road itself ain't even the whole problem, either. It's the huge levee between The River and our farmland that will be heaped up to make the road higher and wider. We figger that road is also gonna cut off *all the water* from The River that all us farmers need for our crops. It'll leave the land dry. *That's* the problem.

"Our crops is what feeds River Kingdom and keeps us all alive, right? The nobles plan to fill up all our levees with their road! *That's* what the problem is. The rich folk don't get it. They just want their road so they can ride on it with their militias and fancy carriages. They ain't thinkin' about nothin but lookin' good and havin' a good time down at Drumsworth Lake."

*Drumsworth Lake...*Willa had almost forgotten about *Drumsworth Lake.* Her family had gone there once before her mother died. It was one of her few happy childhood memories.

All the children of the noble families of River Kingdom went there as often as they could. Of course, it was a day's bumpy ride away. Sometimes it even took two days if they had to detour around levees......*Oh... Farmland and levees... a new road...Oh.*

Now I think I understand at least a part of this, Willa thought. *But a new road sounds like a very good idea to improve a trip to Drumsworth Lake.*

"Do all the landholders the along The River see this as a problem?" Solem was asking seriously.

Note: *A point of history---The Feudal System: Landholders didn't actually 'own' the land that they farmed, but most had occupied their land for years and were responsible for it and what it produced. They were only long-term tenants of small pieces of land that were part of huge swaths of land that belonged to the noblemen or the King. Landholders were, of course, regarded as unimportant by noblemen who could remove them or even do away with them at will. The landholders had no real power but were essential to the agricultural production of the Kingdom. Forcing The River into inland ponds watered their crops and mitigated flood damage downstream.*

Landholders produced food for the nobles and the villagers. They also paid for the use of the land with food and taxes. If the farmers didn't produce for their nobility and the King, they would be evicted from the land and replaced with someone who did. This was often unfair, but it was the system of the day.

Willa was the daughter of a nobleman and had never thought about this system. *Why are the farmers so upset,* she wondered. *Men get so loud and worried over such little things.* She looked over at Mimi. Mimi looked like she was thinking the same thing, Willa could tell. They both rolled their eyes. *The noblemen will do what they want. They always do. This whole argument will be over before dinner,* Willa thought.

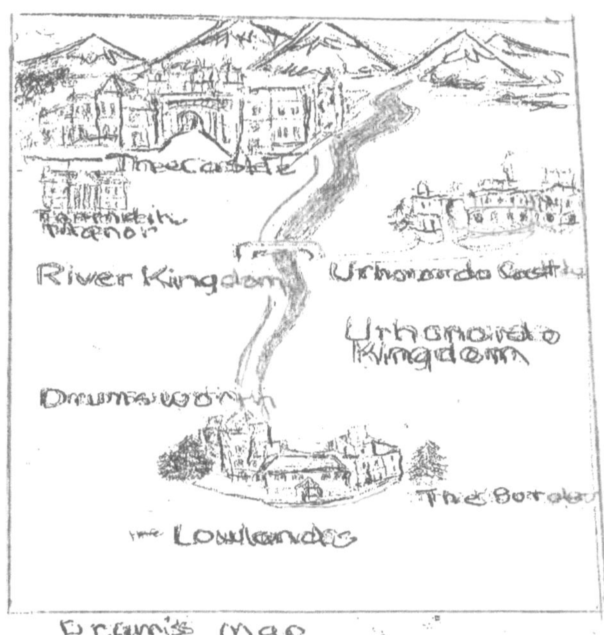

HECTOR'S ARGUMENT

But Solem was still taking the farmers very seriously, Willa noticed. She hadn't figured out exactly what their problem was, but she had decided that her place was here, by Solem's side. She sat back, determined to behave in a queenly fashion and be supportive of her new husband. Cwilla was dozing off.

Again, Solem focused on the curled up 'southern end' of the farmers' map. Unfortunately, the participants from Urhonordo and the Lowlands were interested in it too, and had gathered around it, chattering excitedly. Solem knew what that part of the map was meant to show. It was 'Drumsworth Castle' and the large circle around it that the farmers had stained green was the lake.

They have made a mistake in their drawing, Solem thought. *Their drawing of Drumsworth Lake is big and goes around the house. Must be they've never been there. Oh well. That's just the way they drew it*, Solem reasoned. *They don't claim to be artists.*

"Please finish your presentation, gentlemen," Solem said with respect. "What is it about your lovely map that has made you so angry?"

The farmers jostled each other, trying to push Hector back into his speaking role. Finally, he shuffled forward, bowing again and again as he came. "If we was better at drawin', I think you'd understand right away, Yer Majesty," he murmured. "Do you see that big green space on our map, down by where the Lowlanders are?"

"Yes, of course. That's Drumsworth Lake. You drew it a bit too large, but I see it."

"We didn't draw it too large, Yer Majesty. Maybe we drawed it *too small*. The way yer noblemen want to make it, that lake'll be twice as big as it is now and go all the way around The Castle. It'll even go onto Lowlander and Urhonordo land. *That's* the other part of our problem. *That's* where all *our water* will end up. That River won't water no crops. It'll just fill up a fancy mote around Drum's Castle that'll make *him* feel like a king. Then all his fancy friends can come down there and tell him how wonderful he is."

Hector blinked suddenly, realizing that he might have said too much. He turned and scurried back among his supporters, disappearing altogether in their midst. The group glared ahead stoically but offered no further discussion. They felt they had made their point.

A low rumble of laughter rolled out into the hall. The noblemen nudged each other and chuckled at Hector's hurried departure, assured that *their* presentation would be much more intelligently delivered and would, of course win the day with King Solem.

To Willa's chagrin, the next man to step forward was her father. Lord Thorndike smiled broadly at his daughter and winked. Then he bowed before King Solem. His manner was dignified and polite. He looked smug.

"Your Majesties", Thorndike reluctantly began, bowing slightly. "I thank you both for this opportunity to present *another* map, one that the gentlemen with me and I have had constructed especially for today."

King Solem nodded his approval and stood to accept the scroll. "Thank you, Lord Thorndike for this lovely presentation scroll. I will look it over carefully."

"We have also prepared a real-time explanation to enhance our presentation," Thorndike continued, stepping aside. "I would now like to pass this opportunity to someone more qualified...no, *equally* qualified in these matters. I believe that you know him already, of course. Please allow me to present, Lord Doran Drumsworth, the Third."

The noblemen to Willa's right exploded again with boisterous huzzahs and clapping of hands as Drum strode forward, nodding and accepting accolades as though they were obviously due him. He reached for the scroll.

Willa groaned. Willa's Conscience choked back most of what she wanted to say to Willa but settled quickly on "stay put and be quiet.".

Agnes reached around the decorative trimmings of Willa's throne and grabbed a piece of her skirt to hold it fast, determined to keep the spontaneous young Queen in place. The cheering from the noblemen continued but Willa managed to maintain some composure.

Drum, to Willa's surprise, stepped forward and bowed to Solem with respect that had just been coached by attending Auras. "Would you like me to explain our scroll, Your Majesties?" he inquired over the supportive din.

Solem scowled. *What does all this noise mean? People do not erupt like this in the King's presence. This behavior is rude and against all Royal norms,* he thought angrily. But he was a politician. *No one seems angry, however,* he thought. *These noblemen are, indeed, cheering for something...or someone.*

Since the rude noise makers were men of means and position in his kingdom, Csolem advised Solem to make a politically correct response. He shouldn't use The Royal Castle Guard to quell their rowdy behavior...although he could. He shouldn't yell at them and threaten to send them to the dungeon...although he could. *Remain calm...*

At last, Solem took a deep breath and stood strong on the platform, looking down on Drum as a protocol-loving King should. He raised his hand. "Please tell your supporters to curb their enthusiasm so that I might hear your presentation," he said to Lord Drumsworth.

Drum smiled again and raised his own hand in a King-like fashion. The cheering faded and then stopped altogether, on cue. Drum smiled inwardly at this. He unlocked the scroll's casing and unrolled it proudly across the platform itself, at Solem's feet.

Although the scroll was smaller, it had been artistically drawn and was easier to read. Willa took in the action with fascination. Essentially, it was the same map that the farmers had presented and was still on the floor below. It showed the same thing, although the etching of The River was a lovely shade of blue. River Kingdom Castle showed itself grandly on one end of the map while the other end held a simple rendering of Drumsworth Castle and its current lake.

"This is very impressive," murmured Solem, scrutinizing the map thoughtfully. "But The River flows directly to your lake as it does now. There is no change. Is that true? Am I reading this correctly?"

Drum shrugged his shoulders and nodded. "Indeed," he acknowledged. "The River flow is just as it has been for one hundred years, of course, to a place of leisure and enjoyment for all. Many people here today have taken vacations there, including you and your wife...I mean your former wife and children... in the past. Am I not right in this?"

Solem nodded. "Yes, Lord Drumsworth, I have attended your establishment several times, and this representation looks like it would get a traveler there on a wider, smoother road. Am I wrong in that?"

"Not at all," Drum responded quickly. "But the lake is not the important thing that this map displays. What is really important is 'The Royal River Road', itself, a raised and sculpted structure that will be a smooth and *safe* way of travel that links The Castle in the north to the Lowlands in the south. *All* travel will be safer, and *all* parts of The Kingdom will be easily protected from any enemies due to easier travel of The Royal Guard, an issue that is always a Royal concern, I am sure."

A chant started up in the Thorndike corner... Drum... Drum...Drum...The noblemen were feeling confident about their proposal. Drum had made their presentation well. Who could argue with it? A flat, new road from north to south for Kingdom-wide protection and safe travel would make River Kingdom the envy of all the known kingdoms...and Drum was their man in the lead.

And the nobles can't help but see how regal and confident Drum looks either, thought his personal Auras. *This is set up perfectly.*

The nobles did notice. *Drum is a man who knows what he wants and is going for it,* they thought. *He wouldn't be asking a bunch of ignorant farmers what they thought. King Solem should just step aside and let someone with real understanding take over.* The regular meetings of like-minded nobles at Thorndike Manor had frequently discussed that very possibility.

Solem looked closely at the drawing. *This road would make The Kingdom stronger and more able to protect itself ...a real, defense asset. General transportation is always key, as well,* Solem thought honestly...*and there is always the safety issue.*

"This presentation is very interesting", he stated, looking up from the beautiful scroll before him. "I can find no fault with it."

But the farmers were so set against it, he thought. *Why? Why did they object so much to this road?* What didn't he understand? Was there a flaw in his thinking somewhere?

Willa was thinking the same thing. She hated to think that Drum had a good idea, but it looked like he did. She was shaking her head in disbelief when a complete surprise appeared suddenly in front of the royal platform. It was Princess Amanda of Urhonordo. (Manda)

The Princess stood still, waiting to be recognized, a bronze statue in white, a tiny crown perched on her dark curls. She spread her arms wide and curtsied gracefully. Her head held high, she immediately commanded the attention of the entire room.

"Your Majesty," she began. "My family and I have been studying the map that was presented to you by your landholders. Urhonordo has a few questions. Might we take a moment to present them to you?"

Willa smiled. She had no idea what Manda was going to say, but she admired what she was doing. *How brave she is to stand up in front of everyone and ask questions. She is bold as well as beautiful. I wish I could be like her.*

"You *are* like her," whispered Cwilla. "You just don't know it yet."

When do I find that out? Willa argued back silently.

"You started this afternoon when you stood up to Drum," countered Cwilla, "but you have much to learn. Now is learning time. Listen."

Solem looked down at Manda, also with admiration. *What a lovely child,* he thought.

"She is the same age as your wife," Csolem reminded him quickly.

Solem blinked. He had forgotten that. "What is it that your highness wishes to say, Princess Amanda", Solem asked respectfully. "As you know, I definitely need to hear from Urhonordo and The Lowlands before we even consider something as grand as this proposal. We are all careful protectors of The Great River, together."

Manda motioned to those behind her who had, in fact, been studying the farmer's map with her. An eclectic group of Urhonordorians and Lowlanders stepped forward to surround her, all bowing in one way or another to show their respect to a powerful King.

King Solem was important to these proceedings. Queen Luna and Prince Armando usually handled their issues informally except for these Gatherings. These meetings were necessary for big issues. Now, Manda and the others had already heard and seen something of concern presented by someone from River Kingdom and they could see it would affect *them, as well.* They had concerns that needed to be expressed now. Manda looked intently at Solem. At this point, she would speak for them all.

"King Solem, we Lowlanders and Urhonordorians have studied the map of your landholders. It would appear to be very similar to the scroll that you have on the platform before you. If that is so, we feel that we need to ask you, with this plan, how does the water needed to grow River Kingdom crops inland get to them when water access from The River is blocked by such a large levee, Your Majesty?"

Drum stepped back with a haughty sneer. *Here we go again*, he thought, rolling his eyes. *This young woman is from Urhonordo with a gaggle of brown faced followers. It's ridiculous. This is a River Kingdom issue. These people should have no say in this.*

"The water comes from the mountains in the north and goes where it has always gone, young lady," he snapped. "It goes right past River Kingdom and its farms. What is the problem, girl?"

Willa wanted to jump up and say, "How rude. Amanda is not just a girl. She is a Princess!"

"Sit still," whispered Agnes and Cwilla in unison. Willa sat still.

Manda simply dismissed Drum with a look and returned questioning eyes to Solem. "Your Majesty?" she repeated politely.

King Solem smiled at Manda. "Do you see a problem with the water flow in this ambitious plan that the noblemen have presented, Princess? The water flow looks very similar to that which has always existed except for a lovely road beside it," Solem said seriously. "The road looks exciting, don't you think?"

The contingent around Manda began to all speak at once. They all had answers, opinions, and strong points of view. Manda held up her hand as The King and Drum had done. Her companions stopped talking, apologized and looked embarrassed.

Amanda laughed. "That's okay, my friends. You may definitely speak your mind. I'll just start explaining our concerns and then you can say what I leave out, okay?" Much shuffling of feet and nodding occurred among the Lowlanders, but now their contributions were reigned in.

Manda began again. "Yes, Your Majesty. The road looks exciting, but the first thing I am wondering is where the River Kingdom's farmers' levees are going to be? There are dozens of levees now along that road that divert river water into holding ponds for use, but I don't see them in the farmers' map. That is their concern. How would The Great River's water get *inland* to your farmers' crops? Also. Where would the un-diverted water go, if not to flood Urhonordo in the spring?"

Solem squinted down at the scroll near his feet.

"Also", Amanda continued, "It looks like Drumsworth Lake is larger on the farmers' map and goes all the way around Drumsworth Castle. If that is so, wouldn't that break a legal property line that has existed for years between Urhonordo, The Lowlands and that lake? That would allow Drumsworth Lake to hold much more necessary water away from The Lowlands than it does now and would put the lake on both Lowlander and Urhonordo land." The Lowlanders behind Amanda nodded vigorously.

"Tell the half breed to sit down and mind her own business," growled a loud voice from the nobles section. Willa didn't look around to see who said it. She knew. Solem frowned but didn't respond. (Note: Manda and Queen Luna were both part Lowlander...'half breeds')

Manda continued, unphased. "Does this proposed lake end in front of Drumsworth or does it also go *behind Drumsworth*? Because of these questions and others, we believe that we should all talk about this new road...*together...* before Your Highness considers the concept for its convenience and possible protection, only."

The throne room grew suddenly silent. Drum shook his head and threw up his hands, dismissing the presentation. He folded his arms and stood defiantly in the middle of the room. The noblemen shrugged their shoulders and grumbled. This was a pretty young girl, but she was just a young half-breed girl. What did she know?

Are these real problems that I should consider, Solem thought, a bit sheepishly. *They sound very real. I should probably take time to consider what the farmers and Urhonordo are saying. I should look more closely at both maps. The noblemen are expecting an immediate answer, though.* He looked again at the maps.

I don't understand a word of what Manda just said, Willa thought to herself. *Water goes where it wants to go, doesn't it? But Solem looks perplexed.*

"Water goes where thoughtful people send it." responded Cwilla, reading Willa's thoughts. "Manda is being thoughtful."

The beautiful scroll lay peacefully on the platform as though it somehow conveyed an acceptable answer to everyone's concerns. But quiet persisted. The nobles began to fidget.

This quiet was not one of peace. The Lowlanders had gathered themselves together, glancing suspiciously across the room at the River Kingdom groups... both of them. Prince Armando went to them with quick assurances. The King would listen to their concerns, he told them.

The farmers were tensely nodding to each other, looking at the Urhonordo Princess with appreciation. A*t least someone* understood what they had been saying.

The noblemen now moved to re-huddle around Drum and Thorndike as though they were getting ready to make their 'obviously superior and totally acceptable' plan more 'obviously superior and totally acceptable'. Their arms were waving and their voices were rising.

Now Queen Luna glided gracefully to where her daughter was still facing Solem. She raised both arms in one dramatic salute. "Hail, mighty King of River Kingdom," she declared with respect.

The room became silent again. "If I am not mistaken, we who love and revere The Great River have some very important questions that need to be answered by all of our people." Queen Luna nodded deliberately to the farmers, her daughter and then, with a focused nod and smile, to the noblemen.

"As with all matters that concern The Great River, these matters will need time and thought before they can be decided or implemented. They cannot be decided without study. This day has been long and I believe that the scent of a lovely dinner is wafting to us from The Castle dining room.

"Would you accept my request to 'table' a decision on this large and important matter before us in favor of Your Majesties *'dining table'*? In this way we can all review these maps more carefully together in the coming days and come to appropriate decisions."

The assorted members of The Gathering stopped their respective grumblings with what appeared to be relief. Little by little, they all nodded and began to look toward the great dining hall doorway.

King Solem knew that he was being rescued from a difficult and possibly volatile situation.

"This has always been the reason that you have liked this woman," whispered Csolem. "She has just given you some time to think about what has been proposed and what you might not have thought through. Take advantage of her gift and take charge."

"Thank you and your lovely daughter for your thoughtfulness," Solem said in a way that he hoped was appropriate to quell the unrest. "What an excellent suggestion. Let us all adjourn to the dining hall where we can discuss what we have heard today and schedule a time to meet and consider our respective points of view." He turned to Willa with a trace of relief in his eyes.

Does he see that I am completely in over my head and don't understand any of this situation? she wondered.

"My most gracious Queen," Salem continued clearly to Willa. "Will you be so kind as to join me in leading our guests into the Royal Dining room?" He bowed and put forth a guiding elbow. Protocol.

Willa breathed a sigh of relief and took his arm. *What just happened? Is everything alright? I thought for a minute that folks were going to come to blows,* she thought nervously.

"Manda was thinking today. You should think, too. This 'Royal River Road' issue is going to be more complicated than it looks," Cwilla said quietly.

MORNING REVELATIONS

Willa awoke the next morning with a start. Manda, dressed in boy's clothes and grinning, was sitting on the end of her bed!

Instinctively, Willa jumped and turned to the room next to hers, but her beloved Solem was not in his bed. Thank goodness.

"Wake up, sleepy head. We can't go exploring with you in your night clothes," chattered Manda as she began rummaging through the dresses that were hanging elegantly in Willa's closet. "You can't wear this fancy stuff where we're going. Haven't you got something less 'queenly' to wear?"

"Manda, what in the world are you doing in my room?" Willa whispered. "Are you crazy? Who let you in? And why are you here? No, I don't have 'less queenly' stuff to wear. I'm a queen now. I'm supposed to dress like a queen."

"Indeed, you *are* a Queen," snapped Agnes as she marched in with a fresh wash bowl and pitcher in her arms. "And who is this intruder?" she snarked playfully. "Oh, I see. It's the 'neighbor boy', Amanda, who has snuck in without an invitation, as usual. Hello, Amanda. You used to come all the time, but, after yesterday, I thought you had outgrown that."

Agnes and Manda gave each other a hug, providing clear notice of the fact that they had known each other well for a long time.

"What in the world?" Willa squeaked again.

"I should have warned you," quipped Agnes. "This girl looks pretty when she is all fancied up like yesterday, but she turns into a boy most of the time. She used to show up here and 'bother' Queen Linda and Mimi regularly before Queen Linda passed. I just didn't expect her to start up that visiting routine again with you."

"I'm not just visiting, Agnes," Manda shot back in a horse whisper. "I'm here as a teacher. Please help Willa find an outfit that she can use to ride in a wagon...or can you ride a horse, Willa?"

"Of course, I can ride a horse, Amanda. At home I used to ride often...but teach? You're *'teaching'* somebody something? Who? Why?"

"You, Willa...*I want to teach you.* I could tell yesterday that you had no idea what was going on with the maps and the water and all, so I decided to *show* you what the issues were...or at least some of them. What was suggested yesterday by Drum was potentially a really bad idea for some farmers...and, by the way, call me Manda. 'Amanda' sounds more like my father than me."

"Don't you turn the Queen against our King, Manda," advised Agnes sternly. "I know how you are. You always think people should look at things the way you do."

"If you're right about something, you should share the information, don't you think?" Manda retorted. "At least that's what my Conscience says."

"*My* Conscience and I were both just plain confused yesterday," sighed Willa. "That is certain. And you don't have to worry about me influencing my dear Solem, Agnes. He knows that I don't know anything about the water rights issues."

"That's why I'm here, Willa. I saw the way the King looked at you yesterday. You have won his love already. That means he will ask for your opinion about things so... you have got to develop one. You must learn some important stuff about The River."

Agnes laid out a dress of practical home-spun on Willa's still unmade bed. "Here you are, Queen Willa. It is neat but not fancy. Will it suit for a ride through the countryside...or will you tell Manda to go back to *her* side of The River?"

Willa laughed at her two new friends. She was happy to see them both. But their familiar presence suddenly made her aware of the contrast between her life now and how she had lived in her former life.

She had been so lonely. She had not had many friends, wandering down lonely hall, bothering servants or just reading. She had been alone in Thorndike Manor except for Flo, a servant girl her age... and her Conscience, of course...Flo was nice but not allowed to be chatty.

The atmosphere for a young girl at Thorndike had been isolating. Her mother had died when she was so small. Her father was not fond of social visitors unless they were male and noble. This was the solitary existence she was used to.

How different Solem's Castle was. Yesterday there was a 'Gathering'. Today, she already had two new friends. With enthusiasm, she pulled off her bed clothes. She hadn't been on outings since she was very young. She could hardly remember what it was like.

Willa brushed her still tangled hair from her face and grinned. "Where are Mimi and June? Aren't they going to go with us?" she asked as she pulled her soft garment over her head.

 Manda put her fingers to her lips. "Shh," she giggled. "Mimi is still asleep, of course. She always sleeps in. That's just the way she has always been. She doesn't come alive until noon. June is out with his father, already. They are both early morning people. Actually, I think they may be on their way to your father's house for 'River discussions'. I passed them at your gate on my way to you."

"Really? Solem is out and about already? He must think that I am a lazy sort. He didn't say anything to me before he left."

Manda smiled broadly. "I don't think that he thinks you're lazy, Willa. He probably just thinks that you are *female,* and you wouldn't want to go with him and June on business. That's the way your precious 'family men' think.

"Men do the important stuff and make the decisions in River Kingdom. Women are pretty and good to talk to sometimes, but men do all the interesting things." My mother has been chiding King Solem about that for years."

Willa pulled her boots up over her hose and snickered. "Why should I learn anything, then? If I am just going to be a female ornament to Solem, why do I need to know anything?"

"That's why I am here, Willa. You are going to be different than sweet Queen Linda was...that's why. My mother and I could see that yesterday. You have strong feelings about things and you speak out.

"Queen Linda was a woman of more traditional habits. She was supportive to her husband, and she was a lovely person, but she would never have disagreed with King Solem. She was too loyal."

"I am loyal, too," Willa responded indignantly. "Don't think I'm not."

Agnes stepped forward now and, surprisingly, added carefully to Manda's point of view. "There's a difference, My Queen, between 'loving loyalty' and *'thoughtful* loyalty with love.' You appear to be the thoughtful type, Willa, if I may be so bold as to say so. But I could also see that you were confused by the whole 'Royal River Road' situation yesterday."

"Did I look that stupid?" Willa groused.

 Manda laughed out loud and then realized she was making noise. "Not 'stupid', girl," she whispered. "Just 'uninformed'. I'm going to fix that. I believe you are a daughter of nobility, right? I doubt that *your* father has ever told you anything about water rights or asked for your opinion on anything. Am I right?"

"Indeed. I'm not even sure that he ever thought about me for any reason except that he hoped I would marry well. I think he was planning to auction me off and get something for me from some rich baron and now I'm married to a king. He thinks he arranged that. Can you imagine? I stay out of his way. I didn't really know him much while I was growing up, and I sure didn't like his friends."

"I noticed. I couldn't keep you in your seat," Agnes commented dryly.

Willa sighed. "I may not have looked completely stupid, but I felt that way. I have always known that Lord Drumsworth was a terrible man but then he presented that beautiful, scrolled map. I could see that the noblemen wanted to build a new road and their farmers didn't want them to do that.

"I couldn't figure out why anyone would object to such a lovely thing. I supposed that the farmers didn't like Lord Drumsworth any more than I did, but I thought the scroll was great. I heard *you* say that you weren't sure about the new road, though. Is that right, Manda?"

"That is definitely right," declared Manda with emphasis. "That's the main reason I am here. I *hated Drum's* sketch of the Royal River Road that we saw yesterday. It's just plain ignorant. I know that you don't see it that way, but I *do.* I want you to see what I see. I have friends that are farmers. That's why I want to show you things you don't know anything about yet."

She must think I'm stupid, Willa thought defensively. "You're just a girl like me, too," said Willa, her nose rising in the air. "Why do you think *you* know so much more than I know?"

Manda looked hurt.

"Because she has been riding all over both her kingdom and ours since she was a baby", Agnes contributed. "She has gone everywhere, even when she shouldn't, investigating both sides of The River.

 "Look at her. She has britches on. This is the way she always looks when her mother doesn't make her dress up. She has actually been to and *seen* that whole farming area of River Kingdom that the noblemen want to turn into a road."

Cwilla attempted to clarify. "She knows more because she's been where you haven't been, Willa. That's all. She's trying to be helpful. Don't get snippy. It's not becoming."

 Willa now looked at Manda and hung her head. "I'm sorry, Manda. I don't know why I said that. You're my new friend and I haven't had many friends in the past. Spending too much time alone loses one's perspective, I guess. Will you forgive me?"

"Of course, you will," advised Cmanda.

"Of course," answered Manda, finding an understanding smile. Now she was full of energy. "Let's go," she said briskly and headed out the door. "Are you ready?"

"Indeed," said Willa, pulling a bonnet over her barely combed hair.

What a pair, Agnes thought approvingly as she watched them disappear down the hall. *New friends learning together. This is a good thing.*

EDUCATION

The fall weather was still warm, but a stout breeze caused Manda to pull her curls up under a hat to keep them out of her eyes. She was taking the lead.

"You look like a boy with that hat and britches," Willa laughed.

"Queen Linda always used to say that" Manda called back. They were both on horseback as The Castle had prepared Willa a 'suitable steed' at Manda's request.

Manda had decided to travel at a slow and steady pace. *Willa is a noblewoman,* Manda reminded herself smugly. *She must be used to trained, gentle horseflesh on quiet, supervised rides. I must remember that and be very careful.*

And so it was, for a while. The two were sauntering lazily along when, without warning, Willa's horse leapt into a full gallop, sailing around a curve in the path ahead and out of sight.

Disaster! Manda thought with fear and guilt. She kicked the sides of her horse hard and raced to follow. *I should have kept hold of her reigns. How stupid am I? She's going to be killed and it will be my fault.*

"*I should have warned you,*" shouted Cmanda in Manda's ear. *I wasn't paying attention.*

In panic, Manda hurtled around a bend in the river path, following the clouds of dust Willa was leaving behind.

Abruptly, however, Manda came to an abrupt halt. Willa was waiting for her, sitting quietly on her horse at The River's edge. Her horse was taking the opportunity to drink. Both appeared to be unaware of the panic they had just caused in Manda.

As Manda and her winded steed skidded to a dusty halt, Willa looked up with a smile. "I thought I recognized this horse when I first saw her," she said with happy confidence. "I did. I have ridden her many times before. My father trains horses, you know. Thorndike Stable is the only part of our grounds that I really liked."

"I thought you were going to get thrown off and die," Manda confessed breathlessly. "I thought I had let this horse kill you. And I had no idea you could ride that well...*You* are a Thorndike?"

"I used to be. I had no idea you would be worried about me, though," Willa responded. "I *do* know a few things, even if I *don't* know about The River."

Manda nodded. They were learning about each other and themselves. "I had no idea that your father trained horses, either" she acknowledged. "I guess I *don't* know quite a few things about River Kingdom, even if I *do* know some things about The River."

Willa grinned. She felt better. They were sharing information. "That's okay. My father is all about his militia and horses are a part of that. He has prepared a cavalry just for his militia. He says they must be ready if they are ever asked to confront an enemy as they did long ago. He has been breeding war horses for that purpose for years."

"Really? War horses? Does King Solem know about your father's militia?"

"I'm sure he does. He purchased the horses for Queen Linda's carriage from my father before she died. I don't like the big war horses like those that he sold to the King, though. I like the fast, gentle ones like this one.

"This was one of the best he had in my opinion. I missed her when she was traded back for the big ones after the Queen's accident. Solem said he never wanted to see any of those horses again. I don't blame him. And now I get to ride my favorite again."

The girls guided their horses forward and continued down The River bank at a more sedate pace and in silence. Both held their own thoughts to themselves.

How lucky I am to be riding a horse that was my favorite horse when I was at home, mused Willa happily. *How lucky that she was traded for the big war horses...and to think that my father trained them all. In River Kingdom, all things are connected, I guess.*

Manda's thoughts were different. *Willa is riding a horse that her father trained. It was her favorite. Thorndike also trains big war horses for cavalry. 'Big horses' were pulling Queen Linda's carriage the day she was killed. I don't like Lord Thorndike after the way he acted at The Gathering but I really like Willa. I wonder if Willa has thought about the Thorndike horses and Queen Linda's accident...Probably not.*

Manda turned her mind to more pleasant things and looked around. They were at a place that would be her first chance to show her 'student' what she wanted to show her. She smiled and reigned in.

"Why are we stopping?" Willa asked. " This is only a little ditch with a little water standing in it. It isn't that wide. We can jump it easily. We won't even get wet "

"You're right," Manda agreed, "but this is the first thing I wanted to show you. Do you know what this is?"

"A drainage ditch," Willa stated with assurance. "You must think I am *really* ignorant."

"Nope. I *never* thought that. But you *are* only half right about the ditch. It *is* a draining ditch when the crop and land has too much water, but not now. Now it's an irrigation ditch, dug deep and wide, trying to pull a little more fall water *in* from The River for these last fall crops. That's what is really important about a ditch."

"I don't get it. How can a ditch do two things at the same time?"

"It can't. It changes. What it is used for depends on the time of year and what the crop nearby needs."

Willa scowled. "How does it know which is needed and when?"

"The farmers know," Manda said quietly. "When their crops need water, they dig deeper and let more river water in and store it. When they don't, they dig again and block the extra water from their crops letting it flow past."

"Really? The farmers know how to do all that? They must be smart. But that sounds like a lot of work to make the ditch work right."

"Indeed."

"Who does all that?"

"The farmers and landholders."

"The farmers, themselves?"

"Um hum."

"Why can't they get someone else to do it? My father would never do that, himself."

"Right...Noblemen don't farm, themselves, so they don't know how much work it is. But there is no one but their landholders and farmers to do what needs to be done. They can't hire too many field hands so their wives help. Their kids work, too, when they get old enough, but the farmers and their field hands are the main ones. Their built up, frequently changed levees are what furnishes their crops' water."

"Oh."

Manda's horse jumped the ditch and Willa's horse willingly followed. The horses ambled slowly along The River's edge, carefully stepping over additional ditches.

"I had no idea there were ditches like this along The River...and so many of them. This is what you wanted me to see, right?" said Willa with growing admiration. "Agnes knew that you knew about all this. You have definitely been here before."

Manda smiled. *She is as nice as I thought she would be. How did she get this nice with a father like that Thorndike guy from yesterday?* Manda wondered.

The young women chatted amiably, but Willa became aware of a pattern that was developing under their horses' feet. The farther south they traveled along The River's gentle borders, the more frequent the 'drainage ditches' became. Here and there, the ditches became much wider and logs had been lowered from edge to edge creating makeshift bridges.

"That was pretty smart of someone," Willa observed. "Who built these, do you think?"

"The farmers," Manda responded. "She stopped her horse and allowed her another drink. Willa followed, taking a seat on one of the nearby logs.

Willa was impressed. "I had no idea about all that the farmers do," she mused. "Thanks for teaching me about farmers and the ditches. I wonder if Solem has thought about ditches." She smiled and stretched her legs out daintily in front of her. "You thought I didn't respect the farmers enough yesterday, right? Well, you were right. I never really thought about what farmers do to grow the food that we eat. I just thought they were bad at making maps," she laughed.

Manda stood up straight and put her hands on her hips. With mock sternness, she grinned and playfully shook a finger in Willa's face. "You think I just brought you here to teach you respect for farmers? Maybe you're not as smart as I thought you were," she laughed.

Suddenly a gruff voice boomed out of the underbrush around them. "Is that impudent young ruffian giving you any trouble, Yer Majesty?"

From behind the brush stepped a familiar figure, but one with a very disapproving face. It was Hector, the farmer who had been the landowners' spokesman yesterday.

The girls stared. "Is that you, Hector?" asked Manda, now recognizing the man.

"Yeah. This is me. And who are you, you scamp? I saw you just now. You don't shake your finger in the face of the Queen, scoundrel! Get down on your knees and show some respect before I take you up to The King and have you thrown in the dungeon."

Willa snickered. "Hector thinks you're a boy, Manda. Take your hat off."

Manda did as she was told. Shaking out her long curls, she bowed low, first to Hector and then to Willa. "I am sorry, Hector," she said as seriously as she could. "The Queen and I are friends. I meant no disrespect."

Hector stared and slowly lowered the pitchfork which had been aimed squarely at Manda. He squinted at Manda and then at Willa. Now, his old eyes understood. "Manda, is that you? I haven't seen you since...since yesterday, I guess. I thought you had given up riding all over like a boy like you used to."

Now Hector got down on his knee and bowed low. "Your Majesty," he murmured, quickly converting from rescuer to humble servant.

Willa stood in response to the farmer's respect and returned it. "You honor me with your protective stance, sir," she said plainly. "But it is clearly not needed. Please rise. I should bow to *you*. Manda has just told me about all the work you farmers do. You are a farmer, right? *You* grow all my food and dig all these ditches? You work very hard. I should really tell the King."

The old man stood up carefully and smiled. "You honor me, Your Majesty, but I believe King Solem knows what I do."

"No, Hector. I don't think he really knows," corrected Manda abruptly. "He knows some things, but he doesn't really think about some other things. That's why I brought Willa...I mean, the Queen, here today. I brought her here to see your levees and how they work. The King knows *you* work hard, but he might not know *how hard the levees work* or what the Royal River Road might do to wreck them."

"Well, I'll be darned," Hector said, shaking his head. "I knew you was mad yesterday when you spoke up at The Gathering, but I thought you was just mad about the lake getting bigger and maybe flooding Urhonordo. I didn't know you was standin' up for us farmers' levees, too."

"It all goes together, Hector," Manda said, nodding. "You know that."

"I *do* know that, Manda," Hector responded.

Willa watched and now formed new thoughts that went with what she was watching. Manda and Hector were friends and had known each other since Manda was only a young explorer. They knew each other's problems. They respected each other.

"These two are royalty and farmer friends who *talk together*", Cwilla offered. She had been silent all day. "They speak as equals. You and I have never seen such friendship at Thorndike Manor, have we? These people are very different, but they like each other. Do you think your father has any farmer friends?" *Never,* thought Willa.

The two young women sat down again on their logs. Hector, still in awe of the fact that he was here with a grown princess and now a queen, shuffled self-consciously. "Is there anything else I can do for you ladies?" he asked.

Manda was quick. "Tell the Queen here what you think about the Royal River Road, Hector."

"Really? You want me to say what I really think? I ain't the polite sort, Manda. You know that."

"Go ahead, Hector. That's why we came here. I wanted to let Queen Willa know where the Royal Road would be if we had one...and what all your levees and drainage ditches would do if that happened. Tell her what you think about that."

Hector stuck his pitchfork into the ground for emphasis. "If that Royal Road goes between my land and The River, it'll flatten and plug up all my levees, Yer Majesty. Then I can't get to the water. That'll kill my crops. That's what I think!" he announced fiercely.

"My crops will dry up and die and me and my family will be forced off the land by the noble that really owns all the land around here."

I don't think my father knows about farmers or their levees, Willa thought. *I have never heard him mention them, even though this is where our food was always growing.*

Hector continued. "The nobles don't know about no levees and drainage ditches, so they don't care. When they don't get no product or taxes off their land, they'll get some other fool in here to try to grow somethin' but *that guy* won't be able to do it either. Then the land will be dead and everybody will know that the Royal River Road was the wrong thing to do, but by that time, me and my family'll be dead, too. That's *exactly* what I think and that's exactly what *I know*!" Tears glistened in his eyes. Hector snorted and wiped his nose on his sleeve.

"Thank you, Hector. That's *exactly* what I needed the Queen to hear, today." Manda said simply. "The huge Royal River Road levee will fill in and flatten out all the ditches that we have ridden over and crop water will just dry up. It will be awful for all farming."

Willa was still sitting on a log which was over a water-filled ditch. "Won't a big ditch like this water the crops enough, if they leave it?" she asked hopefully.

Hector raised his eyebrows. *She's actually listening. She's thinking, too,* he thought in amazement.

Hector just shook his head. "Yes, but not enough," he sighed. "A big ditch like this will only take care of the land right next to it. The rest of the land will go dry. We've worked on this system for years. We change it every year...a little more here, a holding pond there... That's what makes it work. The noblemen don't get it. They have never worked the land. Rich nobles just complain about The King, eat the food we grow for them and collect the taxes." Hector shuffled his feet some more. He was sure that he had said too much.

Manda looked closely at Willa. *Does she understand? Have Hector and I done our job?*

"I think Hector may have been a better teacher than you were," snarked Manda's Conscience. "He didn't take all day to get right to the point."

"Thank you for explaining the issue so completely," Manda said to Hector gratefully. "I couldn't have said it better than you did."

"Yes, thank you, Sir Hector. I believe I understand the problem much better, now that I have seen it," said Willa with a smile. "I will be sure that the King knows about your crops and what the Royal River Road would do to them if it is built without thought to your irrigation."

Horace blushed and smiled gratefully. "Me and my family thank you, too, Manda, but you better take Her Majesty home soon," Horace suggested quietly. "Yer almost to The Bridge and this is that time of the year when we lose daylight in the afternoon. You don't want to be lookin' at drainage ditches and levees in the dark."

THE RETURN

Looking downstream from their current spot, the girls saw that Hector was right. The Bridge between Urhonordo and River Kingdom was only a few yards up ahead. They had talked and ridden all day and traveled farther than they thought. The sunlight was beginning to fade.

It was time to go back to the Castle, but the two girls now discovered something else. There were two castles that they should consider. The two were about the same distance from both River Kingdom and Urhonordo Castles.

Manda hadn't realized that they had gone this far. Her Conscience began to comment impatiently. "Nice planning, Manda. You were so intent on showing how much you knew about the farmers' problems that you went too far. Now it's late."

Willa looked at The Bridge and laughed. "It looks like we need to split up, Manda," she announced, casually. "It would be pretty useless for you to go all the way back in my direction and then travel in the dark back to yours."

"Indeed," responded Manda, thinking aloud. "It *will* be a little dangerous at night, but...."

"That's a pretty good idea, Yer Majesty," interrupted Hector hesitantly. "I was watchin' you fer a while this afternoon before I spoke up. Looks like you can ride pretty good...for a girl. Almost as good as Manda here. You could both make it home in jig time if you leave now and push it a bit."

"Is he crazy?" Cmanda shouted in Manda's ear. "You can't let the new Queen go back all the way to her castle *by herself*. *You* talked her into this. What if something happens to her? It would be your fault. *You* have made this mess for yourself. *You* have to fix it."

Willa's Conscience now became involved, as well, however. "If you let Manda take you home, you will be putting *her* in danger, for sure. She's your new friend. Don't let her even consider that. You ride as well as she does and you can both get back to your castles easily by yourselves. Just get going, before she decides she has to go with you."

Willa grabbed the reigns of her horse and was on it in a moment. "I'm on my way, Manda. Don't follow me. That would be silly. I'll send you a message tomorrow. Thanks for the education. I needed it. Bye, Hector, and thanks." She turned her horse north and was gone, plumes of dust billowing behind her.

"Our new little Queen has spunk. I sure have to give her that," Hector said with admiration.

Manda stared at the dust cloud in amazement. Willa was out of sight at a full gallop.

"Yer not goin' after her, are ya?" Hector questioned, seeing Manda's look of concern. "She's good on a horse. You can see that. Don't worry 'bout her. She'll be fine. She's spunky, like you."

"But she's *the Queen* and you aren't taking care of her," Cmanda reminded Manda.

I know, Manda thought miserably. *If anything happens to her...*

"She'll be fine," Hector repeated, still watching Manda's worried face. "Don't worry, girl. Just go home yerself before the sun goes down all the way." He smiled confidently and shuffled off toward his own home.

Willa was enjoying herself, riding fast and free. She hadn't felt so worry-free since she was small. She loved this horse. She was wonderful. They remembered each other. They were two young animals without a care in the world.

Willa's mind began running free as well. *Manda is going to be a good friend... maybe a friend for life. And I have learned so much,* she thought, sailing along The River's edge. *This has been an outing that I will always remember. I need to tell Solem everything I learned.*

She was still riding at a gallop and was beginning to see the outer gates of the Castle when it happened. She heard hoofbeats behind her. They seemed to be coming fast. Glancing behind her, she spotted a rider through her clouds of dust. Someone was following her!

Willa was, for better or worse, her father's daughter. He was a cautious man and she knew to be cautious, as well. Someone was behind her, and it was not Manda! She could now see that it was a man...no, several men. This was not good.

In the few times that Willa's father had spoken to her directly, he had impressed her with fears of 'others' and these early warnings had taken root. Willa dug her heels into the sides of her horse and raced ahead at an even faster pace. She was close to home. She looked back at her pursuers. She could out-run them!

Without warning, Willa found herself slapped hard across the chest as a low-hanging branch in her path caught her and hoisted her into the air! The blow left her without breath. A sharp pain stabbed her from arm pit to arm pit and she was briefly aware that the branch was digging into her flesh as her head landed hard on the rocks below. Daylight faded from view. She was unconscious.

It was only for a moment or two, but Willa's fearful brain pulled itself back to consciousness and she struggled, aware now of two male faces staring down at her. "Get away from me," she tried to order. Only a wheezing whisper escaped and the faces did not pull back.

"Thank goodness... She's hit her head, but she's alive. Let me..." began one.

"No. Don't touch her. Maybe she has broken bones. Don't touch her," said the other curtly. Then another voice came closer. "Willa...Can you hear me?" the voice asked softly.

"Solem?" Willa blinked her eyes, trying to clear her vision and her understanding of what was happening.

"I will take The Guard back with me and we will find the scoundrels that were chasing her," said the first voice. "They will not escape! But be careful. She's had a terrible fall."

"Thank you, son," said the second voice. Indeed, the voices belonged to Solem and June.

"Wait...Willa struggled to sit but gave up as pain made her wince. "Wait. I don't think anyone was following me but you." She slurred weakly.

Solem sat down in the dirt, carefully cradling Willa's painful shoulders in his arms.

My sweet husband is sitting in the dirt with me, she thought briefly as she again drifted off.

June heard nothing from Willa. He was already on a mission. Grabbing the reigns of his panting steed, he raced back down the trail on which he had charged in pursuit only a few minutes earlier.

The Royal Guard that was always with the King followed the dashing Prince. With June in determined lead, they relentlessly plowed through the foliage on either side of the trail, looking for broken branches and evidence of the evil doers June was sure had been chasing the galloping Willa.

Prince June's intensity surprised The Guard. "Why is he so frantic? The Queen is not *his* wife," they commented to each other.

It was getting dark. Gradually The Guard determined that whoever was chasing the Queen had disappeared into a nearby woods and would not be found tonight.

"We will return here tomorrow and find them," Prince June declared with determination. Then he galloped back to where his father still sat on the dirt in a very un-kingly pose, holding a crumpled but weakly smiling Willa.

"I am so glad to see you, June dear," Willa managed to squeak out. "There is no need to go looking for someone chasing me. I was not being chased."

"What?" June gasped, climbing down to hear her and looking at Willa carefully. "You were galloping like fury, Willa. If you were not being chased, why were you flying down the path? Were you afraid of something? Was your horse running away with you? I was sure...I mean, we were sure that you were being chased by someone."

"But she wasn't." Solem said quietly. "Would you believe that Your Queen was just 'having fun,' heading for home on a galloping horse? Would you believe that she was *enjoying* herself before *we* started chasing her?

"*We* were the ones that caused her to look back and not watch out for that tree branch. *We* were the ones that caused her to fall." Solem smiled ruefully and pulled Willa closer.

June scowled and stared at Willa. Then he simply shook his head and walked off.

The King really loves his new Queen, thought many of The Royal Guard who were witnessing the unlikely scene.

So does Prince June, thought a few of the others.

The Call of the Auras

It was morning. Cushioned in her soft bedding, Willa was comfortable while she wasn't moving. That changed when she reached for her covers to get up. Her shoulders didn't seem to want to do anything, and her head throbbed. Willa squeaked faintly.

"She's awake, Sir," Agnes announced quietly.

Solem was up instantly from his chair at Willa's bedside where he had been keeping vigil through the night. "Willa...can you hear me? Do you remember what happened? How do you feel?" he burbled in a rush of pent-up concern.

Tonight was a unique performance for a king, Agnes thought to herself. *He surely does love Willa. No one can doubt that.*

Agnes had cared for two queens now and both had been the lucky wives of a man who loved them. Agnes felt that she had been a part of that in a way as she had dealt with small issues between Solem and his first wife in a somewhat motherly fashion for years.

Among her ladies-in-waiting duties for the Queen, she had become an observer and unofficial assistant to all *her* Royals among the many other duties that she had assumed as needed. Agnes had to admit that she was proud to work for this man who seemed to care for the women in his life, even though he was a bit chauvinistic.

"Now, if you can convince him that women are smart enough to be consulted about important matters, Cagnes added with a snort. "Manda was trying to help Willa with that. I wonder what happened yesterday."

Carefully, Agnes replaced the cool cloth on Willa's forehead. "I think she's going to be fine, Your Majesty. You can leave her now. I will take care of her, don't worry."

Solem did not leave but sat instead on Willa's bed and took her hand. A black eye was visible under the cloth, but Willa managed to smile. "I must be a terrible mess," she told her caretakers.

"You look beautiful, my queen." King Sol choked slightly and looked down.

Agnes heard again from her Conscience. "While you are busy wondering about yesterday, you should take time to recall that the King just lost his first wife to a terrible accident *with horses*. And he found her dead, right there on that same road. He must have been thinking about that and worrying about it all night."

*Of course. How could I have **not** thought of that? Poor Solem.* Agnes shook her head. "I will leave you with the Queen, Your Majesty. I will be just outside the room if you need me."

She quickly pulled together the nursing materials that she had used during the night and hustled into the hall, only to run squarely into the path of a sleeping June.

Agnes felt no shyness in speaking directly to a young man that she had helped to raise. "What in the world are you doing outside the Royal Chambers so early, June?"

"I was just waiting to see how Willa's doing, Agnes. I mean, we brought her in but then...nothing. Nobody said anything. My father went in and didn't come out...I had no idea...I mean, I didn't know if..."

"June lost his mother recently, too, Agnes," reminded Cagnes quickly.

"I understand, June. You lost your mother in an accident, so you are worried now about Willa. I should have come out to say something to you before now. I'm so sorry, dear."

"Well?" June's look was intense. "She hit her head, I think. She was knocked off her horse. She hit a tree branch, for heaven's sake. How is she?"

"She was hurt but not so severely as to endanger her life, I believe...

"How do you know? Has she said anything? We were just trying to catch up with her and... Was she..."

Agnes looked closely at June. His concern was too intense...

"She will be better soon, June. She is not going to die, I can assure you. When the King leaves her room, he will tell you that, too. Try not to worry, dear."

June stood back and looked at Agnes closely. Then he turned and walked quickly away.

June loves Willa just as he loved his mother, she thought tenderly. *Like father, like son.*

"Indeed", observed Cagnes. "June appears to love Willa *exactly* like his father."

June marched down the stairway that led from the Royal Private Quarters to the entrance hall. The door guards were in position and other servants were milling about, accomplishing their daily chores.

Gratefully, June observed that all seemed completely unaware of the trauma that had occurred the night before or that the royal patient lying in her bed upstairs had a problem.

Keeping this information quiet had been deliberate as only The Royal Guard had assisted the King and himself with the transport of Willa to her room through a back entry.

June made his way to the empty conference room and closed the door behind him. He lowered himself into a chair and leaned back with a sigh.

"She is going to be alright," his Conscience contributed helpfully. "She isn't going to die. She is young and healthy. She will probably be up and about tomorrow."

I know, June thought. *My father and Agnes are caring for her. I just wish I could see her and talk to her, myself.*

"*Do* you, now?" questioned Cjune. "Since Agnes told you that she will be alright, why do *you* need to see her *yourself?*"

I don't know, June thought. *I don't know why but I only want to be up in Willa's room to see her in person.*

"Oh. Let me see... Are you still worried that she will die?"

No. That's not it. I just feel like I need to be with her right now.

"I see...I see...Hmm..."

I'll get over it," grumbled June. "*I am not worried any more. Agnes told me that she would be alright. And Father is right there with her...*

Cjune watched June for a minute before he said something more. Then he carefully whispered an idea. "You wish that *you* were there with Willa *instead* of your father, don't you, June?

"Of course not," June said aloud. "She's my *mother* now." His voice echoed in the empty room.

"You really need to keep that in mind, June," Cjune said sternly.

A loud noise penetrated June's thought from outside.

Someone had arrived. June rose to see who it was, happy for a diversion in his thinking. He opened the conference room door and peered into the hall.

"There you are, Prince Solem. You are just the man that I want to see. Where is your father? I have come to have a heart-to-heart conversation with both of you. It's time for us to all think about our futures." It was Drum, of all people.

June took a step back. Drum was there, indeed, complete with a small entourage of noblemen and his daughter, Balynn. The group was gathered impressively in the entryway. Confident in their acceptance, they milled about quietly, awaiting a formal request to come in and be seated.

June shook his head in aggravation. *What are they doing here? Were they invited? Did father know they were coming?... What should I do?*

"It's time to re-focus and act like a man", contributed his Conscience. "Someday you will be King. Start training for the position and acting like it right now."

June straightened his rumpled tunic and stepped forward with what he hoped was a look of confidence. "Lord, Drumsworth. It is so nice to see you and your lovely daughter at our home. And you have brought others with you as well. Welcome, Noble Sirs. Please step with me into the conference room and tell me why you are here."

June motioned quickly to a few of the servants as he escorted the group into the quiet conference room he had just vacated. Quickly he asked one of the servants to summon his father and bring refreshments...while he tried to figure out what these people wanted.

The ponderous group of older, wealthy gentlemen moved as a small herd into the conference room, finding chairs for themselves along the heavy table that was always in position for discussing important matters.

All found appropriate positions except Balynn who simply stood alone in the doorway to the conference room and waited. She smiled at him invitingly.

There is no doubt that she is pretty, June observed. *Not as pretty as Willa, but...*

"Why are you comparing Balynn to Willa, June?" hissed June's Conscience. "What are her father and the nobles here for? That's what is important now...*Ask.*"

June smiled back at Balynn and focused. "Balynn, my friend. I am so pleased to see you. We didn't even get a chance to speak in private at The Gathering. But why have you come here today? Is there a problem of some kind?"

June didn't need to wait for an answer. Drum was not yet seated, although he had claimed a position at the head of the table, the usual position of importance. Now he hustled over to join June and his daughter, exuding a fatherly warmness that Balynn had never noticed before. With a friendly pat on June's shoulder, Drum made an announcement that stopped her cold.

"I brought Balynn here for *you*," he stated to June with a flourish. "She's beautiful, is she not? I know that you have observed that to be true as everyone has. You two have never really had a good chance to get acquainted. I thought, as long as

I was on my way back up here to talk to your father, I would bring Balynn here to meet with *you*. Pretty good idea, right?"

June's Conscience spoke up instantly. "How crude. He doesn't waste time does he? *You* are only important as a possible mate for his daughter. Be careful, June. Don't say what you're thinking. Be smooth."

June swallowed and blinked. He wasn't sure what he was thinking.

Balynn smiled sweetly but rolled her eyes. Her father's little speech was embarrassing, but to the point. He was always like that. He said what he wanted to say when he wanted to say it and seldom thought of others. She was used to it.

Balynn had a Conscience, although she had not spoken with Cbalynn often or recently. In her isolation, there had been no need. They were not in practice. But now Balynn and her Conscience both focused. Neither one was stupid and what her father said should have been expected.

What should a girl say when her father is so pushy? Balynn asked herself.

"Your father is only thinking of himself, as usual," whispered Cbalynn. "That's not your fault, but he *has* made Prince June as uncomfortable as you are, Balynn. See if you can help the Prince out or you may never have a chance to talk to him, again."

Balynn smiled at June and shook her head. "My father just gets these strange ideas in his head," she said, shrugging her shoulders. "I am actually here because *I* am as interested in the Royal River Road as is my father. I think it will be lovely, don't you? I hear that you and your father talked to Lord Thorndike yesterday and suggested that *my* father talk to *your* father, as well, so here we are. By the way, where *is* your father?"

June took a deep breath a smiled at Balynn gratefully. "I sent a servant upstairs to let him know that you are all here, but it appears that he didn't get the message. If you will excuse me, Mistress Drum, I will run up to his room to make sure he knows you are here. Please seat yourself in comfort and I will return to your most gracious presence shortly."

June bowed and escaped up the stairs. "Nice guy," Drum murmured to Balynn as he sauntered toward his seat. "He might make a good partner for you someday. I think he likes you."

June's escape came to an abrupt halt as he met his weary father in the upstairs hall. "She's doing well, June," the King announced, "but she's sleeping. Agnes is with her. Did I hear that something is going on downstairs? Is somebody here?"

"Yes, Father. A group of noblemen are downstairs. Drum even brought his daughter. They're all waiting for you in the conference room."

"You said 'noblemen.' Thorndike isn't there is he? I just saw him yesterday. He would have no reason to be here again. What do *these* men want?"

"Balynn Drumsworth said that Thorndike messaged Drum and the others after he talked to you yesterday and he said that *you* wanted to talk to all of them."

"I did say that, but not immediately, for heaven's sake...and not all of them at once. I prefer to *invite* people to my home, not have them show up in a group unannounced. Don't they have any understanding of proper protocol? These people are too pushy."

That was exactly what June was thinking. He looked more closely at his father. It was easy to tell that the night had been hard on him. His face was creased with worry and his clothes were still covered with the dirt from the road. In fact, June realized, he was nearly in the same condition himself.

June could still hear what his Conscience had said a few minutes ago... "Be a man."

"Don't worry about our visitors," June announced confidently. "I'll handle this for you." Solem was too tired to question. He retired gratefully back to his chambers and Willa as June returned down the stairs.

"You said the right thing to your father," commented Cjune. "I'm proud of you. Now, what are you going to do?"

As June re-entered the conference room, the nobles rose from their seats and bowed. This was startling. He had never been treated with royal deference before. His father had always been present, and June had been of little consequence when his father, The King, was in the room.

"Don't let it go to your head," counseled Cjune. "It's just a reminder that you're growing up. Now act like it."

"Thank you, gentlemen," June began. "I appreciate your presence, but I regret to inform you that my father had no prior understanding of your arrival today and so he is otherwise occupied, I'm afraid."

"Thorndike sent us a note," stated Drum from the position of leadership he had chosen for himself. "My people and I journeyed all the way up here from Drumsworth for this? Who does the King think he is, not speaking with us, when we're all here?"

"I believe he thinks he's The King," June said bluntly.

The Call of the Auras

A giggle emerged from a corner. *That was Balynn,* June thought with an inward chuckle of his own.

June now looked around the room. The group as a whole turned toward Drum. June could tell that Drum was the one that had pushed this meeting for today. They were displaying sheep-like obedience to Drum that was similar to that they had displayed at The Gathering.

They were herded here by Drum, June thought. *He is the one that wanted them here today. They are just trying to please Drum.*

"I am most sorry that my father's schedule today cannot accommodate you, Lord Drum," June continued, "but, as a busy man yourself, I know you understand. If you would allow us to reschedule this meeting with you and the gracious Balynn, River Kingdom Castle will be most pleased to host you. Your contributions to the discussion of the Royal River Road are invaluable, I am sure.

"If these distinguished gentlemen here could possibly allow you to speak *for* them, however, that would save them the time and effort of returning here. What do you say to this thought, Lord Drum?"

The assembled noblemen nudged each other and nodded. They were completely content to let Drum speak for them. They were fans of Drum, not the King. They didn't really understand this whole 'road thing', anyway.

Sheep, June thought.

Drum looked at June suspiciously. He didn't trust this young man. Drum also realized that he had just been 'dismissed' by a Royal that he thought of as an 'underling' after he had driven here with big plans that he and his Aura were ready to put in place. This road was only the start of his even bigger plan.

He and the Auras thought quickly. *Thorndike has pulled together these local noblemen, himself. They are not my problem. I can blame Thorndike for them. If I spend the night at Thorndike's and strategize ...maybe even sharpen him up, I can bring him back to support me. Balynn can stay overnight at Thorndike's as well. She'll do whatever I say.*

Drum decided that he could afford to be gracious, even if this young man who was telling him what to do was only a prince. *I must be here to state my case. I will always be the best spokesperson for my plan,* he concluded. *I can wait. The Auras agreed.*

"This is such an excellent idea, Prince June", Drum responded politely. "Actually, I knew that we should not have arrived un-announced as we did but *Thorndike insisted.* He is very anxious to discuss the new road, as you know. He gets pushy, sometimes. I will talk to him. I, of course, knew that we must not move too quickly on something this important to the Kingdom. Shall we return tomorrow?"

Balynn stood and came forward. *My father is pushing this off on Thorndike,* she thought with chagrin. *He is lying right here in front of everyone...and in front of Prince June. Do the others know that he is lying? My father is the one that has been pushing for this meeting. He's still pushing. Does June know? I know."* *June must sense it,* Balynn thought. *And June looks nervous and tired. This is so embarrassing.*

"Try just being nice," offered Cbalynn without hesitation. This was more action than Balynn's Conscience had ever had. "Don't worry about your father. You can be honest, yourself, and help the Prince out. Think about it."

"This is an uncomfortable situation, Prince June, but your decision is a good one," Balynn offered politely. "It would be an honor to be invited back, but only at your convenience. Did the King give you an idea as to when another time for a visit might be available?"

June breathed a sigh of relief. He was tired and Balynn had just provided him a convenient out. "I believe that my father had time available day after tomorrow", he responded without hesitation.

"You should have asked your father first," Cjune suggested.

I'm too tired to go up there, June thought.

"That will be acceptable, Sir," Drum stated quickly. "Come, Balynn. Come gentlemen. We are wasting the Prince's time here. Thank you all for coming to The Castle on such short notice. I will try to handle things from here. My messengers will keep you informed."

The northern noblemen nodded and hustled out as though they had all made this decision themselves. Quickly, they found their way to their waiting carriages outside. They knew how this would go. They had made an appearance. They had shown their commitment to the new road. Drum was their man, and he would achieve the approval that they (and Drum) wanted. They were more than satisfied.

Drum and Balynn were the last to leave. Drum pulled himself into his carriage with dignity, smiling and signaling to those watching that this was a good solution for today. Balynn turned back and waved to June who was still watching from The Castle doorway.

"You scored with the Prince. He really likes you," chuckled Drum with satisfaction.

The Call of the Auras

The noblemen were being driven back to their manors by their servants, content that the ideas they had intended to discuss with the King would be handled by Drum soon. They were not all that familiar with the details, anyway.

That was because the most important details of why he was pushing for the road were only in Drum's mind. Today's delay was an inconvenience, but perhaps a useful opportunity for Drum to coach Thorndike. Of course, there were important parts of The Plan that Thorndike didn't need to know, but now was not the time to discuss them. It was too complex for him. When it was time, Drum knew Thorndike would fall in line.

Thorndike Manor was not too far from The Castle. Thorndike had declared himself and his militia to be loyal subjects of The Crown, but he and the nobles around him knew that that declaration was just for show. He was a Drum man.

Drum had convinced the nobles felt that they were looked down upon by Royalty, and these Birthers knew the reason well. There were too many Lowlanders in River Kingdom and Drum was going to help the nobles get rid of this problem and bring it under control.

As the Drumsmen drove Drum and Balynn toward Thorndike Manor, the young Aura force riding with them requested clarity from their elders.

The senior Auras were happy to oblige with a history. "One hundred years ago." they began, "a Drum was King for a while, and he was great."

"He was selfish and ego-centric by nature. Auras didn't need to teach him a thing. He threw big self-promoting rallies and spent the kingdom's money on anything that he and his noble friends wanted. The nobility of the time did what they wanted, pushed all the common people around and bullied the Lowlanders."

"Sounds great!" chirped the young Auras.

"Indeed! Lowlanders were put in their place, sometimes even in prison to keep them out of the way."

"The Year of King Drum' was the best year in River Kingdom history!"

"Thorndike was even going to invade Urhonordo for King Drum and he was ready for it."

"Why didn't he? That would have been great!" the young Auras asked breathlessly.

"Because King Sol returned as king and wouldn't allow Thorndike to go through with it. That move is still at the core of Thorndike's dislike of King Solem to this day. He still thinks that Urhonordo Kingdom should have been his then…and might be soon if The Plan goes as Thorndike wishes."

"This Drum has brought all those old stories back to life with the nobility of today. Drum is a natural-born Aura. This Drum works on his plan all the time, and it is a good one."

The young Auras were fascinated. They were so glad that they had decided to be part of an exciting campaign. But they had one more question. "This Drum guy talks about his 'plan' all the time, but what is his plan? What is his goal for now?"

"It's simple," said the senior Auras. "He just wants to be King like his grandfather was, only this time, he wants to be king from the north to the south of River Kingdom…a beautiful Kingdom with *two castles* and a lovely new road for Birthers in between."

AT THORNDIKE'S

The Call of the Auras

As Drum's carriage neared Thorndike Manor tonight, Drum reveled in his position. Wealth and his growing importance with nobility was what his current plan had brought him so far, but it was not enough.

With dignity, he stepped down from his carriage and strode to the entryway of Thorndike Manor without hesitation. Balynn waited to be escorted by her father in vain. Drum was on his way back to see Thorndike and was no longer thinking about her.

I wonder if he even remembers that I am here, she pondered as she climbed to the ground and adjusted the elegant skirts she had worn to impress June. *My father would leave me here all night if I didn't say anything.*

Drum's Conscience was present, however, and watched his behavior. *At least he might respond to a reminder,* Cdrum thought sadly to himself. "Oh, Drum...have you forgotten your daughter?" he murmured in Drum's ear.

It's my stupid Conscience, Drum thought to himself. Hasn't he learned to shut up? *What does he think he is doing here?* But Drum turned.

"Hey, daughter... I am on my way to speak with Thorndike." He waved an arm in a flippant command to a servant nearby. 'Take care of my girl, will you, girl?" he shouted as he sauntered through the doorway.

The servant rolled her eyes and went to help Balynn. This servant girl was a regular in the Thorndike household. She knew Drum and remembered the lack of respect he had always shown toward women.

His daughter looks nice, though, she thought to herself. The young servant seldom saw visiting women anymore. This was a treat.

Balynn and the servant smiled at each other and entered the grand entryway of Thorndike manor. Balynn had never been here before and now she was shocked. The manor was beautiful from the outside, but the grand entry hall was not a normal reception area. It was filled with the glowering suits of armor, standing empty but at complete attention and lining the walls. Each suit of armor had a shield bearing the Thorndike crest and a matching sword. Like a phantom army, they looked ready for battle.

"Very impressive, wouldn't you say?" asked Lord Thorndike as he descended the stairway to the entry hall. "I just set them all up."

Drum stepped forward, accepting the greeting with nonchalance and addressed the man in front of him without preamble.

"Hello there, Thorndike...Yes...impressive...very impressive. I've seen them before. Where can we sit and talk in private?"

"What are you doing here so early, Drum?" asked Lord Thorndike. " I didn't expect you back until this afternoon or evening."

He expected us but not until later, Balynn groused to herself. *My father never said anything about staying the night. I didn't pack for an overnight. And this is Queen Willa's former home...*

'Are you still surprised at your father's lack of consideration?" asked Cbalynn.

"Come upstairs with me, Mistress Drumsworth. I will show you to a room," whispered the servant girl with an understanding smile.

Drum scowled at Thorndike and shrugged. "The King didn't have *us* 'on his schedule'," he growled. "I sent the nobles home and came directly here. I trust you can accommodate an early arrival...oh, and an extra night, as well. We are not 'on the schedule' until day after tomorrow so we'll be here for two nights."

Thorndike rolled his eyes. "Typical," he snarled. "The King shows no respect for worthy nobility. None. That man deserves what happens to him. Come with me to my conference room. I still have your maps laid out. We can review them again. I also have a few new questions." The two men disappeared together into the conference room, pulling the heavy doors of the room tight behind them.

"I wonder what they are discussing and planning exactly," murmured Balynn quietly as she followed the servant up the stairs.

"Nothing good," sighed the servant girl.

Balynn blinked. "Oh dear," she sighed. "I had no idea I was speaking out loud."

"No problem," smiled the young woman in a shy manner as they reached a bedroom door. "I shouldn't have said anything, either. It was not my place. Here is your room. I hope you will find it comfortable. My name is Florence. If you want anything, just ring the bell on the table. I service this room."

"You are very nice, Florence. Thank you. I appreciate it. I am stuck here because of my father, but I have no idea what is happening downstairs. "Do you?"

"Not really, but I fear it won't be peaceful," answered Florence as though talking to herself. "The master wouldn't have all this knight stuff out in the hall if he doesn't intend to use it. He never does anything without a reason. He's up to something big, I think."

"Really?" gasped Balynn. "What kind of something?"

"Oh dear. There I go talking out of turn again. I'm going to get in trouble."

Florence stared closely at Balynn. "You don't know any more than I do, do you?" she asked. "I'm surprised. Your father has been coming here alone for years. I thought *you* would know about what is in 'The Plan' from his side at least."

"My father doesn't consult with females," Balynn admitted.

"That's just like my master. Men have value. Women have children---or *don't*. That's why my poor mistress is out in the east wing by herself. Oh...I shouldn't have said that, either."

"What? What shouldn't you have said?"

"Oh, nothing."

Balynn looked at Florence closely. At Drumsworth, all the servants responded to Balynn's directions without question. "What? she repeated, trying to look forceful.

Florence looked at the floor. "I don't think it matters any more. It's been too long... I don't guess anyone cares, but me, either. And no one will believe *her*, anyway. She's just another female." She was thinking out loud.

Florence stopped murmuring and smiled, starting to leave. Then, responding to a sudden urge, she turned and put her finger to her lips. "Shh. Don't tell anyone... Would you like to visit with the only other female in this manor? *She* might like that. *She* never gets any company."

"I would, indeed," Balynn answered eagerly. This sounded intriguing. "I will be here all day tomorrow with nothing to do."

Florence looked at Balynn again. *Should I or shouldn't I?* she wondered anxiously. She was always alone in the manor and now she had someone to talk to. A young female close to her age was right there in front of her. This was new and different... and an unexpected opportunity.

"I will see you tomorrow, then," she said with a conspiratorial gleam in her eye. "There are extra clothes for you to wear tomorrow in this closet... since you brought none with you. And it's cold. Be sure to make use of them. And Mistress... Please call me Flo. My mother and Willa always did." She disappeared down the stairway.

"I am proud of you," whispered Cbalynn. "You were nice to that servant, much nicer than your father would have been."

"Aren't I always nice?" Balynn retorted.

"No."

BACK AT THE CASTLE

Willa sat up in bed stiffly. There was a painful bruise across her chest and her head ached. "I guess I must have fallen and hit my head," she announced to Agnes. "I must look a fright. Are my eyes black? Let me see a mirror, Agnes. If I look terrible, I don't want to see Solem."

"You look fine, my Queen. Your eyes look like a racoon, but don't worry. Solem has already seen you. He was here all night."

"What? What did you say, Agnes? Speak up a bit."

Agnes laughed and came to stand in front of her damaged charge. "I said...the King has already seen you. He was here all night."

Willa looked closely at Agnes and then brought her hands up to her ears. Agnes continued to laugh and chatter, obviously happy that Willa wasn't hurt any more... but Willa couldn't hear her at all.

"I can't hear you, Agnes," Willa shouted, listening for her own voice. To her, even her own voice sounded muffled.

"No need to shout", warned Cwilla. "Agnes is right here with you."

"But I can't hear," murmured Willa. "I really can't hear."

Agnes *could* hear and finally realized what *she* was hearing. She stared at Willa. "You can't hear at all?"

Willa watched Agnes' mouth move and nodded. "It must have happened when I hit my head," she moaned.

Agnes sat down on Willa's bed, something she didn't usually do without permission. Slowly she looked closely at Willa. Then her motherly feelings took over and she hugged Willa tight in sympathy.

Willa squeaked in pain but hugged Agnes in return. "What should we do?" Agnes asked. Then she pushed back and asked again slowly so that Willa could watch her. "What do you want me to do, Willa? Should I tell the King?'

'Where is he?" Willa responded.

Agnes pointed to the bed only a few feet away where the King was now snoring loudly.

Willa thought for a minute. What *was* she going to do? She was a young queen whose husband needed someone to listen to him and now she couldn't listen to *anyone.* What had she done to herself? And what had she done to her beloved Solem? She hung her head. She would be of no use to him now. She would be failing him.

"The King loves you, you know," Agnes said softly, but Willa couldn't hear her.

"I must tell Solem, Agnes. He must know right away that I am useless as a normal wife. He mustn't be saddled with a wife that can't hear. People will feel sorry for him. I will look stupid. I will shame him as a king. He must talk to the church about ending a useless marriage immediately. They will understand."

Agnes was shocked. Why would poor Willa suggest such a thing? She tried to intercede. "What are you saying, my Queen? The King will never allow that. What makes you think such a ridiculous thing?"

But Willa had had experience with her father. She knew what he would do if he were King. Her father would not tolerate uselessness in a woman. She knew this. Her mother was the perfect example. Not producing an heir was one thing, but not even being able to hear would be worse.

Some of the religious people might even view this sudden disability as a curse on the Kingdom, a judgement from God. For a King, his strength could be challenged because of the evidence that he had a 'cursed wife'. To begin with, no one must know about her condition.

Willa stood up and waited for the room to stop spinning. Her head still hurt. Gradually she got her bearings. She knew what she had to do. Solem would not throw her out, she was sure, but she would be banished to a back room and die there as her mother had.

She would not put her husband in a place of public scorn and doubt. Surely the church would understand and annul their marriage.

"Please help me find a room in the back of the castle where no one will see me, Agnes. I will go there alone and stay. Solem should not have to face this failure of mine after just losing his first wife completely. Please tell Solem that I am sorry." Willa reached out to Agnes, her bruised face determined.

"Where are you going?" came a sleepy voice from the shadows. "You shouldn't be up and about yet. You had a terrible fall." Willa heard nothing.

"Your husband is awake," whispered Cwilla. "You shouldn't just sneak out without talking to him."

"Hurry," Willa whispered to Agnes.

"*You* must tell the King," Cagnes ordered Agnes quickly. "*You must tell him now.* She can't hear him or you."

"The Queen has lost her hearing due to her accident, Sire," Agnes said bluntly." She can't hear you."

Solem was already out of bed and across the room, gathering the wobbling Willa in his arms.

"My brave Willa, "he murmured. "Where do you think you're going? Don't you understand how injured you are? I almost lost you. I refuse to let that happen. You must stay in bed. *Don't move* until you are well. Don't you know how much I love you? I will take care of you."

"I can't hear, Solem," Willa announced, shutting her blackened eyes to his concerned face.

Solem looked at Agnes. "She can't hear? When did this happen? Is this true?"

"Yes, Sire."

"Why didn't you tell me?"

"She and I just discovered it, Sire. She didn't want you to know."

"Why, on earth?"

"I think she believes that she will be a burden, Sire. She wants to annul the marriage so that you don't have to have a 'damaged wife'.

King Solem placed Willa back on her bed and stared down at her at her in disbelief. "How dare you?" he growled. "How dare you think that I would want that, Willa? How dare you misjudge me so? I thought you loved me. I love you. Now you insult me. How dare you?"

"Why are you angry, Solem?" the deaf Willa cried out, searching his face.

Solem buried his face in Willa's chest.

Cagnes spoke quickly. "She didn't hear any of that. Get in her face."

Agnes turned Willa's face to hers. "He's angry because you are hurt, and he loves you. He says he won't let you leave."

"That's not what he said," Cagnes snarked, "but it was a very good lie."

The couple were still in an embrace as Agnes snuck out of the room.

The next morning, Willa was out of bed and tucked in a chair of the sitting room down the hall. Through sign language and funny, scribbled notes, she and Solem had managed to come to terms with her deafness and Solem was back at work downstairs. The day was beautiful, sunlight streaming in the windows of the old sitting room, making it look almost new.

Agnes busied herself with cleaning and dusting, not wanting to stray from *her Queen*. Agnes had emotionally declared 'full ownership' of Willa at this point. The fact that she still had a puffy face and an unsteady gate had made her even more precious to the elderly servant. The fact that she couldn't hear made her dependent on Agnes, if only for a while. Like a mother bear with a cub, only special people were allowed near Willa. June was one of those special people.

He arrived early today as he had both days since Willa had been hurt. Today he brought flowers. He had even picked them himself. "You wait here while I go downstairs for some water," Agnes ordered with a smile.

"Thank you, Agnes," said June politely, sitting where Willa could see his face.

Willa wrinkled her bruised nose. "I feel sorry for you, June, coming up here every day and talking to someone who can't hear you. You have to say things three times sometimes, I know. That must be very boring."

"It's not boring," he answered slowly, laughing, and using exaggerated hand gestures. "When I am up here with you, I don't have to study maps with father. He is still working on recommendations for people who were at the Water Rights Gathering. *That's* what is boring."

Willa laughed, too. Her smile was back, and her eyes were losing some of their blackness. June was happy that she was feeling better, but his stomach was beginning to churn again. He had made a discovery that was difficult and he didn't know what to do about it. He was in love with Willa.

"You have to get ahold of yourself," counseled his Conscience gruffly. "Willa is not available to you. She is your stepmother. She belongs to your father."

I know, June thought miserably. *What should I do?*

"Stop coming up here to see her every day, that's what. You're just making yourself more unhappy."

I know. I know. You're right. I just wish I could tell her how I feel.

"That wouldn't help *her* at all, would it? If you really care about her, you need to stay away from her."

This is true, June agreed miserably. He straightened his back and took a deep breath, moving to stand in front of Willa. Slowly he mouthed the words carefully. "I—am—going--- to stop--- coming up here,-- now that you are better. Try not to-- miss me-- too much," he laughed.

"I understand", Willa laughed back in return. "Thank you for the flowers."

"You are welcome, Willa," said June as he walked toward the door. June nodded and then stopped at the doorway behind her. "I love you, Willa," he whispered softly to her back as she gazed out the window.

I heard you, June, Willa thought suddenly. Willa's hearing had returned.

THE SECRETS OF THORNDIKE MANOR

The following morning, Balynn Drum followed her new servant friend down a dark hall of Thorndike Manor's cold, dusty east wing. They were on their way to visit 'the one other female in the manor' as Flo had put it. Drum was closeted again with Thorndike and Balynn had not seen her father since they arrived yesterday. Actually, she had not seen anyone but Flo.

This wing was very different from the armor-guarded entry hall. It was plain and gray and lit with only a few guttering oil lamps on the wall.

"This hall looks haunted," laughed Balynn as she caught up with Flo. "Don't walk so fast. I can barely keep up with you."

"I'm Sorry...I always walk fast over here", Flo responded. "It's cold. There's no heat. It's terrible."

"Why is it so cold? Doesn't anyone take care of the fireplaces here?"

"Just me mostly. And it's not just here that it's cold. It's everywhere except in the entry way and the conference room... the places that visitor's see."

"Really? I thought Lord Thorndike had a lot of servants and was wealthy. I thought he lived in splendor."

"Oh, he did...before" Florence said. "*Before,* it was really nice everywhere in the manor but here. He loved to show it off. Recently, though, all the fancy stuff...*and the heat...* have been re-allocated to places where preparations for The Plan are being made."

"You talked about 'The Plan' yesterday. What exactly is 'The Plan'?"

Flo stopped in her tracks and stared at Balynn quizzically. They were in front of a thick, heavy door. "You really don't know, do you? That's amazing. You're smart. You live with your father and I'm pretty sure *he* actually *wrote* The Plan. *You must* have heard *something* about it."

Balynn looked at Flo hard. Was she joking or was she serious? Then, with the mention of her father, a familiar thought surfaced. It was a memory of her father at home, railing at the mention of "The Plan". He had been angry and told her to never mention it again and she hadn't. Was this the plan that Flo was talking about?

"Of course, I've *heard* about it." Balynn said with a shrug.

Flo studied Balynn. She was no longer smiling. *Maybe I have said too much,* Flo said to herself.

"Maybe it's past time to say more," said her Conscience bluntly.

Flo was startled. She hadn't heard from *her* Conscience in a long time. With a shrug she reached down to the end of a chain that had been hidden among the folds of her skirt. With a loud clanking of the lock, she unlocked the massive door.

"We're here, Mistress," she called out. "This door is always locked so I must let you in. I won't lock it again, though, until you leave. I'll be waiting outside the door."

"You're not coming in with me?"

"My Lady doesn't like crowds...But I will be here...and we can talk about...'The Plan' then, if you wish."

Flo says there is a lady in this room, but the big door is locked? Why would that be so? Balynn wondered. Cautiously she pushed the door open to view a very strange sight. She was entering a room that was as dark as the hall leading to it had been. It looked at first as if it was abandoned.

There was a small fire burning in the hearth, however, and a tiny female figure was hunched near it, almost swallowed up by a huge, old but elegant chair. Was the figure real?

At Flo's suggestion, Balynn had borrowed some of Willa's old clothes and a shawl against the chill that seemed to envelope the manor. Now she pulled the shawl more closely around her shoulders and tiptoed toward the fire. Yes, the figure was an old lady. Was the lady sleeping or dead?

The hunched creature in the chair suddenly came alive and turned toward Balynn slowly. "Who are you?" The voice was scratchy and dry from lack of use.

But the creature suddenly smiled. She held out a wrinkled hand. Then, squinting through the darkness, she looked up and stared at Balynn.

Like magic, she seemed to grow larger before Balynn's eyes, appearing to almost bloom. She was no longer tiny and hunched. Her head raised and she sat up straight in her chair. Impeccably dressed and perfectly groomed, a lovely older woman emerged.

She looks like Queen Willa, Balynn realized with a jolt.

"Willa...my lovely Willa...is that really you, at last?" the woman asked in her raspy voice.

The Call of the Auras

Balynn gasped. *This is Queen Wilhelmina's mother! It must be. The Queen's mother is locked away in a solitary space. How could this be? And she thinks that I am her daughter. What is happening here?*

Balynn had not had much practice at feeling sympathy, but her little-used Conscience saw a rare opportunity to try for this emotion. "This poor woman is a prisoner in her own home", stated Cbalynn. "The world thinks that she is dead, and here she is, locked away. How terrible is this?"

Balynn stepped forward. "Lady Thorndike...Is that you? . I am so pleased to meet you," she began in a way she hoped sounded more sincere than afraid.

"Lady Thorndike? Who would that be? That would not be me, of course. Lady Thorndike is dead. But I certainly know who *you* are. You are my beloved Wilhelmina, all grown up. How many years has it been since I last saw you? You were only a wee thing then. Why didn't you visit me earlier? You must really allow me to call you Willa as I used to when you were alive. Even if you are now a ghost, you appear real to me, and that is enough for your lonely mother. Thank you so much for finally visiting me."

Balynn couldn't speak. She thinks I am the Queen's ghost. How strange. What should I say?

Cbalynn thought for a minute. "She is happy. You are making her happy. I would wager that it's been a long time since she has been visited by someone she loved. It will not hurt to let her think you are her daughter. Kindness is never a bad thing."

Balynn took the lady's frail hand in hers. "I am so happy to be with you, Mother. I thought you were dead. I had no idea that you were here, or I would have come more often."

"Don't worry about it dear. We are both dead and now and we have found each other. From now on, we should see each other more often, don't you think? Please put another log in the fire, though. Being dead is such a chilling experience. Where is Flo? She usually does this for me."

"I'm here, Lady Thorndike. I am always near. I'll fix the fire for you." Flo scurried to the fireplace out of the darkness by the door. Hoisting a large log onto the flames, she stood back and surveyed the situation.

"I see you have met Lady Thorndike," she said.

"Why didn't you tell me my Willa was coming to visit me, Florence? I would have had you do my hair properly and help me with a nicer dress. This one is so plain.

It's fine when it's just the two of us, but I would have liked to look more special for my baby. Can you see her, too, Florence, or is it just me?"

"I can see her, too, Mistress," Flo replied gently. "I didn't know who she was, though. She looks different than I expected. Are you sure that she is your Willa?"

The woman rose from her chair and reached out for Balynn.

"Stay put," counseled Cbalynn. "She just wants to touch you. You are 'her child'."

Lady Thorndike touched Balynn's arm, her hair, her face. "You feel so warm and alive Willa," she sighed wistfully. "It has been so long...and you were only a toddler, of course. I couldn't believe it when they told me you had died. I can't believe that you grew up even after you passed from my arms...".

The frailty and confusion of the woman began to reclaim her. Flo hurried forward and eased her back into her chair.

"She has loved seeing you, but this is more excitement than she has experienced in years", Flo explained. "She is quite fragile. We should go."

Balynn looked again at Flo and saw more than she had seen before. The servant's gentle kindness was very real but almost guilty. What did this bizarre situation mean to Flo?

Balynn followed Flo through the door and heard her lock it again behind them.

"She's dozing now. I will return later to feed her and put her to bed. This has been a big day for her, I'm sad to say," Flo announced unhappily.

Balynn walked side-by-side with Flo back down the shadowy hall. Neither said a thing.

Finally, Balynn had to say what was on her mind. "You were very kind to Lady Thorndike, Flo. That is who that was, right? Why in the world is she locked away here?"

Flo kept walking, staring straight ahead. "I care for her and love her," she said simply, 'but I have kept her captive here for years. Like my own mother before me, I have taken care of her since the real Willa and I were both children. I am a bit older than Willa, and my lady used to think *I* was Willa. I even used to pretend that I was when she was most distraught."

Balynn stopped and turned toward Flo with wonder. "How long has she been here...and *why?*

Flo didn't look up. There in the shadows of the dark, empty hall, she uttered what now appeared to be a painful confession. "My lady has been here since her baby girl died," she whispered. She stopped walking and began to cry.

Balynn was sorry she had asked. Flo's answer had made her miserable. With a compassion she had not used before, Balynn put her arm around the shoulders of the young woman. She was almost the same age as Balynn and had been left alone to care for a deranged creature in the huge, cold house.

Balynn's mind wandered. "Queen Willa would have had a baby sister, then? That is a shame."

Flo did not look up. "No," she whispered. "It wasn't a baby sister that died. It was Willa."

Balynn stepped back and stared at Flo, who did not meet her gaze. "What do you mean, 'Willa'? Willa's alive and now she is Queen. You must know that. What are you saying?"

Flo looked up and Balynn could now her tears were flowing freely. "I have never told anyone this," she sobbed. "All these years, I never told *anyone*. My Lady was told a lie. When she was told that her daughter died, she became unstrung by the news. She lost her bearings. Evil and misery have kept her mind at bay all these years. She has never recovered. But my mother and I kept her locked away ever since so that she wouldn't be killed. My Master thinks she died.

"When my mother died, *I* took over and became her jailor." Flo continued; her shoulders were shaking with long held sorrow.

"She has been keeping this secret for all these years," Cbalynn whispered in Balynn's ear. With compassion she hadn't known she possessed, Balynn took the servant's limp, damp hand in her own and led her back toward the hallway they had traveled to get to this forgotten wing.

Arriving in her room, Balynn sat Flo on her bed and waited for her to speak. Was her motive compassion of curiosity? Compassion was unfamiliar to Balynn.

Balynn's Conscience noticed something. Auras were fading out the bedroom door. *Flo has been surrounded by Auras of evil that have kept her guilty actions in place,* she realized suddenly. *They have probably maintained Lady Thorndike's insanity in place as well.*

Cbalynn had never seen the result of Auras' actions so clearly before, even though she had been living with them at Drum's all her life. She now realized how accustomed to Auras she had become...and how powerful they were. *Strange that they had never infected Balynn's thinking,* she thought. *Maybe they have been to busy with her father.*

Flo has just shed Auras with her confession, Cbalynn now understood. *Thorndike Manor must be filled with Auras. I should warn Balynn...No, she wouldn't understand. I must simply stand by. She will need protection...Now I know she will need it at home, too. I haven't been paying attention to what Auras can do to good people.*

Flo continued to cry softly but these tears seemed to be tears of relief. At least part of her burden had been lifted.

Balynn patted Flo's arm softly and waited. Flo gradually calmed but continued to look at the floor. "I promised my mother that I would never tell anyone about My Lady, but you are so nice...and your father is so much like my Master...I guess I just felt that you would understand."

Flo looked up cautiously. "You must think I am terrible. Do you? You must...but you won't tell anyone, will you?"

Balynn shook her head. "Don't pry," suggested Cbalynn. "Flo has just released a burden that she has been carrying for years. Give her some time to understand what she has done...and perhaps what she should do in the future."

The young women sat alone in the guest room. The cold seemed to be warming now. The fire that smoldered in the fireplace crackled a bit for attention. Flo responded automatically, poking its embers and adding a log from a nearby pile.

"I bet you have a lot of questions, don't you?" Flo offered quietly. "Do you think I am terrible for keeping my Lady prisoner all these years?"

Balynn was trying to sort out what she *did* feel.

"Does Willa know about her mother?" This question rose up first and begged for an answer.

"Heavens, no," said Flo firmly. "Willa would never have allowed it. She loved her mother...No, that was the other lie that my master told...and so did I. We told Willa that *her* mother had died... She was really young and she believed us. So did everyone else.

My mother and I were the only ones that knew the truth. She knew *she* would die...or *I* would die...or my lady and all of us would be killed if we told anyone she was still alive.

"My Master was always mad because he didn't have a male heir that he could shape into a victorious militia leader as his father shaped him...and he still believes he will be, someday.

"That's why he beat on her in a fit of rage that one night and told her that her daughter was dead...that she now had "no daughter" because she had given him "no son". That is how cruel and heartless he can be. That lie, the 'death' of her daughter, was his evil punishment for her.

"He forced my mother to hide little Willa in town for months to prove to my Lady that her daughter was dead. I think he thought that would *make* her conjure up a male child if she tried hard enough. She became terribly depressed. He abused her and screamed at her and told her she was useless and just not giving him a son on purpose.

"My Lady finally lost reality entirely and my mother thought he was going to kill her, when her depression got so bad. My mother begged my Master to just banish her, instead and when he lost interest, she secretly did and told him she was dead.

"Actually, that is one of the good things that my mother did. She convinced Lord Thorndike that it would be bad for his reputation if word got out that he had killed his wife for no reason.

"So, my Lady was banished, and *my mother* told my father she had died. Then *my mother* was used to secretly bear him two sons, herself. This is why he let *her* stay alive. I think that helped *me* stay alive. too.

"Many servants who knew too much have disappeared, but I have two younger brothers who are now in my master's Thorndike Militia...Yes...brothers. Lord Thorndike is *my* father, as well as Willa's. Willa is my half-sister.

"My Lady has been my secret duty since my own mother died. The terrible years of secrets in this house have been my curse. My Conscience will not even speak to me anymore."

It was now Balynn's turn to sit on the bed. With her mind swirling, she tried to sort it out. Cbalynn was in the same condition...shocked and confused.

And here was a new question that now occurred to Balynn, unbidden. *What is my father doing downstairs with this terrible man. Is my father equally evil?*

"Good question," said Cbalynn reading her thoughts.

And my father likes him, Balynn thought miserably. *Why? Why do they spend so much time together? Is he evil, too?*

Cbalynn was an inexperienced Conscience, but she now thought she had figured it out. *"Your* father must not know about how really awful Thorndike is. *Your* father must be primarily interested in *my* father helping him with *The Plan*, whatever that is."

That must be true, Balynn thought gratefully, *but I have no idea what The Plan is. What in the world is The Plan?*

"Ask Flo."

Florence Thorndike, Flo, was miserable. She had just told a new friend her most terrible family secrets. *What must she think of me,* Flo asked herself. *Balynn must hate me.*

"I think she might be almost as confused as you have been", answered Cflo.

"Mistress Balynn, please forgive me for burdening you with my secrets. I know you must wonder what *your* father sees in *mine,* as awful as he has been in the past. But I hope we can both concentrate on the future now."

A small smile began to creep onto Flo's tear-stained face. "I, for one, am hoping that your father will help my father change for the better and me and my brothers can be proud. After all, they are both working on The Plan. That, I am sure, must be a good thing."

Balynn brightened. This could be true. She hadn't thought of that. But now she had to confess something to Flo. "I hope you are right," she said slowly, "but I don't know as much about The Plan as you know. My father doesn't like to talk about it at all. You think The Plan might be good? What is it exactly? All I know is that it involves some sort of road along The River."

"Indeed," said Flo with a small display of pride. "The part of The Plan that *I* know about *is* about a beautiful road that will wind along The Great River so that *everyone* can travel on it. "It will run from The Castle all the way down to Drumsworth and *everyone, rich and poor,* will be able to travel the whole length of River Kingdom in just a few hours. It sounds grand.

"The only thing I can't figure out is why the suits of armor are collected in the hall. Maybe they are just for my brothers and Father to use to make sure that the Lowlanders are kept under control."

That's curious, wondered Balynn. *Who is worrying about the Lowlanders? Some of them are my servants. Are some of them suddenly out of control?* Balynn tried to figure this out, but the new road sounded wonderful.

She looked forward to finding out more about it tomorrow when they went back to The Castle. And there was another thing to look forward to as well. June would be there.

BACK ON SCHEDULE

The next morning was bright and clear as Manda brought her horse to a halt in the warm autumn sun in front of The Castle.

We'll be into cooler mornings, soon, Manda mused. *I wonder what Willa told King Solem about our adventure. She has had plenty of time to tell him what she learned. I hope, between me and Hector, she got the message about the farmers' levees. Now I just have to be sure that she gave her husband that message.*

Smugly handing her reigns to a familiar servant, she skipped across the colonnade to The Castle door and was ushered into the entry way as she had been many times before.

Her mother, Queen Luna, would be here tomorrow for *her* Water Rights meeting with the King. With a little bit of luck, King Solem had received a good briefing from Willa and Urhonordo's fears about the complications presented by the plan for 'The Royal River Road' would be over. Meanwhile, Manda would just hang out with Willa and Mimi and discuss 'girl things'. Willa was already a friend.

Surprisingly, Mimi called to her from the balcony above. "Yo, Manda! I was wondering when you would get here. Willa is much better today than she was yesterday. She will be glad to see you, even though you almost got her killed. Come on up!"

Better? Better than what? Killed? Mimi is so silly...sweet but silly. Manda shrugged at the youngest in the family and scaled the long stairway to Willa's floor.

She nodded to servants she saw, busy in their daily routine. They all knew her, but some of them were clucking and shaking their heads. The word had now been passed among them that their Queen had been injured on an outing with Princess Amanda, and they were not surprised. This girl was still wild, in their opinion.

Today, however, she was dressed more like a lady. Maybe someone had told her that it was time to wear skirts instead of britches.

Willa was seated by the window and turned to greet her as she entered. Manda stopped and stared. "What happened to you?" she cried. "You're a fright! Were you struck? Did you fall? Are you alright?"

Willa grinned although her face was stiff and sore. "I still have some aches and pains, but I think I'll live."

Willa outlined the story of the fall, the rescue and the recovery. Manda listened with horrified attention while Mimi who had heard it all before, added comments here and there.

"Now...since everybody thought she was going to die...everyone treats Willa like she is made of glass," chirped Mimi with a smirk. "I'm surprised that my father and June are not up here watching her every move like they have been."

Willa laughed, but Manda did not. "You could have been killed and it would have been my fault," she groaned. "Why didn't anybody tell me? You all should blame me and throw me in the dungeon. I deserve it."

"Indeed," hissed Cmanda in Manda's ear. "I told you something like this might happen if you let her go home by herself."

I know. You were right. "I should just go to the dungeon and put myself in it," Manda groaned.

"Hush. You are ridiculous," insisted Willa. "I'm the one that wasn't looking where I was going. And now I am fine. I told Salem everything. He doesn't blame *you*. Actually, I think he blames himself and June for chasing after me. They are ridiculous, too."

"So, Manda...you didn't even know Willa was hurt?" Mimi was now fixing Willa's hair. "What brings you here this morning, then? And you're wearing a dress. It must be something serious."

"Well, I just wanted to check in with Willa. I wanted to see what she thought about the levees and what Solem said about them. *I* wanted to talk to him and to you, too, Mimi, now that you are out of bed and awake," she added snidely.

"At least I didn't get hanged from a tree and dropped on the ground during an adventure with *you*," Mimi snickered. "Do you think she got more out of her ride than bruises? What do you think Willa should have learned from her amazing trip to the country?"

"Didn't you tell her anything yet, Willa?" asked Manda. "Didn't you tell people about meeting Hector and hearing what he said?"

Willa hung her head. "I guess I didn't," she said sadly. "I was intending to tell Salem everything today, but now you're here. Now *you* can help me do it."

Cwilla moved about lightly on Willa's sore shoulders. "You and I were preoccupied with jumbled thinking and I forgot to remind you," her Conscience whispered. "I am still thinking about June and what *he* said to you."

Me, too, thought Willa sadly, *but this is more important. I should have said something to Solem right away.*

"Say it now, then", said Cwilla. "This is *really* important to the Kingdom...and to Manda. She came all this way here again to talk about it."

"I'm sorry that I didn't bring our message to Solem yet" said Willa. "He and June have been busy down in the Conference Room. I think they're working on Water Rights stuff from The Gathering. Maybe, if we ask Solem, he will spare us a moment so that we can tell him what we...I mean, I... learned. You already knew it."

Manda looked at Willa's bruised face and *her* Conscience spoke up, quickly. "You are really being pushy with your friend, Manda. Willa was hurt on an outing that *you* took her on. You should ease up."

You're right, Manda thought. "I can wait, Willa. The River will not stop flowing if we don't bother the King today. I will just stay up here and talk to you and Mimi while you mend. I just get too anxious for my own good...and yours."

"This is all very touching," snarked Mimi with a laugh, "but I don't think you could talk to my father today even if you wanted to. He's expecting company. I can already hear them. That noise outside should be Lord Drum and his daughter arriving, 'per schedule.'

"June messed up and invited them back right away when they were here the other day. Father doesn't have a choice, even though Drum is not his favorite person."

"You're kidding. Drum and his daughter are coming here today?...and they're here already? That's terrible, Mimi. Are you sure?"

"Of course, I'm sure. Drum was here before with a whole passle of old Birther nobles gathered around to make him look good. I don't think they even knew what they were in favor of. They were just here to support Drum, like usual."

"You're right," came a voice from the doorway. It was June.

"I just wanted to tell you what Amanda and Mimi just told you," he said with a brief nod to Manda. "Drum and his daughter are on the colonnade and will be in the conference room shortly.

"I know that Drum is not your favorite person, Willa, and I don't know about his daughter, but are you well enough to see either of them?"

"Does her face look like she wants to entertain guests," snorted Mimi. "You are the dumbest man I know."

June lowered his head.

"He just came up here for an excuse to see you," whispered Cwilla.

I know, Willa thought. "Thanks, June. Thanks for asking, but I think Mimi is right. I don't need to speak to anyone or answer questions today. I'm sure you and Solem can handle guests and make excuses for me just fine."

June nodded and turned away.

Manda watched him leave with a surprising twinge of sadness. *He didn't even say hello,* she thought. *I was right here, but he didn't say a thing. And he still calls me Amanda. Maybe he's still mad at me for beating him up when we were kids.*

Suddenly June was back at the bedroom door. "Hi, Amanda. I forgot to tell you. Father said that since you were here, you would be welcome to meet with Drum and his daughter. You have some opinions about the new road, too, don't you?"

"Yes, I do, June. I have a lot of opinions about that road. Nice of you to at least say hello."

"You always have opinions about everything, from what Mimi tells me," June retorted with a grin as he headed back down the stairs.

"He made that up, Manda. I don't tell him *anything* about what you and I talk about. He's too dumb to understand it, anyway." The girls were back to being girls.

THE DELAYED MEETING

Downstairs, things were much more serious. Drum and Balynn entered The Castle. Drum strolled ahead with an air of confidence and a bevy of Auras. Their numbers had grown and become more enthusiastic.

And beside Drum was Thorndike himself. The two were focused on obtaining the official support that they would need to begin the road that had caused so much ruckus at The Gathering. This was a day for which they had plotted...and recently re-plotted.

Following the pair at a distance was Balynn, determined to make a good impression but not nearly as confident as the men. Yesterday had been an unnerving day. She had made a servant friend, something new for a young noblewoman who had always looked down on servants as did her father. She had also learned disgusting facts about Thorndike who seemed to be her father's best friend and co-conspirator on the mysterious 'plan'.

But Balynn was her father's daughter. Stubbornly, she decided to concentrate on something she was more interested in... June.

And there he was on the colonnade that graced The Castle entryway. Beside him stood the King with a friendly look and beside the King stood ...a beautiful woman?

To Balynn, this was not a good thing.

Is that Queen Willa? Balynn thought in confusion. *She certainly looks different than she did at the Water Rights Gathering...taller...darker...she almost looks like a Lowlander.*

"No, idiot," snickered Baylyn's Conscience. "That's not Queen Willa. That's Princess Amanda. Remember her? And she *is* part Lowlander you know, considering who her father is."

Balynn stopped in her tracks. She *did* remember her. She backed up. *What is **she** doing here? What right does **she** have to be here when my father is here for an important meeting, and **I** am here to see June?* she thought, her father's prejudice showing strong.

"That's not your call, Balynn," whispered Cbalynn. "You sound like your father. Just smile. June is looking at you."

Balynn smiled and June smiled back. For a moment, Balynn relaxed and moved forward, but a lifetime of seeing Lowlanders as 'lowly servants' was inside her. Lowlanders were beneath her. Her father had told her that this was true...unless they were needed for something. They had no status and, according to Flo, 'needed to be controlled' by The Plan.

Flo was a servant but at least she was not a Lowlander. In fact, she was actually a Thorndike. Balynn followed her father proudly into The Castle.

In the conference room, an elderly servant was setting the table and preparing for guests. Horace was not as spry as he had once been and so this light duty matched his creaking abilities well. Besides, his brother Hector had been upset at the Water Rights Gathering because of something to do with a road of some kind. Hector had asked Horace to keep his eyes open if anything about that road came up in The Castle. If Horace wasn't mistaken, the new road was what this meeting today was all about.

Horace prepared the table elegantly with glasses and pitchers of ale and comfortable chairs replacing the more utilitarian ones that had been there before. He stood back in the shadows as King Solem led the guests and took his place at the head of the table, the primary location that Drum had chosen for himself two days before. To the amazement of Drum, seating had also been arranged for women...Balynn and Amanda...at the far end of the table...as observers, of course.

"Amanda has been invited to our meeting as a representative of Urhonordo. She was here on a visit, and I asked her to stay," Solem announced matter of factly. "I trust no one objects to her lovely presence."

This is a man's meeting, growled Drum to himself. *Women should not be here... unless they are my daughter who is here for decoration and potential meeting later with the Prince.*

Thorndike, seated directly across from him, had the same thought. *This is just another indication that this King does not know what he is doing...or what he is up against. Maybe my daughter will finally turn out to be useful, though.*

"No problem," said Drum.

"You know that woman's half Lowlander, right?" growled one of Drum's nearby Auras. Drum sat up straight. This young woman, although lovely to look at, was both female *and* part Lowlander. This made her presence here inappropriate on two counts. His Auras were obviously warning him to watch what he said. The King was not a Birther. Drum would need to be politically correct today to accomplish his first goal...the road.

Drum cleared his throat...and his thinking. Quickly, he reviewed the major talking points that he and Thorndike had discussed. The surrounding Auras had counseled that many of these ideas were not for either Lowlander or Urhonordo consumption. Hopefully, Thorndike would let him do most of the talking.

"As you know", Drum began, "The Royal River Road is designed to be a marvelous addition to the life of the citizens of River Kingdom. It will stretch from The Castle to the Lowlander border and reduce days of travel time to hours. It will be smooth, allowing for *safe* horse and carriage travel, free from ruts and obstructions. I know safety is important to your majesty." Thorndike had suggested to Drum that he emphasize this point considering the Queen's accident.

 King Solem winced. Drum knew Thorndike's advice had struck a nerve in a monarch. Thorndike had stressed that the idea would be a strong one. Sometimes Thorndike had good ideas, Drum acknowledged.

Solem nodded sadly. "I understand the basic concept, Lord Drum, and it is a good one. Everyone will benefit from safer, quicker travel between north and south."

Manda stood. "Excuse me, Lord Drum, but what will happen when a big levee road flattens the small levees of the farmers? Won't a big *road* levee block all the farmers' *irrigation* levees and cut off the irrigation water for the crops inland?" she asked calmly.

From the shadows, Horace listened carefully. Amanda was sticking up for the farmers.

Thorndike stood and glared. He was not about to listen to a woman, especially one with dark skin and hair. She was the very picture of Lowlander 'mixes' that he and the nobles were against. *She* was the enemy. She shouldn't be here. He knew his rights.

"This was supposed to be a River Kingdom meeting," he snarled in immediate disgust. "Who are *you* to be asking questions, woman? You are not a citizen of River Kingdom. You are nothing more than a foreign blot on these proceedings! And *you* do not even farm. You do not harvest. Who are *you* to be asking stupid questions of the white men who oversee such things?" He stepped toward Manda threateningly.

King Solem stood immediately, himself. "Beware, Thorndike. You are a guest in this Castle. You will not be rude to my other guests."

Manda didn't flinch. She simply turned and stared at Thorndike, unruffled and determined.

June's admiration of Manda at The Gathering surfaced. He suddenly also remembered her 'reaction' to him at the wall when they were young. Surprising then...beautiful now. Even though he knew she was tough, he suppressed an overwhelming desire to punch Thorndike on her behalf as he towered over her like a bear ready to devour what it had cornered.

Balynn shrunk back in terror. She was remembering the cruelty that Flo said Thorndike had shown the woman in his life. *He was a monster then*, she silently proclaimed. *And he still is one now. How can my father tolerate him?*

Manda remained silent for a moment. Then, amazingly, she spoke directly to Thorndike, this time with a tinge of disdain. "I can see that you haven't the self-discipline for a sensible discussion, my lord. That is a pity as we both might have learned something.

"Since you are unable to control yourself, discussion will be impossible. My mother, father and I will be here tomorrow. I will save my questions and comments for then when this table will be more civil, I am sure." She turned and walked out the door.

Thorndike plowed after her, waving his arms and cursing. "*You* are the one who needs to be controlled! *You* and your low-life Lowlander family!" he shouted.

The Royal Guard, always stationed at the door, blocked Thorndike as Manda departed. Thorndike glared at them, breathing heavily.

Balynn was relieved. Thorndike had been bested...and not only that. *Her* female competition for June's attention was gone. It didn't matter if that girl was part

Lowlander. She was beautiful and impressive. No one could deny that. Balynn had *some* faith in her own beauty and charm, but she didn't want to compete for the Prince with a Lowlander. The Auras that had drifted in with her father curled around her feet. She might have promise, the Auras observed.

June rose and approached Thorndike with determination, still marveling at Manda's dignity. She had demonstrated that this was the time for adult behavior so he would give it a try. "Would you like to take your seat at the table again, Sir?" he asked. Then he stepped aside and nodded purposefully toward the table.

Drum shook his head and gave his partner a look. "Come and sit down, Thorny." Drum added with a familiar but definite tone. "She's gone. We can do *what we came here to do*... without interference. Come... Sit down."

Horace was fascinated. He pulled back Thorndike's chair for him. It was habit. The man is unbalanced, Horace decided.

It took a moment, but Drum's words sunk in. Thorndike composed himself. He and Drum were on a mission. They had confirmed their plans last night. The Urhonordo wench only confirmed the importance of what they needed to accomplish today. "Sorry, Your Majesty. She rubbed me the wrong way. Let's get down to business." Thorndike squared his shoulders and nodded to the King, resuming his position with what he hoped resembled confidence. The surrounding Auras rose to calm and bolster him. It looked like this was going to be part of their job, today.

For a moment, the room was silent, but Drum and *his* Auras were primed. He would not sit still for long. The Plan was at stake. "Your Majesty," Drum began again, "as we were saying before we were interrupted, The Royal River Road will make *safe* travel from here to the Lowlander Border easy and quick for the first time in our history. It will be a smooth and dry levee road, rising high above The River.

"On one side of this levee road, the traveler will see the crops of River Kingdom displayed. On the other side, the traveler will look down on The Great River, itself. It will truly be a Royal Road." The Auras billowed from beneath Drum's chair, rolling and circulating under the massive table, on their way to King Solem.

Solem chose to move on. "I must admit, this road sounds wonderful, Lord Drum, but our departed guest *did* raise a good question. How will the crops be watered if this large levee road blocks their water from The River? Isn't that a real problem?"

Thorndike had regrouped. He stood again, but this time with studied control. Carefully, he pulled a smaller map from his tunic and spread it before the king. He stepped back with pride and let the story told by his map sink in. "Here is your answer to that Urhonordo mut's question, Your Majesty. Feast your eyes on the 'Royal Road Irrigation System'."

The Call of the Auras

Solem and June leaned over the new map. Clearly, they could see two small rivers that flowed through the crop land on the east and cut through the large levee on its way to The River. One of them was in the north, passing through the holdings of Lord Thorndike. The other was in the South, in Drum territory. These small inland rivers were covered by beautiful bridges, connecting to banks of the road on either side. "Behold the inland waterways that will irrigate the farmlands," Thorndike boasted.

Now all at the table stood and peered down at Thorndike's map. Auras wafted back and forth unnoticed." It looks perfect" the Auras whispered to those assembled. "The road is high and dry with two smaller rivers that will water the crops. It looks like everything has been considered. We should just do it."

Slowly, one by one, everyone sat down. They liked the idea. They could see no flaw. The Auras flowed around King Solem, inconspicuously, continuing to whisper positive thoughts.

King Solem was impressed with the completeness of the presentation. It had obviously been planned well. He was not fond of either Drum or Thorndike, but they had thought this through. Solem was a King so he also needed to think things through himself. He had a question.

"This presentation is admirable in terms of the road and irrigation, but two additional issues haven't been discussed yet," he mused aloud. "I am wondering about cost and labor. This road will take a lot of work and I have no idea how much it will cost the Royal Exchequer. What say you on these issues, gentlemen?"

Drum sneered. The Auras were ready for this. "I don't think there will be a problem with that, Sir," Drum replied. "River Kingdom is ready. You may have noticed the enthusiasm that took place at the Water Right Gathering when I walked in."

Solem rolled his eyes at the recollection. "Indeed. I did notice that ruckus."

"Well, I apologize for the disruption, Your Highness. It was rude and I must admit that I encouraged noble enthusiasm. But that noise was not for me. It was for the Royal River Road. The nobles of the countryside are aware of the new road idea, Sire. They want it. They are even volunteering to have their landholders help build it."

A chilling gust sent an Aura swirling from Drum to Solem, himself. King Solem shivered a little but continued to listen with fascination. "Is that so?" he commented incredulously. "I had no idea."

"It *is* so, Your Majesty. All the nobles are with me on this. They are insisting on it." Wisps of Auras curled around The King legs like friendly kittens.

"I had no idea," King Solem repeated quietly, realizing that he had not been out among the people as much since his first wife had died. Csolem thought about this, too and agreed. The Auras continued to curl near The King under the table. Csolem, unfortunately, didn't see them.

"And the cost will be next to nothing," Drum continued enthusiastically. (He was making progress and he knew it.) "And just think about it, Sire. What is a levee made of? Dirt and rocks. We certainly have enough of that in this kingdom. There is no cost there. The bridges over the two small inland rivers will be built from trees grown on our nearby land. There is no cost there, either...and the nobility, riding smoothly in their elegant carriages, will love you."

Drum sat back in confidence. He had been thinking about this moment for years. An image of himself riding in a royal carriage rose up before him. Thorndike leaned back next to him. He, as Commander of the Militia saw himself riding high on one of his war horses as it pranced along on a smooth, new rode toward victory. They shared this dream now, but no one would know of it but them. The Auras crept across the table and snuck quietly onto The King's shoulder.

King Solem rose at the head of the table and smiled. The presentation looked complete.

June leaned in and pulled the map toward himself for more study.

"This presentation is too smooth, whispered CJune.

June agreed. He had spent time with the common people. He still delivered extra food to the poor. He had heard nothing about the new road being of interest to anyone in the town.

"This looks like something that Drum is encouraging the nobles want, whether they want it or not," Cjune offered. "Something is just not right."

"Lord Drum," June began politely. "If I remember correctly, not everyone was cheering for you at the Water Rights Gathering. The landholders at the meeting were *not* cheering nor were many of the townspeople or Lowlanders. They didn't seem to like your map as much as the nobility. Do you know why that was?"

Horace almost gave himself away. June was talking about his brother. The faithful servant didn't move but hoped his old ears would serve him well.

Drum looked around as though unconcerned, but Thorndike was ready for this question. "I remember that, too, Your Majesty. That was embarrassing. A lot of the men in that rowdy group were *my* landholders and Drum's. They are a lazy lot, the whole bunch of them. Digging these levees will be hard work and none of them want more work, you can be sure. *That's* their whole objection.

"The landholders and their families on my land will dig the big levees that are near my land and Drum's workers will dig the ones on his. The ones between our holdings will do the same. It's that simple. They will come up with all kinds of excuses, but that's what 'the labor issue' will boil down to. And it won't be a problem. They will either do what we say or they'll lose their livelihood."

Horace cringed and shrunk further back into the shadows. He knew how hard his brother worked. That was why Horace had chosen to work in The Castle instead. The work inside was easier and the Royals were definitely better to work for than Thorndike. Horace knew Hector would not be surprised by what Thorndike had said.

But Horace *was* going to be surprised by what happened next.

King Solem nodded and rubbed his chin in thought. The Auras swarmed him and whispered carefully chosen facts. "You can decide now easily. Everything that these men have said today makes sense. The dirt for the levees won't cost anything and it is undoubtedly true that the landholders and their field hands will do the work, even if they were not in favor of it," they whispered.

ROYAL RIVER ROAD

Drum - Thorndike Map

By the look of the map on the table, the two small rivers that flow down toward The River might even irrigation easier, Solem reasoned with his Aura-assisted reasoning. Drum and Thorndike drew their chairs up closer to the King to discuss details.

Csolem suddenly thought he felt the rumble of Auras. *Are there Auras in this room?* he wondered. *Of course not,* he concluded. *This is The Castle. This is just a preliminary meeting about a road in The Kingdom along The Great River. Solem's protocol will prevail. No decisions will be made today. There is nothing important for Auras to be interested in.* Csolem relaxed.

June was not having the same thoughts as his father, however. Surprisingly, his thoughts had jumped to Manda. She had spoken up for the map of the farmers and landholders at The Gathering. Her objection had nothing to do with the hard work that digging the levees would require.

It was more about the dozens of small levees that coaxed the water *into* the crops now, many of which were *between* the Drum and Thorndike properties in The Central part of The Kingdom. June didn't know much about the smaller irrigation systems, but he somehow thought Manda did. When she had mentioned it today, it had caused Thorndike to lose composure, completely.

I shouldn't have let her go without asking her to explain her question, even if Thorndike was being a bully, he thought. *Maybe I can get that clear tomorrow. I know there is a reason she was mentioning this twice. She's smart. She will still think it's a real problem even if it makes Thorndike mad.* June laughed at the memory of her sticking up her nose and walking out on Thorndike's bluster.

Now, without noticing, another woman came into his thoughts and vision. Balynn was standing beside him. Curiously, she was also looking at the map.

"Can I look at the map with you, June? I feel so ignorant about The Plan. I know this road is something important, but I have no idea why."

June had forgotten she was there. "Everybody has really ignored Drum's daughter", Cjune whispered to June. "Be nice. It's the most polite thing to do."

Balynn smiled at June and then focused on the map.

She wants me to notice her and like her, June thought impatiently. *She's pretty but she is not who I want to talk to right now. I wish Manda was here. This road decision is more important to me and my father than this girl.*

"You are extremely full of yourself," chided Cjune. "As a guest, she deserves at least a *little* attention."

Balynn was still looking at the map. "Look, June," she whispered. "Daddy's lake is nowhere on this map. Our name is there but there is no lake. Shouldn't these men be talking about the lake, too, and where the water will go when the small levees are blocked? Shouldn't they think about that before they make up their minds about the big road levees?"

June looked again at the map on the table before him. How strange. The name of Drumsworth was clearly spelled out on the southern end of the map but Drumsworth Lake was nowhere on the map at all.

June looked again at Balynn. Cjune snickered. "She's smart as well as pretty, June. You were ready to blow her off as just a flirt. She came up with something you didn't even notice.... Like I said. You are extremely full of yourself."

"Thanks, Balynn. You are so right," June acknowledged with respect. "I didn't notice, and I think my father didn't either. Let's tell him before he makes up his mind." June and Balynn hurried up to the men at the other end of the table, ready to bring up their issue before anyone left.

To June's amazement, however, Thorndike and Drum were on their feet and shaking hands with his father. All were smiling. What had he missed?

"It looks like we're about to get a beautiful new addition to River Kingdom, Son," announced Solem, smiling broadly. "No more drainage ditch ruts in the road. No more low hanging branches. These two men will leave all that to our memories, while *their* men construct the levees for our safe, new Royal River Road. In a week or two we will put it all in writing and make it official."

Thorndike stood to the side, hands on hips, looking very much like the Commander he wished to be. Drum clapped his hands as he had at The Gathering. "This will be a day that you will always remember, Your Highness," he chortled, his eyes gleaming with his success. "This is the day on which *the* agreement was made that will change the future of River Kingdom, forever."

June couldn't believe his eyes. In just the few moments that he had been contemplating the missing Drumsworth Lake, his father had made a decision, a huge decision that should have required more thought...more questions, a meeting with Queen Luna, at least. This road would affect both sides of The River. How could this have happened so quickly?

The Auras in the room were ecstatic. Their leader would be proud of them. This was a major coup. As they had already decided, Drum was going to be their champion...and probably King.

Horace, still observing from the shadows, wondered what had happened, too. Everything had seemed to be decided in just the last few minutes. He couldn't believe it. He couldn't understand it.

Only Chorace, Horace's Conscience, understood what was really happening. It was just as had been predicted during The Conscience Alert. The Auras that had been gathering and strengthening were now in full combat mode. Something important and *bad* had occurred today...and Auras had made it happen.

And Csolem hadn't even noticed. Horace would tell his brother and he, Chorace would inform Chector. Today had been a serious of River Kingdom Goodness and could mean that worse was yet to come.

BALYNN AND THE AURAS

Horace hurried to clear the remains of the meeting while the other men in the room discussed the future as though he didn't exist. This was common. He was a servant. He wasn't listening to the jovial men anymore, anyway. He was, however, very aware of his Conscience.

"Did you see that?" whispered Chorace. "See what?" mumbled Horace, gathering the silverware.

"Did you see that 'Aura Take-over'? That was outrageous! I've never seen anything like it. One minute they were just discussing the road project and the next minute the King decided to go for it, just like that. It was amazing! "

The King did make up his mind fast, didn't he? Horace mused as he worked. *I thought he would wait until he talked to Queen Luna tomorrow. I don't think she'll be happy that the King has already decided something this big without her. And I'm going to need to talk to my brother soon. I know this is not what he wanted. By the way...what's an Aura?*

Drum was still busy, telling Solem that they should get started on construction right away. Thorndike was assuring the King that he would set his workers on the project immediately. He wanted to get the road built quickly. He had plans for it.

Balynn and June looked at each questioningly. Something was very odd. The King's decision was too quick. Only one good thing had come from their discovery about the lake. June had discovered a nice, smart person that he had not known before. Slowly, the two young people left their elders to enjoy a bit of their new friendship.

Balynn had discovered something new, just as June had. June was not just the marriage target that her father wanted him to be. He was nice...and thoughtful...

and terribly handsome. Quietly, she realized that she had time now to be with him without her father, even if it was for only a minute.

The carriage was waiting in the driveway. They would be leaving right away. Balynn made an unexpected decision. She wanted to impress June as her father wanted her to do, but she wanted to know more about him as a person, too. Her Conscience approved of this added thoughtfulness.

She thought back to Flo and her revelations about men and relationships. She had never thought about these things before. Her mother had left (or been dismissed) too early for Balynn to ask her any of these kinds of questions.

"You aren't happy about your father's decision, are you?" she asked June sincerely. "Why is that?"

June looked at Balynn cautiously from the side. "She really wants to know," whispered Cjune. "Answer her."

"I guess my main problem is that I think he's making the decision too fast. A new road will be riding on a big levee that runs the whole length of The Great River. I'm sure it's going to make changes in growing and planting things that he hasn't thought about yet...and then there's Drumsworth Lake. I think a big idea like this needed more study with Urhonordo."

"...and Princess Manda?" Cjune added. June didn't answer.

"Too bad we didn't get to talk to the men about Drumsworth Lake and things like that sooner. Do you think the lake will change?" asked Balynn as she leaned up against a column and sighed. "I hope it won't change the lake too much. It's beautiful now and I know that Daddy is proud of that lake."

June laughed. "If the new road does anything to the lake it will make it better, Balynn. Your father seems to know how this road will help him and I know *he* has thought the lake part through, if nothing else. I think it might even make Drumsworth Lake bigger...maybe too big. That would cause problems with Urhonordo and the Lowlands. I trust Queen Luna will see that that doesn't happen."

"And she will be here tomorrow?"

"Yes...so I'm not worried. She will mention the lake. The ones I'm worried about are the landholders. They won't be here tomorrow to hear what Thorndike and your father promised they would do. They are going to be expected to work very hard for something they *don't* want. You heard them at The Gathering. They were definitely against the road, although I'm not sure why."

"Too much work, do you think?" asked Balynn sincerely.

"No... Farmers work hard anyway. I don't think work was the main issue. I think the concern was for the big road levee blocking their irrigation."

Balynn shrugged. "Well, Queen Luna and her daughter will stick up for Urhonordo interests, I'm sure, and no one cares about what the Lowlanders think so it will probably all work out, right?" She smiled at June confidently.

June raised his eyebrows. He hadn't thought about the Lowlanders. He hadn't thought about them at all. He didn't know any. "What would happen to the Lowlanders if there was too much water down on their land?" he began, but it was too late for more conversation.

Drum, Thorndike and the King had arrived next to them on the colonnade, still smiling and chatting about the pleasure that they would enjoy from a carriage ride on their new road.

Drum, however, now noticed the couple in front of him. *Marvelous,* he thought to himself. *My beautiful daughter is holding up her part of my Plan with June. She and June can manage the southern castle at Drumsworth while I rule all from River Kingdom Castle...a kingdom with two castles...it has never been done before.*

Balynn sighed again. Her father had arrived here on the colonnade. Any further personal discussion with June was impossible now...and just when she wanted to ask him more about the 'lazy landholders' as her father referred to them. But at least she had discovered that June was nice. She would see him again, soon. The new road would give them more chances to meet, she was sure.

With a wave and a smile to June, Balynn climbed into the elegant carriage behind her father and Thorndike.

Prince *June wouldn't be so rude as to make me climb in behind them without assistance,* she thought to herself. *I know Father doesn't care about women. Sometimes I think he's as bad as Thorndike.* But Balynn was used to being ignored. It had always been that way.

For a while, the carriage bumped along in silence. The Drums had a long way to go. They would drop Thorndike off at his manor but it would take hours of bumping and thumping and stopping and starting over irrigation ditches before they arrived at Drumsworth Castle. And her father would force his driver to drive straight through with no stops.

He always does that, Balynn thought miserably. *And then we will arrive home where there is nothing to do and no one to talk to except Lowlander servants.*

Balynn scrunched down into her seat beside her father. Thorndike began snoring loudly as he relaxed into a sprawl.

Balynn shivered. The coach felt as cold and damp as the outside. The triumphant Auras had followed them and her father was in a good mood. Baylyn suddenly had a brilliant idea.

"Father," she whispered sweetly. "It will be such a long ride home and I'm already shivering. Wouldn't it be lovely to stop for a while at Thorndike Manor before going all the way home? It is so chilly and I, for one, would love to have a cup of something warm before that long journey."

"We'll go straight through, like always," grumbled Drum.

"But Daddy, it's such a long way home after we pass Thorndike Manor and this is already our second trip today."

"Get used to it, girl. You're going to be doing a lot of this kind of travel before you finally land Prince June. I've already arranged for you and me to come back up here in a week or two to sign official papers for our deal with the King. You were sweet enough with June today so he'll be glad to see you if you come back with me. By the way. What did he talk to you about when you two were together? Did he say anything interesting?"

Balynn sulked. "He didn't get a chance to say hardly anything before you made us leave," she pouted. "He was worried about the landholders, that's all. I didn't get a chance to ask him why."

Drum sat up straight. "June was worried about the landholders? *That* is a problem. *He's* a problem. I had big plans for him but now I'm worried about him, already."

He nudged the snoring Thorndike. "Wake up. Thorny. Maybe we are being too smug. We left a 'negative force' in The Castle that will be aimed directly at our Plan. And I didn't even think about it until my daughter brought it up."

"Don't worry, Drum. We've got this Plan all sewed up. Let me get some sleep before I get home and have to explain what happened today to my militia. They will want to hear about the big decision."

"Your militia can wait. *This* problem is more immediate, Thorndike. I just remembered that Prince June is the idiot that started the so-called food sharing program in River Kingdom a few years back."

"So?"

"So, *the Prince* talked his father into doing it. June is soft hearted and soft headed, but he can obviously be persuasive."

"What has that got to do with us?"

"So now I just realized that we left 'soft hearted June' alone with his father and June is worried about how the landholders feel about the new River Road. He will probably discuss that with his father."

"How did you find that out?"

"I told him," Balynn said proudly. She was at least part of the discussion, for once.

"How did *you* find that out," sneered Thorndike.

"June himself told me," retorted Balynn with her nose in the air. "You men don't know everything." The Aura that had fallen asleep in her hair earlier wiggled loose and joined the other Auras that were dozing in the carriage. The already happy Auras now saw the opportunity for a new advance. This day had gone better than they expected but it might have even more negative potential.

"Hmm...Interesting...What else did the Prince say, daughter?" asked Drum. He looked at her closely.

Balynn was silent. She thought for a minute. The Auras circled her with the loneliness and irrelevance she had always felt with her father. "You are becoming important" they whispered. "How does it feel? Your father loves you when you are on his side. Stay *focused on that*. He is your father. Please him and make points for yourself with him."

Balynn had talked to June because she wanted to get to know him. That had been a uniquely important personal step for her, but something even more uniquely rare was happening to her right now. *Her father was paying attention to her.* He was calling her 'daughter' instead of 'girl.' He was asking her a question that he actually wanted an answer to.

"Your father is interested in you for the first time. If you help him, he may love you for it", the wiley Auras continued.

Balynn's still inexperienced Conscience struggled through the barrage of the Auras. "If you tell your father what June said, won't you be betraying June's friendship with you?" Cbalynn asked.

Balynn closed her eyes. Her father was right here. He was looking at her with interest. He was actually *seeing* her. "Your father's Plan is the most important thing in his life and you have a chance to be part of it right now," the Auras whispered intensely. "Just think how nice it will be to be *part of The Plan* instead of simply 'marriage bait'.

"*You* will be of value for what *you* can contribute as a person. If you play your cards right, he will continue to value you. Think about it." The Auras knew just what to say.

Balynn looked her father in the eye. "I have much more I could tell you, but it is very cold in here," she announced with new-found strength. "If we can stop to rest for a while at Thorndike's, I will tell you what else I know...and what else I think I can find out."

The Auras were on a roll. They had not expected this much good fortune again today. The Plan had become real in the eyes of Drum and Thorndike. These men were a pair they could count on as they both had very active greed and resentment to build on. Now Balynn was joining their plot. *She* was lonely and ignored. *She did* have potential. They could feel it.

Thorndike was no longer sleeping. He was wide awake. The fact that Balynn might have more useful information was tantalizing. And the Auras that were now winding around all of them, had given him another idea, something that even Drum had not thought of. *Maybe I won't bother to let him think of it,* either, Thorndike thought with a burst of new Aura-inspired bravado. *Balynn is beautiful and I have two grown sons in my militia. One of them may need a bride. Drum's daughter could become a useful bride. She would probably love to wed a 'Junior Commander'... Hmm...*

"I think your daughter has the right idea," offered Thorndike in a surprisingly friendly manner. "Why don't the two of you spend another night at my place, Drum. I would like to hear what you have to say, Balynn. It could be very interesting and helpful. What do you say, Drum?"

Drum was shocked by his partner's sudden change of heart...and, frankly, a little suspicious. Thorndike was never friendly. This was not his style.

Suspicion was one of the Auras' primary tools and when the Auras were active, they didn't even need to choose sides. They could put evil cohorts *against each other* as well as against the known enemy. Right now, the Auras had a functioning grip on everyone in the carriage.

"Watch Thorndike," one of Drum's Auras warned him. "He may be up to something."

But Drum was still intent on what Balynn had to offer in terms of useful *Royal* information. "Thank you so much for the offer, Thorny," Drum responded readily. "Balynn would like that."

Balynn smiled at her new-found importance. As they drew up to the Manor, the Auras settled themselves snugly on all three, determined to keep or enhance the momentum they currently had. This was not a time to ease back.

Balynn's poor Conscience was still with her, of course, but confused. She was inexperienced. A direct attack of the Auras on *her* human was something she had not confronted before. And she had not attended The Conscience Alert. The only thing she could think of to do was to stay with Balynn and hope to figure it out.

Florence Thorndike (Flo) heard the carriage pull up to the manor before she saw it. With dread, she straightened her skirts and apron and placed a smile on her face. This was the 'armor' she wore to please her father.

Her Conscience spoke up. Cflo was now back on speaking terms with the lonely young woman after hearing her story as she had told it to Balynn yesterday.

"I see now that you are not a bad person," Cflo had told her. "I should have seen that earlier. "You are just a person who is trying to survive. Be brave. *I* will be a braver Conscience. I am learning, too."

Flo rushed to the driveway to wave at Balynn. She was sure that the visitors wouldn't be staying long, and she considered Balynn a friend now. But she was shocked to see Balynn not only stepping down from Drum's carriage *but being assisted by her father as though she mattered.*

Hmm...With survivor's intuition, Flo stepped back. Something had changed. Now everyone was walking slowly *together* toward the front of the manor, smiling and chatting as though they were friends coming home from a party.

Flo turned and rushed back into the manor. She had not expected this. Her two brothers were at the dining room table. She invited them in often when no one else was in the manor to see them. They were her brothers, after all.

Thorndike would not be surprised to see his sons inside, but unrelated guests had never seen these boys treated as sons. The world did not know about Thorndike's children. The militia was the militia and not a related family group. His daughter, the servant, was even less significant.

What *was* true was the fact that Flo's brothers ate like...militia men. They were messy. The table was piled with all the things that big men ate and drank. Pieces of some of these things had landed on the floor and dribbled from their chins. They were not expecting company. In fact, they had *never had company*. Flo panicked.

"You can do this," coached Cflo. "Your brothers are not neat, but they love you. They will do what you say. Tell them that today they are *militia only, not sons or brothers... definitely not Thorndike's sons...militia only.*

Flo nodded and went to work on the young men who, because their father had drilled it into them, minded her orders without question. They also loved their sister and didn't want her to get in trouble. Within minutes, the untidy mess around them was nearly cleared and the boys had mopped their chins.

Without any thought to what was going on inside, Thorndike ushered Drum and Balynn into the entry hall. "Florence! Are you here, girl?" he yelled rudely. "The Drums are back. Where are you? They'll want to eat. Serve us without delay." He turned and marched toward the dining room.

 Florence quickly appeared at the door of the dining room, harried but with a clean apron. "I invited some of your militia to eat inside, today, milord," she tried to explain calmly. "There were only two, but the table is not ready for company. Perhaps, your guests would prefer to eat in the conference room?" She curtsied and smiled nervously.

Balynn smiled back. "It's good to see you, Flo. Have you met my father? Daddy, this is Florence, but she goes by Flo. The Aura slid away. This was not the time or place to be active. They were strategic.

At least Balynn's not changed, Flo thought gratefully. "I am happy you are all back, but I am afraid I was not prepared. If you would like to go to the conference room, I will fetch you something to eat and drink." The Aura continued to hover silently.

Thorndike, his new idea about 'the informative Balynn' still active, had a slightly different idea. Carefully he opened the dining room door and peered in. His boys didn't look that bad. Flo had made them straighten up. He stuck his head in further. "You are my 'most favored militia, right'?" he said pointedly.

"Yes, sir!" came the prompt answer in deep throated unison.

They understand, Thorndike thought with satisfaction. He opened the door wide and motioned proudly to Drum and Balynn. "I would like you to see two of my finest militiamen. I have these two at the ready *inside* from time to time," he announced.

Drum and Balynn had not expected anything like this, but they were impressed. Before them, seated at the far end of the huge table were two large, handsome young men. Together, they rose and turned to face Thorndike and his visitors. They slapped their chests in salute. "At your service, Sir," they bellowed, also in unison.

Drum was impressed. Balynn was enthralled. She had never seen this much fine male presence aimed in her direction before. And they were beautiful...strong... manly...and staring at her. And they were exactly alike. They were twins.

She almost fainted away. She wanted to squeeze Flo to show appreciation of these wonders that she knew to be Flo's brothers. But she couldn't. Flo's eyes held a plea for her silence. Balynn tried to gather herself, while her father thankfully commented on how fine and 'ready for anything' they looked.

Thorndike stood back with fatherly pride. "I have trained these men, myself," he announced. "These two are definitely my finest. If we have any trouble with the landholders or anyone else, these men will end it in a flash."

He surveyed the men (his boys) for a moment. *Which one will I choose to make a move on this wench beside me?* he thought as he reviewed his male inventory.

"Either one of them or both," the Auras responded happily.

Thorndike was ready to focus on his new target, Balynn. With a cordial look that he hadn't used forever, he escorted Balynn and Drum into the conference room. Flo watched her father's behavior and made a mental note. Something had definitely changed.

The conference room held the trappings of a war room. Maps hung on the walls. Swords and shields leaned in the corners. The drapes at the windows were of quality but gray at the folds, showing they had not been drawn back in years. This was now a secluded, secret planning place. Balynn knew that she had been granted access to a location that few women would ever see. She was thrilled.

Balynn sat up straight but fiddled nervously with her fingers. "What are you going to tell them that keeps you in this favored position?" the Auras coached. "They expect you to say something interesting and useful to them. What do you know that they would like to hear?

"They probably want to know everything that the Prince said," reminded the hovering Auras. "That's what they'll want to know about first."

He didn't say that much that was important, Balynn thought sadly. *Daddy and Thorndike will be disappointed.*

"Then use your imagination," the Auras directed impatiently. "Embellish. Make it exciting. Get creative if you want to stay relevant here tonight."

'You mean lie?" The Auras laughed like the new conspiracy comrades they had decided to be.

Drum sat across from Balynn and looked at her closely. "Well, let's hear the rest of what the soft-hearted Prince Salem had to say," her father said as he grabbed for a biscuit that Flo had just brought to them. He leaned back and munched, grinning at her expectantly.

"Indeed," Thorndike seconded enthusiastically. "Let us in on what the boy was thinking when we left, girl. We already know that he is worried about the poor, shiftless landholders. I wanted to throttle the whole lot of them at The Gathering. Who do they think they are? They live and work on *our* land. We can get rid of the lazy fools any time we want to. They had no right to speak up at The Gathering."

Balynn nodded sagely. "I know," she agreed, trying to appear wise, "but June thinks that they work hard. He thinks that the new road is going to give them more work on something they don't want and they already work hard. That's what *he* said, anyway."

"Any fool knows that the lazy landholders are just too stupid to appreciate what they have. They should be more grateful for the land we let them work on," sneered Thorndike. "The next thing you know, they are going to want a trip down to Drumsworth Lake for a vacation." He slapped his leg and laughed heartily, biscuit spitting from his mouth on both sides.

"This man is a disgusting boor," whispered Balynn's Conscience. "Stop talking to him. How can you and your father work with a crude, ogre like this?" Balynn heard her Conscience but she was in a tough spot. This was her father paying attention to her, maybe for the first time.

"You are so right," said Drum nodding to Thorndike. "All servants will say that they are overworked even when they have twenty Lowlanders under their control to do the hard stuff.

"You should hear what *my* Lowlanders say when they think I can't hear. They think I should pay them, feed them and even house them just because they take care of my manor and make my food. I tell them, 'Find your own housing. I am not here to take care *of you*. You are here to take care of *me and mine*. If you could do better, you would. You don't deserve anything but scraps and a lean-to.'"

"Listen to your father," coached the Auras. "Listen to your father. He knows."

Balynn was listening. *That's right. They do think what Daddy said they think,* she thought. *I used to feel sorry for them when they were hungry and it was cold. But is Daddy right? Are all servants just lazy?...especially Lowlanders?*

Now she remembered something else that the Prince had said. Eagerly she brought forth her new contribution. "Prince Solem was also worried about the

Lowlanders and Urhonordo and how they would deal with Drumsworth Lake if it changed size because of the new road," she announced proudly. "I thought that you should know that, Daddy," she added with a look of purpose and confidence.

Thorndike and Drum snapped to attention. Drum threw his biscuit to the floor and drew up close to his daughter. "The lake?" he hissed pointedly. "He was worried about Drumsworth Lake? We didn't put the lake on our map yesterday so that no one would bring it up, and June worried about it anyway? See, Thorny? I knew that boy would be a problem."

Balynn nodded. Without a doubt, her father didn't need to know that *she* was the one that brought up the fact that it was missing on their map.

Drum began pacing and ranting about the lake. "That is none of their concern! My lake is my property. They have no right to interfere with what I do down south when they are way up here!"

Balynn thought desperately. What could she say next? Her attention from her father seemed to be slipping away as his anger rose. *What can I say?* she thought. *I am going to lose his attention if I don't tell him something important.*

"It's not *what* you say so much as *how* you say it," advised an Auras.

Balynn leaned forward and looked her father in the eye. "June didn't seem worried about the size of the lake, though," she informed him darkly. "He knew that Queen Luna would be coming to The Castle tomorrow. He figured that *she* would point out the problems with the lake from *her* point of view. I agreed, of course. She will do that and then she and June will pair up against the road. That's the way I see it, Daddy."

"Well said," whispered the Auras.

Thorndike was listening intently. Now he rose and grabbed one of the swords from a corner. Like a general in combat, he strode to the maps on the wall, swinging his sword toward one of them. Balynn hadn't noticed it before. It was like the map that the nobles had presented to The King at The Gathering but very different in one place.

Drumsworth Manor was clearly bigger than it had been before, and the lake around it on this map was huge, more than twice the size of the current lake.

Balynn stared at it. In size, the lake now resembled the size of the lake that had been shown inexpertly on the landholders' map and it wound all the way around Drumsworth Castle.

"Is this still about the way you see it when it's finished, Drum?" Thorndike asked, pointing at the lake with his sword.

"That's it, alright", Drum answered with pride. "That mote' ll be a beauty. River Kingdom will have two Castles to be proud of and a beautiful road to run between them." He smiled a conspiratorial smile. "What do *you* think, Balynn? Do you think that you would enjoy pleasure trips on a new road between *two* castles?"

Balynn stared at the map with a new understanding. For years she had looked at a similar map that was on her father's wall at home. She always thought that her father's map of River Kingdom was a thing of beauty, but she had heard him curse it more and more as the years had gone by. He had cursed The River. He had cursed the size of his manor. But, most of all, he had cursed the size of Drumsworth Lake

Over the years, he had tripled the size of Drumsworth Castle and now he was going to triple the lake into a mote. "A mote is a symbol of power," Drum now growled. "It will be three times the size that has been. If it weren't for the stupid Lowlander and Urhonordo objections, Drumsworth Lake *would be that big already.* And Drumsworth Castle is bigger and more beautiful that River Kingdom Castle, right now. This is the way it was always meant to be," ..."and *will* be as befitting a king," he added.

Balynn was beginning to understand the real importance of the 'access to The Plan' that she was being granted. She was now privy to new and amazing information. Not only did the big, new road stretch from The Castle to the Lowlander border, but Drumsworth Lake skirted around Drumsworth Castle, surrounding it like a mote...like the protective mote *of a king.*

Balynn pulled her cape around her and frowned, but the Auras were not deterred. "This is your father's dream, Balynn," the Auras persisted. "Now you know what your father has been working on over the years. It is what the Birther population of River Kingdom cheers for. It is what your great grandfather had and what your father deserves. Drumsworth will truly be the southern version of 'The Castle' and your father will be the triumphant king of two castles with a wonderful new road to join them."

Balynn's eyes grew wide with the enormity of what she was seeing and hearing. Even with no experience, she knew that something this big could not be done peacefully. *This isn't just about a road. This could mean a war. Do they really want to do this?*

The Aura had answers. "It's simple. Your father dreams of the kingdom he lost when his grandfather relinquished it to King Sol back in the day. Thorndike wants to be the Commander of 'King Drum's' Royal Castle Militia and *all militias* as his ancestor was. They both hate Lowlanders because they and the nobles are afraid that Lowlander darkness will ruin their whiteness. That's why the nobles are your father's biggest fans."

Balynn was stunned. Now she knew what she knew, but what should she do with the information? This was her father's dream versus right and wrong. Cbalynn was overwhelmed as well.

<u>PROTOCOL ERROR</u>

The Castle visitors from yesterday were gone, but Horace was at his duties again in the Conference Room of The Castle. This time he took special pains to make the table look welcoming. Queen Luna and Princess Amanda were expected. He knew them and liked them both and yesterday had been very strange.

Upstairs in the Royal Bed Chambers, preparations for Queen Luna's visit were also underway. Willa was determined to attend this meeting, even if she still bore the signs of her encounter with a tree branch.

Solem had asked her to attend. Prince Armando would be there along with Queen Luna and Manda. It would be a friendly meeting of royals who were used to working together, he said, and Solem was looking forward to it.

Willa was still a bit concerned. She had not yet had a real discussion with Solem about what she had learned on her ride with Manda. She had, of course, explained that Manda had invited her on an outing, but she hadn't told him what she had learned. Manda would be disappointed in her, but she knew Manda would understand. Willa would just talk to them both about it downstairs at today's meeting.

Solem had said nothing to Willa about the surprising road approval at yesterday's meeting with Drum, but he was not at all concerned. He had made a decision with ease yesterday. He had not had a chance to tell Willa about it, but she was very new at Water Rights issues and this only involved River Kingdom's side of The River. He would tell her about it with everyone else at the meeting.

Agnes was fussing with Willa's appearance. She was clucking, as usual. " I know that you want to attend this meeting with Solem today. I understand that. But you are still early in your recovery. You shouldn't exert yourself. Just because you are young, there is no need for you to rush about."

Agnes stared down at Willa and reviewed her handywork. The bruising and swelling were almost indistinguishable. She had colored Willa's lips and cheeks and arranged her hair carefully over the gash on her head. The young Queen was dressed in lavender, a color that Agnes thought was becoming to her fair-haired mistress.

Willa wrinkled her nose at the scrutiny and the advice. "Do I look alright for a 'hot-house plant', Agnes?" she quipped. "Don't worry about me. I am fine. I can hear again. I am strong enough to join my father's militia, for goodness' sake."

"You are too young to know your own limitations," Agnes responded, but she smiled.

"Maybe so, but Solem would like me to be present in his meeting with Urhonordo today and I don't want to let him down. Besides, I want to get to know Queen

Luna more. Now that I know Manda, I am sure I will like her mother. I hear that she and Queen Linda were friends."

"Indeed. They were friends for years. You will like her, but Luna is very different from her daughter. She is a highly educated *lady,* not a tomboy. And then, of course, the Crown Prince Armando will be there. *He* is a gentleman. You will like him. too."

The upstairs chit chat continued, with Agnes and Willa while only a few feet away, Solem studied the map that had been given him by the nobles. It was spread out on his bed which Agnes had straightened for him for that purpose. Studying in his room or the upstairs sitting room was his practice before meetings when issues were of particular importance.

Now he scratched his chin in thought as his eyes traced the path of the new road. "This road looks like a thing of beauty," he mused quietly. "I know that Princess Amanda had some concerns at The Gathering yesterday, but I'm sure that we can work out a solution for whatever her concern is. What do you think, Willa?"

Willa felt trapped by what she had learned and had not told Solem yet. Slowly she walked over to the bed where the map was spread from side to side. The picture showed the road to be wide, smoothly topping all the farmers' inlets and smaller levees. It sat gracefully on its tall, winding levee, overlooking green fields that represented crops while a lovely, blue river flowed peacefully on the other side. It was truly a beautiful sight. Strangely, there wasn't even a representation of Drumsworth Lake anywhere on it.

Willa's Conscience snickered in her ear. "You are a scaredy cat. You're not going to say anything about what you learned with Manda, are you? You know what's wrong with this picture and you aren't going to say anything...Ha! I thought you had pledged to help your new husband be a good king. Being a coward now isn't going to help him, you know."

"I know," Willa whispered, not realizing she was speaking out loud.

"What do you know, my love?" Solem responded.

Willa stood up straight. Her Conscience was right. "I know that there is more to this road and the problems it might cause than shows on this map," she said clearly.

Solem turned around and looked closely. "Really? You don't say. Do you have opinions about the new Road that you haven't shared with me yet?" his tone was congenial and interested.

The Call of the Auras

"Yes, Solem. I *do* have some thoughts. I should have said something earlier, but..."

Solem hugged her carefully and smiled. "Don't worry about it now, Willa. You can speak your mind at the meeting. We are all friends here today. You can say whatever you wish, but I think I hear our guests' carriage arriving on the colonnade already."

He rolled up the map and tucked it into his vest. "Would you like me to help you down the stairs? You may still be weak. I don't want you to fall." He opened the door to the hall. June was right there.

"I was just coming to get you two", said June shuffling his feet. Then he nodded to his father. "There are two lovely ladies in our front hall that are asking for Willa. Would you like me to escort you down, Willa?"

Solem raised his eyebrows at his overly helpful son. "Thank you, son, but I believe I will take care of my lady, myself. Please go down and escort our guests into the Conference Room and we will join you." June disappeared down the stairs as Solem held out an arm.

"I'm not that delicate, Solem, dear," Willa whispered as they descended the stairway carefully together.

Horace had arranged the seating for the Royals in positions of appropriate protocol. After cordial greetings between the Royal adults and careful hugs between Manda and Willa, the meeting began. Deliberately, Solem removed the map from his vest and spread it on the table with the lovely representation of The Castle in front to him and the position of Drumsworth at the other end. This was the lake-less map Drum had left with him yesterday. "I thought it would help if we take care of this item on our agenda first," Solem said with a smile.

To his surprise, Luna pulled out a map as well and Manda unrolled it in the opposite direction so that Drumsworth Manor and the lake were in front of the King. "I suppose we can look at two maps if we want to," Solem laughed. "They are the same, are they not?"

"They are not," said Prince Armando quietly. "They are very different, Solem. Look closely." He stood and walked to Solem's end of the table. "There are two major differences. They are both very important, but the one that is most important to us right now is the one right in front of you. Do you see Drumsworth Castle?"

"I do. It's right under my nose."

"Indeed. And do you see Drumsworth Lake?'

"Yes, of course, although it looks to be overly large. It looks more like the one that the farmers drew...but more carefully rendered," he added with a smile.

"Thank you, Your Highness," chirped Manda with a grin. "I drew it, myself."

Those around the table chuckled briefly but the King continued. "Why did you draw the lake this large, Amanda? Drum's lake in your rendition looks like it flows around his manor and into the Lowlands and even Urhonordo. That makes it much bigger than it is, doesn't it?"

"Yes, it does," answered Prince Armando. "It makes it bigger than it is now, but it makes it as large as Luna and I think it *will be* soon."

June and Willa leaned in to see the drawing more closely. Drumsworth Lake was huge and blossoming out into both Lowland and Urhonordo territory. Willa was surprised that this was their issue. She hadn't even thought of this big lake as a problem.

Now June was up and examining the two maps. Slowly he walked down the table's length. "It looks like you think you think that a lot of water is going to end up in the lake," he said as he peered at the map. "Both maps looks the same to me except Drum's map doesn't even show his lake.

"I noticed that yesterday," June continued. "I just thought that was because the lake wouldn't be affected much by the new road. The River is The River. It has flowed the same way for a hundred years. We have to tinker with the outlets to Drumsworth Lake every spring to keep it from overflowing and give the Lowlanders their share of the water, but we know how to do that. We'll just do it again. Why did you make this version of the lake so big, Manda?"

Manda stood up and looked June in the eye. "Because that's how big the lake is going to be. And look again, June," she responded with a friendly nudge of her elbow and a challenging smirk. "Look closely. Don't you see some scratch marks along the side of the new road?"

"Yeah. I see 'em," he answered, returning the smirk and standing back with confidence. "I wasn't going to say anything. I just figured that you didn't know how to draw maps that well. You should take lessons."

Everyone laughed but Willa. Willa knew what the scratches on the new road were. "Actually", she said slowly, "I think Manda drew the road pretty well, June. Those aren't scratches. Those are small levees as they are now, before The Royal River Road fills them in. I think that's what they are. Am I right, Manda?"

Solem looked at Willa. What did *she* know about this?

Manda sat back down in her chair and folded her hands in her lap. "You are exactly right, Queen Willa," she agreed stoutly. She looked smug. "These scratches are the thousands of small, landholder levees and inlets that keep The River in check and water our crops now. They also help to deter flooding in the south."

June grinned. "They don't look like anything but scratches to me."

It was now time for Queen Luna to take control of the discussion. This was too important for the young people to dicker about. "King Solem, I understand that Drum was here yesterday and made a special presentation to you about the new road. I believe that his 'special presentation' without mutual consultation was not proper protocol, however. Something as large as this road should have been presented to both our Kingdoms and The Lowlands together at the same time as it will affect us all."

Solem stopped short. He was a firm believer in protocol. Being told by another monarch that he was not following protocol was something that had never happened to him before. He was insulted.

"What I do with the people that I entertain in *my* castle is not up for review, Your Majesty. Lord Drum is a subject of *my* kingdom. He requested an audience with me and it was granted." Solem sat stiffly and stared straight ahead...end of discussion.

It was Luna's turn to stop short. King Solem was a good man, but protocol was important to him as she knew. Her Conscience was in her ear in a second. "You insulted the man, Luna," whispered Cluna. "You know how he feels about protocol. You spoke too fast and too harshly. You had better fix it."

The meeting was silent. Everyone simply sat in their chairs. Horace was dumbfounded. He had watched many of these meetings from the sidelines for years. They were always friendly and cordial, but not today. He didn't like the change.

Now *his* Conscience got active. "These are good people who are at odds with each other for some reason. Small acts of kindness sometimes help," whispered Chorace.

Horace walked to the overly quiet table with a tea pot. "Would anyone like some tea?" he said into the silence.

"Yes, please," said King Solem. "Would you care for some tea, Queen Luna?"

"Indeed, if you please, Your Majesty," Luna replied. The negative spell was broken.

"Your turn, Luna," whispered Cluna.

"I apologize, King Solem, "Luna said softly. "I was too harsh in my statement. Of course, you are the ruler of *your* kingdom and lord of *your* castle. Of course, you can entertain anyone you wish. I spoke out of turn."

"So did I, Queen Luna." Solem admitted as he sipped his tea.

"Let me start again, please." Luna continued. "What I *should* have said is that I was concerned when my daughter told me that Lords Drum and Thorndike were here to pitch their 'Royal River Road' to you, alone. I have looked closely at the situation, and I believe that this road may cause harm in several areas."

Solem smiled. "I understand what you are saying now, Your Majesty. But I had no idea that you were concerned about the road at all. I know that Amanda voiced some concerns at The Gathering. She left the meeting I was having with Drum and Thorndike early as I am sure you know. I don't blame her for that. Thorndike is a crude sort. He is not my favorite person."

"He is disgusting," contributed Manda. "And he is up to no good."

"We need to discuss this on an impersonal basis, dear," said Luna with a smile. "You are talking about the Queen's father."

Willa sighed but said nothing.

June was now ready to contribute. " I brought up your issue about the irrigation of the crops, Manda," he said with pride. "I think they have that issue worked out well, though. You see, they have these two small rivers that cut inland from The Great River. They will be set up to water the crops. Pretty smart, right?"

Manda shook her head. "You don't really believe that those two streams will water the entire west bank of River Kingdom, do you, June?"

"Indeed, I do. Why won't they?"

"Because those small canals are on Drum land and Thorndike land alone, of course. They won't water any of the Central land in between. Nor will they hold water as inland holding pools do. They don't flow up. Those rivers will flow down, causing possible floods. The planned new road will completely plug up the levees and holding ponds of the Central lands between them so that land will go dry and so will their crops...while Urhonordo floods. Surely you can see that."

"I trust they will figure that out," snorted June.

"'Trust' is one of the main things we need to talk about," Prince Armando now offered. "I have no problem with Thorndike, yet. I don't know him. I do, however, have many concerns about trusting Lord Drum."

"You don't trust Drum, Armando?" Solem raised an eyebrow. Armando was usually very positive.

"Drum is not a person that I trust at all," Armando stated flatly. "He is harsh with his Lowlander servants and cruel to his farmers. All are Lowlanders. They tell me that he does not respect any of them...or pay them...or feed them. Many of those that work for him are afraid of him. And he cheats the rules that the Water Rights Gathering has set up for governing his lake unless forced to comply. I would not take his word for anything without proof."

Solem was stunned. "I had no idea that you felt that way," he said. "I had no idea, at all. You have never told me this in our previous discussions."

"I regret that now," Armando admitted. "I should have said something before. But our work up here in the north has always been good and I didn't want to burden you with issues that primarily affected only Urhonordo and The Lowlands."

"I told you that you should have said something about Drum before," whispered Carmando.

Solem was beginning to have a very uneasy feeling. He looked around him. These were all people he liked and respected and they didn't seem happy at all. Was the new road really such a big issue? Had he said yes to the road too quickly? He had been so sure.

"You were smug, yesterday," chided Csolem. "You thought you knew it all and you made a snap decision without your usual protocol. You didn't consult with Luna, did you?"

You're right. I didn't. I jumped at Drum's idea. Now that I think about it, why did I do that? Solem thought to himself.

"You were overcome by some Auras," conceded Csolem. "Me too...That's exactly what happened. I slipped up in my Conscience duties at the same time. The idea of Auras didn't seem relevant at the time. Under normal circumstances like today, we don't have Auras here. I take the blame. It was my fault. I should have had my guard up. But I have it up now. I won't let it happen again."

I am the one that made the bad decision, Solem thought. *But now what do I do?*

"A big man admits when he is wrong," counseled Csolem, trying to get back on track.

Queen Luna was looking at Solem and smiled at her fellow monarch. "I think I understand, Solem," she said quietly. "You appreciated the look of the grand new road and I agree with you that it looks grand, but it cannot be built the way it is currently drawn. The water of The Great River is too important to the farmers. As it is now, although The River is mighty, it is coaxed into our fields and orchards by thousands of levees dug by individual farmers and landholders on both sides. That is what grows our food and has for over one hundred years. "

Solem studied the huge lake drawn around Drumsworth Castle in front of him. "Will Drumsworth flood, too?" he asked.

"Indeed, it will," said Prince Armando with a knowing nod. "It will flood right into an enlarged pool around Drumsworth Castle. That enlarged pool will find its way onto Urhonordo and Lowlander land. I believe the digging onto Lowland territory has already begun to get ready for that."

King Solem walked the length of the table. Armando walked with him, pointing out the dry areas on either side of the canals that were far north and far south leaving the wide distance between. Amanda rose to emphasize her hundreds of 'scratches' to the map and pointing out how the water from the small levees currently travelled into holding ponds on the various farms.

"Surely there can't be that many small levees," Solem mused.

Willa spontaneously jumped to her feet. "Yes, there can, dear" she added with a smile. This was something she knew. "There are hundreds of them."

All conversations stopped. Everyone turned and looked at Willa. The new Queen had an opinion? What did Willa know about any of this?

But she had jumped to her feet too quickly. Now, her face turned white, and her scars glowed under the powder that Agnes had used to disguise them. She felt suddenly faint. *This was not the right way to tell him what I learned,* she thought grabbing the side of the table. She sat down weakly. *When am I going to learn to hold my tongue*? She was dizzy. Agnes had been right, as usual.

Amanda ran to Willa's side. "Willa knows the truth but it's my fault she got hurt. I made her go with me!" She steadied Willa carefully and looked at Solem with defiance.

Solem just stared at the two young women. June, however, understood the whole story at last. "Willa went with you so you could *show her* the little levees, didn't she, Manda? That's where she was coming from when she ran into the tree, right?"

"Is this true, Princess?" asked Solem.

Manda stood up straight. "Yes, Your Majesty. Willa went with me so that I could show her the levees. And everything that happened after that is my fault... everything. If I hadn't asked her to go...and then let her ride home on her own, she would never have been hurt. It was all my fault."

"Are you going to let Manda take the blame?" hissed Cwilla.

Willa's strength was coming back. "Be still, Manda. You speak of me as though I am a brainless child. I am not. I went with you that morning because *I wanted to learn*. And I did. I rode home by myself because *I wanted to and was able*. I fell because I didn't look where I was going."

She turned to Solem with sadness. "You should be mad at *me* for not telling you what I learned sooner but don't be mad at Manda. She was just trying to make me less stupid about The River. And what I learned was worth it, Solem,"

Willa now spoke with conviction. "I saw that all those levees are necessary. Hector explained everything to me. He was very worried. I think we really need to hold off on approval of the new road for a while until we understand all the problems it might cause. What do you think, dear?"

Solem sat down slowly in his suddenly uncomfortable 'position of authority'. His gaze focused on Willa first. He understood her feelings about what she had 'learned'. June and Amanda had asked about the levees before. Queen Luna had mapped out the possible problems with the road and flooding...but he had already approved it.

"Nice move, Solem," snarked Csolem. "Everyone you love and respect thinks that the road is a bad idea and *you...we...* approved it without consulting any of them. We blew it."

Solem knew this was true. He had not thought this through well enough, but what could he do now?

"To begin with, you have to admit what you did," advised Csolem. "Especially your own, precious new wife deserves the truth."

Solem took a deep breath. "Please take your seat, Princess Amanda," he said quietly. "You did nothing wrong. You took my lovely wife with you to show her what you thought she would want to know."

Solem took Willa's hand. "I know you would have shared your findings with me if you hadn't fallen," he murmured to her with a quick smile in her direction.

Willa squeezed his hand.

"You have all been honest with me," Solem continued. "Now I must be honest with you in return." He took a deep breath. "I have already approved the Royal River Road. I understand now that I was probably too hasty, but I did so yesterday."

The entire table of assembled royalty inhaled at once. Those who were not already sitting, sat. There was a stunned silence.

Queen Luna's look was incredulous. "Why? Why did you do that, Solem? Of all the times that we needed to work together, this time was the most important. Why did you decide this on your own? That is so unlike the way you have worked in the past."

Slowly, Solem rose from his chair and walked to the back of the room, avoiding their stares and hoping he would somehow find an answer there.

Csolem was also unhappy. Silently, he accepted his share of the blame for the spot Solem was in. But Solem couldn't undo what was done now...or could he?

Csolem suddenly remembered a remark that Solem had made to Drum. *This might save Solem today,* he thought.

"There *is* a possible way to stop Drum before he goes too far with your approval." Csolem whispered in Solem's ear. "The road cannot be begun until legal paperwork is completed. When Drum comes to do that, you will have a chance to 'modify the plan', 're-work it' and 'include others'...in other words, do then what we should have done yesterday."

Solem stopped his pacing. *The paperwork hasn't been signed yet! My Conscience is brilliant. Why didn't I remember this? The hasty decision I made is not official yet.*

Solem returned to the silent table. These people were important to him. They were all still watching him with disappointment on their faces. King or not, he didn't want to fail them.

Solem struck a kingly pose. He told them of the solution he had remembered and implied that he had had this idea in mind all along. (Csolem knew that wasn't true but decided not to say anything. They both needed to make up for their mistake together.)

The people around him raised their eyebrows but smiled in understanding. They gratefully accepted his information even though they understood that he had just discovered this solution for what he had done. Now they could work together. These Royals and their Consciences were all an understanding lot. Slowly the tension among them faded.

No one was entirely happy, but all agreed that another meeting with Drum and Thorndike was the best thing to do next. Getting *all* of them together was essential to work out the details of such a huge project.

Suddenly, June remembered something that Armando had mentioned. The Prince had said he thought work on enlarging Drumsworth Lake had already started. "Do you really think that's true?" June asked the monarch.

"It is," shrugged Armando. "I have seen it. It almost looks like a mote and the back half of it is on Lowland territory. As I said before, Drum cannot be trusted. Even if we agree on a plan with Drum for the road, we are going to need to watch him closely on his changes in Drumsworth Lake."

June didn't hesitate. "I would like to see what he has done already, Your Highness. Father was worried about giving Drum approval to start the road but if he has already started on enlarging his lake *without Water Rights approval*. He doesn't seem to think that the rules and laws apply to him."

"Excellent idea, son," agreed King Solem.

Armando and June made plans to travel together tomorrow to a spot where they could see what was happening at Drumsworth Castle. The Royals completed the remaining Water Rights business that their citizens were awaiting and concluded the rest of the day's meeting early.

Solem and Willa waved their farewell to the Urhonordo Royals and June from the colonnade. June decided that he needed to take a more active and informed part in this Royal River Road decision. He would return to Urhonordo with them so that he could continue down toward Drumsworth in the morning. Drumsworth Lake sounded like it could be an issue...and this would also give him a few days to stop thinking about Willa.

Solem and Willa watched them depart and breathed sighs of relief as did their Consciences.

DRUMSWORTH FLASHBACKS AND REVELATIONS

Balynn and her father had returned yesterday to Drumsworth Castle after a trip that had been eventful to say the least. Balynn finally knew The Plan. That was huge, of course, but now she had many things to think about, many questions to ask and no one available to answer them.

"*I am here,*" whispered Cbalynn. "Don't forget about me. As your Conscience, I will always be with you when you have a lot of thinking to do, especially after that evening at Thorndike's."

"You're right", sighed Balynn. "And I know you will try to help, but I wish I had a human to talk to, too. What do you know about maps and water rights?"

"Not much" Cbalynn admitted, "and motes are not my specialty, either."

Balynn stared out the front windows. Now she knew secrets. Was that a good thing or a bad thing? Until now, she had only known one important secret. She smiled. That secret was the one she had always loved...the secret of The Crown Room. Her mind drifted.

Balynn couldn't remember her mother. Lowlander servants had cared for Balynn's basic needs, but that was about all at first. Her father didn't want her to talk to them, so Balynn had wandered about the manor by herself. The newer parts of Drumsworth were locked off, however, as though being held in reserve. Only the older part of the mansion, her room and adjacent bedrooms were available to explore. *What's in the new towers and the ramparts of the rest of the space,* she wondered. *And there always seems to be building going on.*

And even in the common area, one room was always locked. That room was called "The Crown Room" and had always been a place of mystery to young Balynn, something she wanted to see but had been afraid to ask about.

But she had finally discovered its secret back while still only a small child. Her father seemed to use the room for his 'alone time'. She had often stood near when he was inside, waiting to be invited in but that had never happened. This day, however, the wandering Balynn had discovered The Crown Room mistakenly left open. Drum was in conference with nobles as usual.

She tiptoed in. The room was very fancy, even fancier than the rest of the richly decorated mansion. Keeping a watchful eye on the door of the room, young Balynn began her exploration. She opened drawers. She slid books around on bookshelves. She opened cabinets...nothing.

Then she noticed an imposing painting over the fireplace. It was strange and a little creepy. It was the portrait of a man that looked very much like her father, but he was dressed in old fashioned, fancy garb. The man was smiling down on the room as though he was superior to all and was seated on an elegant chair that almost looked like a throne.

In fascination, Balynn approached the opulent portrait and ran her fingers over the rich curves of its frame. *Father must love this picture. It must be very special to him. This must be the reason he keeps it hidden away,* she had reasoned.

That was when it had happened...Slowly, silently, the frame moved forward on its own. The frame was a door in the wall!

As though pulled by a magic hand, the huge frame swung toward the tiny Balynn. For a moment she saw what was inside. Gleaming in gold and rich, purple velvet was a *real* crown, the very crown on the head of the man in the portrait! Little Baylyn ran from the room, afraid of being detected in the forbidden room. Her father had never suspected.

Balynn remembered that day of The Crown Room discovery well. *She had always treasured her secret.* Today, after the amazing trip to The Castle and the revelations of the Thorndike maps, she thought she understood at last the reason that the room and the secret crown hidden inside were so important. The crown was her father's prophecy. It was... The Plan.

Today, the secret of The Crown Room didn't give her the same comfort that it had in the past. She was unsettled. If the hidden crown was part of the plan as she now thought, it seemed sinister and foreboding. How and when would it fit in?

And she had so many questions about so many *other* things. She thought about Lady Thorndike who was still locked in a dark wing of the Thorndike Manor. She thought about the lonely young servant, Flo, who was still keeping her locked there while she served Thorndike and his sons.

And Balynn thought about the map on Thorndike's wall and the Auras' whispers that had circulated in that room. Had those whispers been real or just her imagination?

"I finally know one thing about that," announced Cbalynn proudly, reading her thoughts. "I can now tell you that those whispers you and I thought we heard were coming from something called 'Auras of evil'. I was told about them once, but I never knew they were real until we were there at Thorndike's and I heard them. Could you hear what they were saying? They seemed sort of pushy and I hear that they are definitely evil. What did they want out of you?."

I think they wanted me to join the plan that Father and Thorndike are excited about, thought Balynn to herself. *Now that I think about it, I've heard them say things like that to visiting nobles here at home, sometimes, too. I think they were doing the same thing, supporting my father.*

"I think maybe you're right," mused Cbalynn, thinking back. "If that's true, that means there have been Auras of evil running around loose here at Drumsworth for a long time without me noticing," she said thoughtfully. "I must go on alert and keep watching for Auras for you. They are always a bad thing, and other Consciences say that they can lead people to do awful things without their even knowing."

I think I am smart enough to figure that out and stay free of evil whisperers, thought Balynn with a smirk.

"Don't get smug," warned Cbalynn.

Balynn wandered on. The servants were all working hard somewhere in the manor. There was only her inexperienced Conscience to talk to and Balynn had come home filled with such mixed emotions. The vision of Thorndike's map kept popping up in her mind. The map showed Drumsworth Lake extending all the way around Drumsworth Castle. Balynn knew that wasn't a part of the discussion with the King and June. Did that mean that the new version of Drumsworth Lake was part of The King's approval or just her father's. Maybe her father was keeping that part of his Plan secret. Maybe it was a surprise. If so, why?

The autumn sun was still shining *outside* and Balynn was *inside* as usual. *That is really dumb,* she thought, now feeling more liberated and inquisitive since her experience at Thorndikes. *What I need to do is go outside and get some fresh air. I can think things through out there as well as in here...better maybe.* She gathered up a shawl and slipped out the front door.

The air was cool and the lake before her was glistening. It was still beautiful, bushes trimmed and lawn benches still available, patiently waiting for visitors that would not be here until next summer.

There is no reason why this lake should be made larger, she thought, remembering Thorndike's massive map. "That would just be more for Daddy's Lowlanders to keep neat."

The front of the manor was kept immaculate by the hard-working Lowlanders, but there appeared to be an untamed jungle around the sides of the manor. No one ever went there. Balynn, as house bound as she had been in the past, had never even ventured into that ignored area. "Probably no one has ever made their way around the sides to the back of this place", said Balynn to herself, "much less build a mote. Why in the world would anyone want a mote back there, anyway?"

"For protection, of course," came a deep voice at her side. Balynn jumped. Then there was laughter.

Glancing quickly to her left, Balynn noticed a small boat that was moored at the edge of the lake. More startling, there was a man standing beside her and a second man was jogging toward her. There were happy, almost familiar grins on their faces.

"Hello, Balynn," the closest one of them said in a friendly fashion. "It's good to see you again. We had hoped for weeks that we would meet up with you someday and I hoped, after yesterday, that you would come out and talk to us. Now, here you are. Your father doesn't have you hidden inside like usual."

Balynn was dumbfounded. Who were these two huge, friendly men? They obviously knew who she was. With a wave, the second of the men dashed off to the right of the mansion, darting around the corner and into the brush.

"We have been coming down here for weeks, but we never got to see you, before", the man next to her explained. "Jack fancies chatting with one of the Lowlander ladies. He's on his way behind the house to meet up with her but don't tell your father. He would have a fit if he knew about that."

Suddenly Balynn knew who she was talking to. These two were Thorndike's two sons, the ones that she had seen for only a minute in their dining room. And they were just as good looking here as she had thought at the time...big and young and almost as handsome as June. She smiled. The day was definitely looking less lonely now.

"My name is Jorge," the one near her announced. "Jackson is my twin brother, but you can tell us apart. *I* am the handsome one." He laughed his deep, rich laugh again.

"I am pleased to meet you," said Balynn demurely. "I didn't recognize you at first. You are a long way from home. What are you two doing down here?"

"Supervising the digging on the mote, of course," sighed Jorge. "We've been at it for weeks. I would rather stand here and talk to you, but at least I got to see you and say hello. I'd better chase Jack down so that we can get the Lowlanders started on the digging. We don't want to spend time in the prison again."

"Prison? What prison?" Balynn laughed. "We are all the way down-River from The Castle Prison. I wouldn't worry about it."

"We're not too far from *your father's prison,* though," grinned Jorge as he sauntered off in Jackson's direction. "We both got thrown in *that* prison once for trespassing when we were kids. We don't want to go there again, that's for sure."

"What are you talking about?" Balynn called after him, but he was gone around the corner. Balynn was miffed. She finally had someone new and interesting to talk to and he and his brother were somewhere "digging a mote?"

It's time for some investigating, she decided. She marched after the disappearing Jorge with determination. The little inquisitive girl that found The Crown Room and the new inquiring mind of a woman joined and came out of hiding. *I have no idea what he and his brother are really doing,* she thought, *but I'm going to find out.*

The sides and back of the huge Drumsworth Manor were places that Balynn had never seen. And, in truth, she had never wanted to see them. She had seen the servants throw buckets of garbage and dirty water out windows in these unexplored directions and she was sure that the bramble covered parts of her father's property were not nearly as welcoming as the landscaped front.

But she was following an interesting man who knew something she didn't know. She knew this was not lady-like behavior for a young woman of breeding and she didn't want to be caught doing it. She dropped her delicate shawl to the ground so that it wouldn't get snagged by the bushes and ducked behind the tangle that skirted the side of the building. She crept forward through the underbrush cautiously. It was a longer distance to the back of the manor than she expected.

Daddy's manor is bigger than I ever realized, she thought. *I guess I have been living in the middle, the original, homey part all these years. This side must be part of what was always locked.*

She looked up. Although the part of the house where she lived looked like a welcoming home, the side where she was now towered above her with slabs of unwelcoming masonry and iron around the windows. It looked more like a fortress. *Amazing,* she thought. *Why didn't I know this?*

She continued to creep toward the back along the wall, brushing aside the thorny branches and frightening the small animals that were hidden among them. *This is an adventure,* she thought with excitement. *I should have done this sooner. I have been living in this house for years and now I am scouting 'a fort'. This is great. Who knows what I will find?*

Now Balynn heard some digging sounds indicating that the young men might have been telling the truth. *Someone is 'digging something'. Let's see what their 'mote' looks like,* she thought, grinning.

She reached the corner of 'the fort' and there it was. It was actually a huge ditch that had been dug from twenty yards beyond the rear of the manor far out into what the Thorndike map had shown to be Lowlander territory! It ran the whole width of Drumsworth Castle, which was now two or three times as large as she had envisioned it to be from the drawing.

Positioned at the far end of the ditch were dozens of Lowlanders, digging and hauling away heavy loads of dirt while Jorge and Jack walked among them, yelling orders as they went. The workers were obviously extending the ditch to where it would eventually curve around the west side of the mansion and create an intake with The River and the original lake. It really was a mote!

Balynn could now see with shock that the entire back of her father's castle overlooked this long, deep canal which, though massive, was currently completely hidden from any view from the front.

Not only is there a mote, but there are Lowlanders that are digging it like slaves with the twins as their slave masters! Slaves? Could that be true? Was she really looking at slavery? Balynn asked herself in horror.

She pulled back behind the bush at the corner of the castle and huddled in its cover. Again, what she had seen long ago in The Crown Room united with what she had seen and heard in Thorndike's map room. Now there *was* a mote. It was all coming together.

The Auras had revealed a plan that she might have brushed aside until today. But now it all made sense. Her father was not only dreaming of becoming a king. He was preparing to *actually become one* and this mote might someday be here to protect him as a king when his plan actually took place.

"This is not a good thing, is it?" asked Balynn's Conscience.

He's my father, Why not? thought Balynn defensively.

"He appears to be using Lowlanders as slaves, Balynn. Are you in favor of that?"

"Maybe they're not really slaves. Maybe they just look like that from here."

"Perhaps," answered Cbalynn, "but honesty and fairness are keys to goodness. Your father hasn't been honest with the King about what he is planning has he? And he is using Lowlanders in a way that 'looks like slavery' and that's not 'good behavior', either is it? *And,* if your father is planning to be a king as you suspect, what country do you know of that needs a king right now? Would he be replacing someone? If so, who might that be? It's time for you and I to start really thinking."

Balynn's head was spinning. Was she going crazy? These questions were almost too big to comprehend. She must be missing something. Her father was a good man. He was always meeting with leaders of River Kingdom and saying that his plan was "the people's plan" so it must be good.

And what about the Lowlanders? Lowlanders were friendly, helpful people. and so were the beautiful twins that were now yelling at them. They couldn't be slaves and slave masters, could they? That couldn't be true, but Flo had said that 'controlling Lowlanders' was part of The Plan.

"But what about your Father not mentioning any of this to King Solem?" Cbalynn continued. "Think about that,"

Suddenly there was a rise in the noise level out back. Cautiously, Balynn peeked out again from behind her bush. It was then that she noticed something toward the rear of the property that she hadn't noticed before. Jorge and Jack were now struggling with a male Lowlander who was screaming in protest. With his arms flailing and legs kicking, the twins were hauling a Lowlander man toward a dark, evil looking structure. Jorge opened a heavy metal door and Jack threw the man into a dark opening within.

"No need to throw him," grumbled Jorge. The door clanked shut. The work of the Lowlanders that had stopped during the kerfuffle resumed in sullen silence.

"That will teach you to lag about rather than working as you should," growled Jackson.

Balynn knew what she was seeing. She was looking at 'the prison' that Jorge had mentioned. Now, she was also looking at the reason that the Lowlander servants in her home had always operated in fear. They undoubtedly knew about the prison, some of them firsthand. She shuddered.

"Mistress Balynn," whispered a soft voice. "You mustn't be here. Come with me." It was one of her Lowlander servant friends from inside. "Your father is looking for you. I saw you talking to Jorge. Then I found your shawl and followed you. Come with me. I know a secret entrance."

"Go with her if you don't want to be in serious trouble," Cbalynn added. "You can figure all this out, later.

RECKONING WITH ARMANDO

June was on an early-morning fact-finding mission with Manda and Prince Armando to see exactly what was happening down at Drumsworth Lake. Was there really a mote being built as Armando thought? Were Lowlanders being abused?

At last, they arrived on a bluff of Urhonordo land that overlooked the valley and The River below. From here they could look down easily on Drumsworth and the view that Prince Armando had brought them there to see together.

"Well, what do you think, June?" the Prince asked with a sad smile. "Is it as I told you?" He climbed down from his horse and found a familiar rock for seating. Armando had been here before. He had studied his perceptions with care. Pulling a flask from his hip, he took a sip and held it out to his companions, watching them closely. "It's only water. I wish I had better," he said after another swig.

"It's a mote, for sure" said June in astonishment. "It stretches from one side of the castle to the other. And look, Manda. Dozens of Lowlanders are working on it, completely out of sight from the front. No one can see it from the front at all. Drum has kept this secret from everyone...especially my father."

June took the offered flask and chugged. "So, you were right, Your Highness. It looks like they got started a long time ago and well before Drum had permission from Father or The Water Rights Board for anything. He has never even mentioned this excavation. Very sneaky and against all River Kingdom Water Rights protocol."

"That was my point," Armando said slowly. "Your father said he was worried that he had given Drum and the nobles permission to start on their road project and would feel bad about reneging once he had given Drum his okay. But, obviously that didn't matter. They had begun what they wanted to do already."

"My father will not be pleased. He has strong feelings about protocol, as you know."

"I *do* know that", agreed Armando with a nod. "But breaking protocol rules is not the main reason that I wanted you to come down here with me and my daughter today. More importantly, please look at who is doing the work and how they are being treated. If I am not mistaken, the workers on the mote are slaves."

June stared at the men who were working hard below. "No... You don't really think they're slaves and not just workmen, do you?", he croaked, taking another gulp of water.

"Slaves, Father?" questioned Manda. "You don't believe that those Lowlanders down there are really slaves, do you? I spotted this work two days ago, myself, but it didn't occur to me that the Lowlanders were slaves." Manda stepped closer to the edge of the bluff and peered down through the grasses. "

Why do you think that, Father?"

"Check out the men standing over them, Manda. What do you see?" asked her father.

"I see what you mean, Father. The men standing over them are yelling at them and showing no respect. They're being awful. Those poor workers really *are being treated like slaves... Lowlander slaves!* They're doing that to *your* countrymen, Father...*our* friends! We must do something, don't you think...I mean now...right away!"

"That's what *I've* been saying to you since yesterday," growled Armando's Conscience. "Now your daughter is telling you. *Your* Lowlander brothers are being *enslaved* and *you* are just sitting here on a rock, drinking water."

Armando stood but lowered his head. "I just discovered the slave issue yesterday," he moaned. "The first thing I wanted to do was storm the crew, free the poor guys down there and open the prison. I was pretty sure that I was looking at slavery, but I needed to be completely sure before I acted. This is why I brought both of you here... to verify what I was seeing."

"Prison? You said 'prison'. What prison?" June jumped up and peered down more intently.

"Look farther south," said Armando. "The prisons are actually back in Lowland territory, behind that row of bushes."

June and Manda craned to the south. There was no disputing what they were seeing. A Lowlander man was being hauled into a large, dark cavern and locked in...by two very big, very *white* men.

"You knew what was happening," hissed Carmando. "*You knew* it was just a matter of time before you knew it for sure. And what have you done to fix it? Nothing... nothing at all!"

June stared in disbelief. "It *is* a prison...a *real* prison... Lowlander slaves are being thrown into a prison that is on their own territory... And look! Those men that are putting them there are Thorndike men. I see the markings on their tunics. What are Thorndike militia men doing down there? What have they got against Lowlanders?"

"Their slaves have dark skin. They look like me," sighed Armando. "Those Thorndike men are probably Birthers, just like Drum." He began to pace. "I should have knocked that prison down, years ago," he said sadly. That old prison was rotted and falling to the ground but it wasn't in anyone's way so I never bothered with it. Now it has been rebuilt...and *now* it is being filled *again* with Lowlanders, just like it was one hundred years ago. River Kingdom noblemen want control now, just like then."

Armando's Conscience was more than active. He was incensed.

Manda was no more charitable than Carmando. "You should have done something the minute you saw it, Father" she fumed. "There are only two guys down there. You could have beat them easily. The Lowlanders would have helped you. It would be all over by now."

June had said little. "No, Manda. Your father was right to get 'more eyes on the issue'. What we are looking at is more than a breach of protocol. It's *slavery*. And it's Lowlander slaves ruled by *River Kingdom* citizens. That makes it my Father's responsibility, as King. Closing this operation down is essential but it will be complicated. Punishing those responsible will also be complicated...and it could be dangerous. It will involve everyone on both sides of The River." He began to pace.

Manda watched June closely. *He looks like his father,* she mused. *I wager he's thinking about protocol like King Solem always does...but maybe not.*

"He's thinking about the need to do something this big in the right way," contributed Cmanda, reading her thoughts. "Be still and think. About that, yourself."

"Now all three of us have seen this horror," June continued. "We need to tell The King and the Castle Guard, immediately. This isn't a matter for just one man to solve, even if that man is an esteemed Prince."

June bowed to Armando. "I stand by what you have done here, Prince Armando." June's use of this formality was meant to show what it *did* show. This was a serious situation and young June respected Crown Prince Armando.

"I thought you would see this the way I do, June," the Prince said with a nod of satisfaction. "That is why I brought both of you here. We can't afford to act too quickly without knowing exactly what is happening."

He's right. This prison may be for more people than Lowlander slaves, Manda thought suddenly. *Slavery is bad enough, but maybe something even bigger is happening. The ditch they are digging looks more like a mote than a simple enlargement of a lake. Motes are built for protection ...of a king.*

"We need to talk to mother right away," announced Manda, going to her horse.

"And my father, too", added June. The three were on their horses immediately and headed north toward Urhonordo Castle. The original concern of Drum's apparently selfish desire for a new road to his beautiful, enlarged lake now seemed insignificant to all.

*****

At Urhonordo Castle, Queen Luna was studying her own maps as she waited impatiently for the three 'investigators to return'. There were maps of all kinds hanging on the walls while two large ones were spread on the table before her.

"Mother... are you nearby?" hollered Manda as the travelers entered a side entrance. Groomsmen took their horses for care as they scurried toward the room where Luna had been spending much of her time since the meeting with the King.

"I am in here," she called to them. "I have a fire going in the hearth. I hear that it is getting chilly now at night. Come in and get warm. How did your travels go? I've been trying to figure out how River Kingdom landholders can get water to their crops with that new road. So... what did you all discover?"

"They saw what I brought them to see, Luna," Armando sighed as he sank into a nearby chair."

"Mother...did Father tell you what he thought was happening down at Drumsworth?" asked Manda anxiously.

Luna looked up from her maps, coming around the table to hug her family members. This was her style. "Of course," she responded sadly. "And now you have both seen the atrocity for yourselves. Slaves... Slaves in this day and age! Who would ever have thought that there could be such a thing in River Kingdom? Did you both agree that slavery was what you were seeing?"

"Indeed! Father will be horrified," June added. "*I* am horrified. Father needs to know about this immediately and I am sure he will want to take immediate action."

Luna motioned June to a chair and sat down with the three, herself. *What a wonderful group of people I have before me,* she thought gratefully.

"A wonderful group? Yes. Now they need direction ...leadership. You must be up to it," counseled Cluna sternly.

I know my role, Luna thought in agreement, *and I must be prudent. This is a very serious matter. The first thing I must do is listen, right?*

"Right...but time is important here," grumbled the impatient Cluna.

"Tell me everything you saw," said Luna, leaning back for a moment to watch and listen.

"It was terrible, Mother," Manda began. It was awful." Manda was up out of her chair. She couldn't sit still. "The whole back side of Drumsworth Castle was covered by a deep, ugly trench that stretched far out into The Lowlands."

June needed to be more formal than Manda. "Your Majesty, it is just as Manda says and as you and Prince Armando suspected. Not only does a deep ditch that is hidden behind the manor show evidence of having been worked on for weeks, but it is, indeed, being dug by a Lowlander workforce of *slaves on Lowlander land.* And there are prisons there to house the slaves...hidden, dank prisons.! Evil such as this has never been seen before in River Kingdom!"

Luna lowered her eyes and stared at her hands. "I hear you, June, and technically, it's on Lowland territory, but it's being run by River Kingdom men *and,* I regret to say, it *has* been seen before. My family's history is a sad addition to your observation. Slavery and imprisonment, both, happened one hundred years ago, in the same place. I can only hope that it is not for the same reason, but this looks like history repeating itself."

"That's what father said on the way home," blurted Manda. "I had no idea. Was that when Urhonordo won against River Kingdom?"

Luna rose and went back to her table. "That is exactly when something like this happened. We didn't actually *win* at that time, Manda. King Sol arrived and called off the Thorndike Militia raid. You need to be accurate. And Drumsworth wasn't as huge then. Invading nobles of River Kingdom just imprisoned Lowlanders so that they wouldn't warn us here in Urhonordo that River Kingdom wanted to take over our Kingdom."

"Really?" croaked June. "You mean to tell me that River Kingdom nobles wanted to rule Urhonordo? Really? I never knew that."

"You don't read enough history, June," Luna commented with a sly smile. "And tell me...was any of the Thorndike clan visible with the slaves today?"

June nodded. "Yes. There were two of them. They were the slave masters. We saw them, plain as day. How did you know that?"

"That's their leader's style," answered Armando. "It was then, and it is now. It was the Thorndike Militia that wanted to invade Urhonordo back in the day with acting 'King Drum' in full support. Thorndike would probably love to do that again, right now."

Manda sat quietly for a moment. Then she looked at June. "Drum and Thorndike were a team for expansion of River Kingdom back then, right? What if they still are? "What if that new part of Drumsworth Lake *is actually* a mote and Drumsworth

Castle becomes a real castle? Then Thorndike can be this Drum's Royal Militia and they can invade Urhonordo all over again."

June was smart and intuitive, but he had not read much. He was more a man of action. He looked at Manda in amazement. "Don't you think you might be putting 'slavery' and 'pushy white guys' together and reading too much into this?" he asked.

Queen Luna raised an eyebrow. She had been having the same thoughts as her daughter. Their reading of history provided them with some insight. "You might be right," said Luna thoughtfully.

"Drum does want his grandfather's power back again for this generation," snarfed Armando from his comfortable chair. "That is what I hear from the nobles. The Birthers at The Gathering spoke of "the good old days" when Drum threw a lot of rallies for the locals and convinced them that *they* should oversee both sides of The Great River. Of course, none of today's nobles were alive back then. They just hear stories, and they *are* Birthers."

"They wouldn't dare think of such things now. My father and his Royal Guard would wipe them out in an instant," smirked June with a prideful grin.

"Thorndike has his own militia again, you know," countered Manda. "And they even have a cavalry."

"I know, daughter. I know, but they are pledged to River Kingdom," Prince Armando contributed.

"They are loyal to my father," added June with pride.

"How do *you* know?" asked Manda. "Have you talked to any of them recently? Have you even met even *one* Thorndike Militia man? Everybody knows that Thorndike is no friend of your father. What if he decides to go against the King *with* Drum?"

"He would never do that, even though he doesn't like my father," June countered. "That would put his daughter in danger. Willa is our Queen now."

Manda looked at the floor. *He likes her,* she thought instantly. *I see that now. He really likes her.*

June's mind snapped to thoughts of Willa. *Willa is her father's daughter, and she's my father's Queen,* June thought to himself. *I shouldn't be thinking about her at all...but I wonder how long it will take me to get home to see her.*

Cjune jumped into action. "Long enough for you to know that you need to talk to your father about possible slavery and an over-active Drum in his kingdom... *not* about the beautiful Willa, you lovesick child," his Conscience hissed. "Grow up. Take a look at Manda, if you want to think of a lovely lady...or is she too smart for you?" snarked Cjune.

Manda's Conscience heard June's Conscience. *June has fallen for Willa. That's too bad,* Cmanda thought to herself. *June is a very good man, but it sounds like he is in love with his father's wife, for some reason. That's not good. He should fancy my Manda, instead. I know that she likes him.I should let Manda know that she needs to go after June...but that wouldn't be right...or would it?*

The silence didn't last long. The situation was too important. Decisions needed to be made.

Queen Luna's Conscience was strong. She was ready for this. "You must put everyone's best thinking to use on this," Cluna declared to Luna. "But first, you must clearly separate what everyone *knows* from what we all just *think* we know."

"Indeed," said Luna firmly to herself and the others. "Let's think about this carefully. What do we know for sure? We know that a hole is being dug in secret behind Drumsworth Manor without consultation with the Water Rights Gathering or the King. We know that Lowlanders who are being treated like slaves are being used to dig that hole. We know that the new hole does not yet hold water. We know that Thorndike Militia men are acting as bosses of those who are digging the hole. *These are the things we know.*

"We only think we know about slaves and a mote, but I think *these things that we know for sure* are the things that King Solem should also know about right away. What do the rest of you think?"

June listened and nodded. He stepped forward with a determined look. His father needed to know all of this...and so did Willa, especially Willa. He needed to talk to Willa. He had been on the road to and from Drumsworth since dawn, and he missed her. He decided now. He would leave immediately. "I am on my way to tell my father of all of this," he announced with determination.

Manda's Conscience jumped into Manda's thoughts without hesitation. This was a chance to help Manda do something that she had been wanting to do but she needed to do it in the right way...now.

"June is excited by what *he believes* to be true," counseled Cmanda. "He may get his father all riled up without proof and cause trouble for everyone, including himself. If you care about him, Manda, it might be helpful to slow him down a bit." Cmanda hoped she had said this right.

The Call of the Auras

Manda listened with a lonely and loving heart. *I hadn't thought of that,* she suddenly thought. *June is only my friend, of course, but he is very excited and he could stir things up too fast. I am not really that worried about him...or am I interested in another way?*

She didn't think twice. Manda placed her hand on June's arm, stopping him just as he turned to leave. "Please wait a minute, June," she said, looking into his eyes with concern.

June stopped and looked back, impatiently.

"Will you do me a favor, June? I was just thinking... Just as my father asked *you and me* to verify what *he* saw, wouldn't it be smart to use *me and my family* to verify what *you* saw?

"This situation is too important to be misunderstood or acted upon without *everyone* verifying what is really true. If you are returning tonight, please arrange for *all of us* to present our findings to your family *together, tomorrow*. We must all speak up but be careful not to say too much too fast."

Sometimes there is something magical in a touch. June felt the small hand on his arm. Then he saw the sincere, thoughtful look in Manda's eyes.

His thoughts stumbled a bit at new feelings arising at just that moment. Why hadn't he noticed those eyes before? Her touch was warm. He returned Manda's look for only one meaningful second... or was it more?

"You're right, Manda. You are *so* right," he responded slowly. "Sometimes I get ahead of myself. Thank you."

He turned to the King and Queen of Urhonordo again with a more controlled stance. "What do *you* say, Your Majesties? What would *you* have me do?

Queen Luna watched carefully and understood both the two young people in front of her and the situation that was before all of them. She nodded with dignity and responded, "Please ask your father and his Queen to receive us tomorrow to... "discuss matters of grave urgency" only... nothing more. If you would do us a great honor to deliver only this message, we can all discuss the many issues in this situation together calmly and think carefully.

"If we do not hear anything to the contrary, we will leave Urhonordo Castle at dawn tomorrow. If you can do this for us, we will be grateful."

"As you wish, Your Majesty", June replied. "I understand." He nodded to Manda who had not yet dropped her gaze or her hand. "I will see you tomorrow then?" he murmured, looking at her intently. She nodded and he was gone.

LOWLANDER REVEAL

The Thorndike twins poled their way slowly up stream along the dusky shallows of The River. They had finished a long day's work on Drum's mote and they were exhausted. Poling against the current was hard work, even now when The River was slower.

"Next time we should take the horses", grumbled Jorge. "This is too much hard work after all day with ornery Lowlanders."

"Not all the Lowlanders are ornery", grinned Jackson. "Have you taken a good look at Paulina? She is definitely *not* ornery. Weve been talking...hanging out a bit. She's not bad...for a Lowlander."

"Wonderful," snickered Jorge. "Shall I tell Father to arrange a wedding date? I'm sure he will love the idea of watching you produce 'half-breed Lowlander babies.'

The boys chuckled as the mist of their companion Auras circled around the floor of their small craft. "No need for a wedding, Bro. Maybe I'll just claim Paulina as my 'perk' for the work I'm doing on that stupid mote. If Father wants us to handle the slaves for Drum, he will have to understand that there needs to be something in it for us. I spotted her as my 'bonus' the first day we started on the mote. How about you? I saw you shining up to Balynn Drum. That looked like a connection that might have some promise."

"That would be more than a 'perk' for me if she likes me," answered Jorge. "Balynn is classy. She is a white girl. I think she liked me a bit, too. We might turn out to be something real."

"Horse pucky," Jackson hooted. "That girl is the daughter of a rich Birther whose blood lines go back for centuries. You and I are mongrels with no bloodlines that anyone will admit. We both know that she would never have a real interest in you. Get serious. Besides, you couldn't afford her slippers, let alone a wedding." Jackson gave the little craft under them a strong push with his pole as they at last reached the shore. They were close to home and they were hungry.

What was left in their journey was only a short walk to a bricked road that was maintained by their father's servants. They could see Thorndike Manor ahead of them. Maybe their sister would have dinner waiting for them. If so, they would eat inside in the kitchen where the rest of the militia couldn't see them and be jealous.

The twins had not been raised inside the Manor. Surrogate mothers and older sister, Flo had raised the boys that were the sons of Thorndike since their mother died. Their heritage was not acknowledged in public by the rest of the militia or Thorndike, himself.

"I still think we should go down to Drumsworth on horses next time we go", groused Jorge. "Poling *with* the current is fine on the way down, but poling is bad at this time of the day. "And we would look good, arriving on fine, Thorndike cavalry horses, don't you think? The ladies along the way would be impressed."

"And the farmers and landholders along the Central Riverbank would see us and give away the whole construction of the mote before Father and Drum are ready to let it be known, idiot. "

"Yeah. You're right. Drum is a pain in the neck, but he's smart and good at strategic planning. This is his call. And Father is totally focused on The Plan. If we blew The Plan, we would soon find ourselves floating down The River."

"Sad, but true. Sad, but true."

The exhausted young men circled on foot to the back of the manor. Characteristically, their sister was waiting for them and motioned them quickly into the kitchen. "Father is working in the map room with some nobles so you won't bother him at all," said Flo with a grin.

Flo loved her brothers even though none of them could acknowledge how, when or with whom they had been conceived. She ushered them into her kitchen and began to ladle out a good portion of stew.

"So", she began as usual. "How was your day? What did you do down at Drum's? Did you work hard?"

"We worked like dogs," Jackson began, reaching for a spoon. "It's always a long day when the slaves are just a bunch of lazy Lowlanders. We had to throw one of 'em into the prison again. Some of them want to talk back as though they think they are equal to us."

Florence shook her head and winced, pouring another ladle of soup into Jorge's bowl and handing it across the table.

"Ask him who is his favorite Lowlander, Flo," snickered Jorge.

"Do you have a favorite Lowlander, Jackson? They are all hard workers, are they not? Why do you have a favorite?"

"Because this one is a 'she', Flo," offered Jorge. "She is an inside worker at Drum's. Her name is Paulina. She *is* a beauty...a little dark for my taste, but still pretty. I think Jackson has a crush on her."

"You can keep all your opinions about *my* Lowlander friend to yourself, big mouth," mumbled Jackson, his mouth full of stew. "A Lowlander is a slave, male or female. Flo doesn't need to be bothered with the details."

"You mustn't call the Lowlanders 'slaves'," grumbled Flo as she added warm bread to the savory fair. "They are Lowlanders from The Lowlands and deserve to be treated with respect."

"They're Lowlanders, that's for sure, but they're also slaves. I thought you understood that, girl. There is no need to respect slaves." Jackson took a swig of wine that Flo had laid out for him.

She stood back and frowned, hands on hips and eyebrows raised. "I don't like *you* to call them 'slaves'. *We* don't believe in slavery in this house," she announced staunchly.

"Maybe you don't, but Drum certainly does," Jorge countered. "The Lowlanders down at Drums have been slaves forever and more so now since he and Father rounded up so many more to build the mote. You knew that, right?"

Flo sat down across from the boys and frowned at them. "But why call them slaves, boys? Aren't they just hard-working Lowlanders, like always?"

"Yeah, you could say that," Jorge agreed slowly.

"Then why call them 'slaves',?"

Jorge took a swig of wine and sadly delivered the obvious. "When people are kidnapped from where they live, forced to sleep in prisons at night and forced to work hard all day with no pay, what would *you* call them, know-it-all? Yes, they are Lowlanders. Yes, they work very hard, but yes...they are slaves, a whole bunch of Lowlander slaves...but they're only Lowlanders. "

Flo closed her eyes. She folded her hands in her lap and lowered her head.

For a few moments, the boys continued to eat their stew, the sounds of chewing and gulping continuing to bounce off the plaster walls of the kitchen with no competition. Jorge was the first to realize that his sister was not just resting. She was crying.

"Are you hurt, Flo? Are you in pain? You're crying? What's wrong?"

Jackson looked up and squinted at Flo. "You have something in your eye, right? Want me to see if I can get it out? I have pretty good close-up eyesight."

Flo lowered her head into her hands. "I knew I would need to say something soon," she sobbed. "My Conscience told me I would."

The boys put down their spoons and stared at their sister. She was still crying. They had never seen that before. She was always in control, always so stoic. What was happening here? They had no idea.

Flo's Conscience crawled up to Flo's shoulder and snuggled under her ear. "I'm here for you, Flo. You needed this day to come. There is no shame in what you have to say. You have beautiful brothers. They need to know what you know."

Conscience activity was rare at Thorndike Manor. The Auras were a constant there and pervasive, but Flo's Conscience had become stronger since she had spoken with Balynn. Flo's honesty had brought her Conscience to life.

"You made your sister cry, Jorge!" growled Jackson. "Look at that, Bro. Flo is crying. What did he say, Flo? If he makes you cry again, I will beat him until he can't stand. What did he say?"

Flo looked up at her two ridiculous brothers. They were so foolish...so adorable... so unaware. She had been dreading their finding out this one thing since she had figured it out herself but she could not ignore it any longer. Her brothers had become slave-masters. Slave-masters of Lowlanders! She dried her eyes with her apron. "I need to tell you something, boys. Have you finished your supper?"

These 'boys' that were already two large men nodded dumbly, looking very much like overgrown children. Flo gradually became 'Florence Thorndike', their big sister. She drew up close to the table and stared across at her brothers, patting their bearded faces and began.

"You two know that I am your big sister. You also know that Thorndike is our father, right? Now here is an important question. Who was *our* mother?"

Jorge and Jackson looked at each other for the answer. None came. The question had never even been considered before. They vaguely remembered their mother but she had died when they were very young. Actually, big sister Flo had been more of a mother to them than anyone.

Flo smiled. "Let me tell you what *I* know now, my brothers. Our mother was a beautiful woman who also worked for the Thorndikes. She loved us and had a decent life, working as I do here in the manor but she was also Thorndike's servant... his *favorite* servant.

"Everyone noticed our mother when she was young because she was beautiful, and the next thing you know, our father took advantage and soon...we became his children. After three children and hard work, our mother died."

"So? We sort of knew that," grinned Jackson. "This is nothing new. That was a long time ago."

"You knew about our mother," Flo continued slowly, "but I have more to tell you," she said. "Now I want to tell you about *our mother's* mother. She was very beautiful, too."

"Of course," snickered Jackson. "I already thought that would be true. That's why I am attracted to beautiful women. *You* are beautiful, *our* mother was beautiful and *her* mother was beautiful. It's all in the family line." Jackson's Auras waited, suspecting an opportunity for action.

Flo sat up straight. "Indeed. Your grandmother *was* beautiful and...your grandmother and our mother were both... *Lowlanders.*"

No one moved. Even the Auras swirled to a stop on the floor. The Auras' opportunity for action was coming.

Jorge stared at his sister while her words sank in and began processing. His inexperienced Conscience sensed danger and climbed into Jorge's hair.

The excited Auras leapt into action, not waiting for Jackson to respond. "What do you mean, 'a Lowlander'?" Jackson snorted. "Like you said before, we don't have any slaves up here. We don't have any Lowlanders, either. The only place I know that has Lowlander slaves is Drumsworth and Drum didn't even have many either 'til so many were brought in to build the mote. What kind of story are you telling us now?"

"She's talking about *our* history, Jackson," Jorge said slowly. "Yours and mine. What she's saying, if I'm not wrong, is that our mother was a Lowlander."

"Yes," said Flo softly. "The people who you are now calling slaves, are related to you. The people that you might have treated badly today may actually be our family. You two are half Lowlander because our mother was a Lowlander."

Seeing increased opportunity for mayhem, more Auras below the table began to stir. Slowly, methodically, they wound their way up Jackson's legs, around his body and into his hair. Jackson leaned across the table until his sneering face nearly touched his sister's. "What you are telling me is that we three are the off springs of a line of Lowlander whores. Is that what you are saying?"

Flo drew back in horror. She had expected that her brothers might be upset by what she had to say, but she had no idea that their level of reaction would be so fierce. She didn't know about the Auras under the table, either.

I should have noticed the Auras, Cflo thought sadly. *I recognize now that there are always Auras around this house. They are so common that I just forgot about them. Now they are getting busy. Is it too late to fix this?*

Jorge's Conscience was also anxious. He had never really been tested. Jorge's Conscience had never talked to Jorge much before. Jorge didn't even know he had a Conscience. Jorge's ancestors had never been discussed before, either. Now Cjorge knew why. This could be a challenge.

"I am your Conscience, Jorge. I know we have never really spoken before, but I am a friend. Also, I am here to tell you that your sister was very brave to tell you these things," Cjorge whispered. "She is a good person. You should be proud of her. Your brother is sort of a jackass', though. Looks like the Auras got to him before his Conscience did. See what you can do to help your sister. Later, maybe we can talk more about slavery."

Jorge shook his head. He hadn't spoken with this thing called 'Conscience' before, but his words made sense right now. "Ease up, Jackson. We need to process this whole thing. It's something we've never had to think of before," offered Jorge, smiling at his sister as though seeing her for the first time.

"I don't want to think about it now, either," huffed Jackson. "Shut up, Jorge, you Lowlander. You're not even fit to be my brother anymore. You're a Lowlander."

"You're a Lowlander, too, dim wit. Don't forget that." Jorge slid his chair next to Flo and put his arm around her drooping shoulders. "Don't listen to your idiot brother over there. He doesn't think deep thoughts very much so he's not used to it."

"I'm proud of you," whispered Cjorge.

Flo accepted the hug from her brother gratefully. What should she do now? She had one brother that looked to be accepting of her announcement and another that was calling her names.

"Just tell the truth and see who will listen," suggested Cflo. "Maybe Jackson is just more like your father than Jorge. It may take him longer to digest this."

Jackson has always been more like father, thought Cflo, thinking back. She took a deep breath. What did these boys need to hear that they had not already heard. "I am sorry I waited so long to tell you," she offered. "I wanted to tell you about the Lowlander part of our heritage when mother told me, but there didn't ever seem to be a good time. And I was sure you wouldn't be happy, especially when you got so active in Father's militia.

"The Lowlanders who have lived up here in the north have been such sweet, nice people, that I really hoped that Lowlander blood wasn't that much of a problem and you both wouldn't feel bad when I finally told you. Then Drum started coming here regularly with his Birther nonsense. He is even more of a Birther than Father. I'm sorry I waited so long."

"Father has always been a Birther, Flo!" growled Jackson. "Don't you understand that? He doesn't like half breeds, either. Now he has three of them for family members. And *we* are Birthers, too. We're white! We look as white as father! Now we find out that we are part *'trash'*. How did you *think* we would feel?"

"I was hoping that you had grown up enough and were smart enough to understand. Lowlanders are good people like white people, Jackson, but they are not fighters. They never have been. That's why some people like Father and Lord Drum take advantage of them," sighed Flo.

"Does Father know that we are part Lowlander? If he did, he wouldn't have married mother, I'm sure," Jackson stated with assurance. His Auras continued to swirl around him. His inexperienced Conscience tried in vain to find an opening.

"He may not have thought about it. The Lowlander side of our family has always been lighter skinned. But that didn't matter to father, anyway. Mother was only a 'useful woman'. And he didn't ever marry her, either, Jackson. That is another piece of our secret. I have always felt that both you boys should know *all* of your heritage and that you were almost old enough to understand it. But I couldn't stand by anymore and let you abuse good Lowlanders as though you are better than they are."

"We *are* better!" shouted Jackson. "We're white! Look at us!" The Auras were already in action, forming a hood over his head and scowling face.

"You are part Lowlander and you should be proud," pressed Cjackson, his frantic (but also inexperienced) Conscience. "Listen to me. I know we haven't been close up until now, but I am here for you! I always have been."

But it was too late. The hood of the Auras was now in control. His Conscience's truth could not penetrate when the Auras were at this level of strength. Jackson's Auras were now in charge.

"No one knows anything about this 'part Lowlander lie' except you, Florence... right?" hissed Jackson. "Even if what you say is true, we're better than 'pure Lowlander', any day! And, as far as I'm concerned, what you *think* you know about our heritage no longer counts. We are all Thorndikes. You weren't around when our so-called 'grandmother' was alive so you don't really *know* anything, for sure.

"You can believe anything you want, Jorge, but *I'm* going to go outside where the *white men* of the militia are getting ready for our father's call to combat. When the time comes for choosing sides, I know where I'll be." Slamming his fist on the table for emphasis, Jackson stormed from the kitchen in a cloud of Auras.

THE AURAS' PLAN

"What's all this shouting and pounding in here, woman?" Thorndike blustered as he burst through the door of the kitchen, *his* Auras swirling around him like a glorious cape. "Do you want the nobles in my map room to think I have no control in my own house?"

"Beware and be smart," counseled Cflo. "Watch what you say."

"Of course not, milord," Flo answered quickly. "It was just a spirited discussion between the boys. You know how boys are."

Thorndike squinted at his son and daughter as he and his Auras came closer. "So, you were in a discussion with both brothers, right? Where did *the other one* go? What was the discussion about? Was there a problem? Speak up. Speak up. You boys were down at Drum's all day. Did you work hard? Did you get a lot done? What is the status? When will the mote be ready? Were there any problems with the low-lifes? Speak up."

Jorge's Conscience was still on alert. "Your father needs to impress his friends in the other room. This is not a time for a 'family discussion'. Be careful. Truth only."

Jorge stood immediately, coming nearly to attention. "Everything went well down at Drum's, Father," he reported in a formal way. "The mote is nearly complete. Urhonordo will be cut off by tomorrow. We only await your order to join the mote to the rest of the lake and fill it, Sir."

Flo busied herself silently with the dishes.

"So. The mote is nearly ready, eh? That's what I was waiting to hear. The nobles will be glad to hear that. They're getting restless...but we won't fill it with water yet, son. We have to finish putting together the hardware for the drawbridge, don't we?

"The ironsmiths from town say they are almost finished with that, too, but we don't want a mote full of water before we have drawbridge to get across and a way to pull it up, right?" Thorndike laughed heartily. "But the Nobles will love to hear this much." He laughed some more.

Flo smiled and nodded and Jorge nodded, as well. Neither one of them knew what the joke was.

"Well, so much for the mote. The other reason I came back here, girl, is to tell you that you are *late* in your duties in the conference room. The nobles are waiting. Bring out the cask of ale I ordered and some flasks... now! Everything should be out there already. Don't you know to tend to *my orders* before wasting time on the boys, woman?"

Flo blanched and looked at the floor.

"And prepare some rooms and beds, too, girl. The nobles will probably want to stay here tonight for more strategic planning in the morning when I tell them the big news." He glowered at Flo and marched back through the door in a cloud of Auras. Flo and Jorge got to work on the ale.

The occupants of the conference room were shifting uncomfortably in their chairs and fiddling with the drawings on the tables as Thorndike returned. These men were nobility. Waiting for someone who was no more important than themselves was irritating. Even waiting for something they wanted   to hear was not their frequent practice, especially in a chilly conference room with Auras under foot.

This evening had been billed by messengers to the nobles as 'an important planning session because Thorndike and Drum had just returned from a meeting with the King'. A progress report from 'the mote build' down at Drumsworth would also be announced. So far tonight, however, only Thorndike was here and the discussions had not begun at all. And where were the refreshments? At least there should be refreshments.

Thorndike could feel the impatience of the nobles in the room as he entered. He knew he must act with confidence tonight and show progress toward their goal. Drum was not here but would insist on a report from this meeting, a fact that irritated Thorndike. Fortunately, his boys had just arrived with 'positive mote information' on the mote. That was the information he needed to top off tonight's presentation. Now he was ready.

"Well, men, the ale is on its way," he announced with gusto. "My people will have it here shortly. Before it gets here, however, I can fill you in on the latest information from Drumsworth Castle. *My* men have been down there every day and have nothing but good news for us tonight."

Now there was a noise outside the door of the map room. Refreshments had arrived. Thorndike smiled and backed off. 'Refreshments' would enhance his presentation greatly.

The nobles, now seeing the one thing they knew they wanted, began to gather around the barrel. Flo entered behind Jorge with flasks. With efficiency born out of years of practice, she filled the flasks, one by one, passing them to outstretched hands and deftly avoiding the fondling that could accompany their leers and thirst.

The ale barrel was delivered and served. The servants were dismissed. Jorge escorted his sister quickly through the gaggle of men toward the exit.

"Hey, boy," shouted one of the nobles, taking a huge slug of ale from his flask. The overflow washed down his vest, unnoticed. "I hear that some of you boys were down *supervising* slaves at the mote today. Were you one of 'em? Some Lowland women are as good looking as this girl here and ready for a quick role, I hear. Did you get any action? There ought to be something in it for you white guys down there with all the pretty slave girls." He laughed at his wisdom as did his friends around him.

The Auras danced and swirled across the floor. This meeting was becoming more interesting. Drinking noblemen without anyone to limit their behavior were great fun for Auras to work with.

Jorge tensed but maneuvered Flo gently out the door. He smiled and waved but didn't answer. "Good job," whispered Cjorge.

Thorndike relaxed for a moment while his fellow noblemen helped themselves to refills from the ale tap. The more 'content' they were, the more supportive they would be, he figured, and he definitely wanted that support. Each one of these men controlled the livelihood of their landholders and fieldhands. This was 'labor' for the new road.

Each of these ale-swilling nobles also possessed personal militias of their own, militias which, when joined with the Thorndike Militia would create a fighting force that would rival if not surpass the size of The Royal Castle Guard. This was Thorndike's main interest in these nobles.

No other nobleman controlled as great a number of men as Thorndike, himself, but he wanted control of all of them. Drum didn't have a militia of his own. He was dependent on Thorndike or would be when the time came. Together with the militias of the men in this room, their combined numbers were enough to present an unbeatable force. Thorndike knew that he and the men these noblemen controlled would be the most important part of The Plan...and he would be their Commander.

The noblemen now jollied their way back to their chairs, their flasks re-filled and their lubricated minds ready for the exciting ideas they had come here to receive. Thorndike took his place at the head if the table. He purposefully positioned himself to demonstrate *his* authority. *Drum isn't here and I am now their leader,* he thought.

Then he thought about that again.

A warm glow began to rise from somewhere deep within Thorndike's fragile, 'second man' soul. Why had he not realized this before. *I am as smart as Drum,* he now announced to himself. *I am as much of a leader as he is. I am a Commander.*

The Auras came to attention. They could feel jealousy bubbling up from Thorndike's memory, jealousy from past slights at the hands of Drum. The Auras loved jealousy. They would take advantage.

"My friends", Thorndike began, "The next phase of our plan is about ready to begin. I know you are all here for an update on the progress down at the mote, but first, let me tell you my most important news , news that is even bigger than our mote news."

The nobles lifted eyebrows skeptically and drank some more.

"For years we have spoken with enthusiasm about the way things used to be back in the days of King Drum," Thorndike continued quickly. "Your forefathers and mine lived well. They could do what they wanted to do and go where they wanted to go. Their King was on their side and never wishy-washy like our king today, right? River Kingdom was the greatest it had ever been."

The noblemen all nodded and took another chug of ale.

"We are here tonight to put a plan in motion that will see that greatness achieved again. What we have needed was a plan to return River Kingdom to the way things were and the way they always should be for Birthers who know how to run a profitable, successful kingdom. And who is going to lead River Kingdom back into the position it once held and should hold again?"

"Drum, Drum, Drum!" the noblemen shouted in unison.

"Right you are! Drum! The old King Drum's time was a time when people could build what they wanted to build where they wanted to build it."

"Here, here!"

"...when a man who owned land could do what he wanted to do with it..."

"Here, here!"

"...when nobody had to have a 'Gathering' just to figure out where a levee should go."

"Right."

"...or ask whether or not a road should be built."

"That's right!"

"We want things back the way they used to be and better, don't we? Of course, we do. That's what The Royal River Road and the mote are all about. It's about those of us who *own the land* being able to say what happens to *our own land* without having to ask a king...and *especially* not having to ask Urhonordo or *any Lowlanders* about it! Am I right?"

The noblemen stood up and cheered, clinking their flasks together in a sloshy salute. Thorndike had their attention now. They were all Birthers. Birther control was what they wanted more than anything. They had grumbled about Royal rules and too much protocol, but the influence of 'others' on their lives was their main concern.

"The increasing influence of Lowlanders in The Kingdom, even the mere presence of Lowlanders makes Royal rules and protocol even more unbearable," Thorndike continued. "Tonight, my friends, I am here to announce that we're actually going to *do* something about Lowlanders."

The Aura whipped itself into a happy froth. The noblemen slapped each other on the back and cheered. They were ready to hear what they had come here to hear.

"Two days ago," Thorndike continued with authority, "You may not believe this, but me and Drum met with our lackluster king and laid out our demands on behalf of all of *you*. We told him that *you* needed The Royal River Road for protection and safe travel *and removal of riffraff,* and he approved our proposal...all of it!

In spite of the fact that they were slightly drunk, the noblemen were as thrilled as Thorndike had hoped they would be. "Anyone going against our plan now will be run out of the kingdom. Nobility should have rights to control River Kingdom's future and those rights are returning to us, at last!"

Thorndike took a big swig of ale himself and enjoyed the reaction to his announcement.

The noblemen were ecstatic. They cheered. This was even more important than good news about the mote. This news was more than what they had been waiting for. They couldn't congratulate each other enough.

From the kitchen doorway, Flo and her brother heard the cheering. Cheering nobles were not a common thing. Cheering, *drinking* nobles could be a frightening thing, however. They, as servants, had seen such happenings before and it was best to be prepared. They snuck back down the hall to investigate but were ready to hide behind Thorndike's hall-guarding armor if necessary.

Thorndike seated himself at the head of the table and leaned back, pointing dramatically over the heads of the nobles to the maps on the wall, ready for his second revelation. The Auras encouraged his relaxed look of superiority. Drum was not here tonight. This allowed *him, Thorndike,* to be the most important person in the room. His palms began to sweat. A small tick appeared in his left eye. Power was intoxicating.

"Gentlemen," he began with pride, "you will note the presence of Drumsworth Lake as it has existed until now on Map I. On Map II, you see how it will look when the lake is increased and extended around the back of Drumsworth Castle and across the entire southern tip of River Kingdom and Urhonordo.

"This increase of the lake's size will mean that it, alone will form the complete boundary between any part of Urhonordo and River Kingdom thus insuring that River Kingdom will no longer need to ask Urhonordo for permission to do *anything* on The River. This mote will mean River Kingdom independence for the first time in one hundred years!"

The room erupted in cheers. More ale was consumed.

"That's really fine," rumbled a portly gentleman, studying the map closely. "When we get the mote in place, our man Drum can rule from there and be protected in Drumsworth Castle, right? That sounds great for Drum, but what about our troops? Where will our troops be during the rest of The Plan? Who will go where when the battles start?"

Thorndike raised his eyebrows. The Auras gathered around him closely. This was his moment to shine. "You don't need Drum, the Auras whispered. "Take the lead."

They're actually paying attention to me, Thorndike now realized. *This must be what it was like in my grandfather's time. Grandfather Thorndike knew how to achieve a goal and so do I. Drum is an idea and money man, but I am the military leader. These nobles now see the value in what I was born to do.*

"When the lake becomes a huge mote, gentlemen, it will be accessed only by a guarded drawbridge and our troops will guard that drawbridge and that mote as they have been trained to do by me, their Commander.

"Some will be guarding the mote from the outside, and some will guard the mote from the inside, protected with Drum. The grounds are large and Drumsworth Castle is huge. Have you seen how big it has grown in recent years? This was by design, my friends. Inside the mote or out, there's plenty of room to house troops at Drumsworth during any battle.

"Drumsworth is built like a fort. These guarded and defended positions will do what we Birthers have been trying to do for years...keep the Lowlanders in their place and out of our country!

"And now I repeat the news that I just received from my militia: the mote is nearly completed. We will soon be ready to install the drawbridge! It is now time for you and yours and me and mine to start *our* part of The Plan...the building of The Royal River Road for us Birthers and our combined militias that will put our great plan for River Kingdom into action! Do I have your commitment on this?

The nobility cheered again lustily.

"The Royal River Road, as you know, will take our combined troops from our farms to River Kingdom's Southern Border and back. A constant troop presence will maintain River Kingdom Borders and our sovereignty under Drum's rule and my command. We Birthers are finally going to take our kingdom back!"

Another roar of approval rose and rustled the maps on the wall. Thorndike was up again on his feet. "With our mote filled and the new road built, we can put The Plan in place. With all of our united militia, Lowlanders will at last be sent back to the 'sand lot' where they belong and will finally be blocked from ever returning to River Kingdom... *our* birthright!"

Flo and her brother flattened up against the walls of the frigid, Aura-filled hall. The siblings stared at each other in amazement at what they had heard. Inside the room they had just left, Lord Thorndike and his noblemen were discussing more than a road and an enlarged lake. They were also plotting to push *all* Lowlanders out of The Kingdom forever! Was the King in favor of this, too? What about Urhonordo?

An old curmudgeon in the back of the conference room asked the questions that Flo and Jorge wanted to ask. "Does the King approve of all of this, Thorndike? This doesn't sound like something Urhonordo would approve of, either. Prince Armando is a Lowlander, you know. You say that the King approved the road but what about the Lowlander part of this?"

Thorndike bristled. The details involved in this question were not ready for discussion yet. Maybe he had said too much. A cold sweat rose on his brow.

The Auras were ready, however. "Stand strong," they whispered as they wound around his head like the helmet of the warrior he wanted to be. "Be the leader you know you are. You are in charge now. Do not hesitate. Tell them all 'how it will be'."

"I can tell you this." Thorndike said with renewed confidence. "All objections to our plans will vanish soon. Why will that happen? It will happen because we Birthers of River Kingdom will start building our new road at the crack of dawn tomorrow. With the labor of all of your landholders and mine, we will make it happen! Do I have your commitment on this?"

The noblemen stood as one and shouted their approval. It was about time that someone showed leadership and determination. They were thrilled.

Quietly, Flo and Jorge scurried back to their kitchen in fear to serve the nobles their dinner but they could tell. Their entire world was about to changed, and Thorndike was changing it.

THE MORNING NEWS

Agnes heard June as he ran up the stairs and ran immediately from her room in the maid's quarters. It was the middle of the night. She would try to convince June that waking his father at this time of night was not a good idea. Quietly, Agnes provided some needed advice. "June, dear, why don't you go to your room and change after your long ride. Perhaps your father will be awake enough by then so that you two can talk. Could you do that, dear?" she asked June pointedly.

"And I will bring you some tea," she added respectfully. "You can leave your dusty clothes by the door for cleaning." Agnes turned and descended the stairway toward the kitchen wing of The Castle.

June stood alone in the hall that was the upstairs walkway between the bedrooms. The hall was still dark, with only a few wall lamps guttering in vain to provide sufficient light to the large space. Slowly he walked down the hall toward his room. His mind began to clear and become more thoughtful. At the door to his room, he arrived at his first realization. Agnes was smart. He needed to relax for a minute and think.

When he entered the room, another thought emerged. He had told Queen Luna that he would wait until she and her family arrived before he told his father about the horrible slavery situation he had seen. If Agnes hadn't stopped him, he would have blurted it out just now. June stripped off his dusty shirt and trousers and splashed himself with the frigid water in his bathing bowl.

June sat on his bed. Suddenly he was aware of its welcoming softness, plump and clean and ready for him. His exhausted body responded as he also became aware of the fact that he had not slept since the day before.

"Don't even look at your pillow," warned Cjune. "You have work to do and problems to solve. Get to work on them. Think."

June leaned back among the pillows and thought. At least, that was his intention, but he was too tired. Questions of what needed to be said to whom... about what ...and who was coming today...tumbled through his head... and Manda's warm hand was on his arm...He was soon asleep.

The soft knock on the door a minute later was Agnes with a tray of tea. Carefully, she placed it beside the sleeping prince and tiptoed out.

*****

The Urhonordorians were almost to The Castle. The front gate was open. Servants hustled out to the colonnade to greet them. Armando brought the carriage to a halt and handed the reigns to a waiting groomsman. The man smiled. He had seen Armando many times before and he was also a Lowlander who had been working for the King for many years.

"I'm glad *you're* here, Your Majesty," the groomsman said warmly. "I know the King is upstairs preparing for a meeting, but I don't think he was expecting *you*."

"He was not expecting *us?* Who was he expecting then?" asked Armando quizzically.

"We're all expecting Drum, of course," answered the groomsman. "We have no idea how long it will take him to get here, but we are ready when he arrives." The man leaned forward and whispered, "but we are happy that it is you instead."

Armando grinned. "Thanks for that," he whispered back. Now he remembered. Of course... Solem had sent a note for Drum to come back and discuss The Royal River Road and slow down the approval until everyone could discuss it. But Solem hadn't anticipated that the Urhonordorians would be here today, had he? Obviously not.

Amanda was already out of the carriage and helping her mother. Armando hurried to them both, but, as usual, the ladies in his life were taking care of their own exits. With one on each arm, he proceeded up the colonnade toward the front entrance of The Castle. But where was King Solem? Why wasn't he here to greet them, his usual protocol.

Agnes spied the carriage and the Royals from an upstairs window. At once, she realized what June had been running up the stairs to tell his father when she had let him go to sleep.

"We thought we were so smart, letting June sleep without speaking to his father, didn't we?" moaned Cagnes. "Guess what. Now there are Royals from Urhonordo on the colonnade and the King isn't even up yet. You better get a quick message to his majesty...now!"

Agnes hurried to the Royal Chamber Door and knocked firmly. No answer.

She knocked again. Gratefully she saw it open with a fully dressed, smiling king standing in the opening. "Good morning, Agnes. Why are you knocking?

Agnes swallowed hard and said simply, "I believe we have guests downstairs, Your Majesty. I believe June was trying to tell you that they were coming last night, but I let him fall asleep. That was my mistake. It is all my fault. Please forgive me."

King Sol smiled. "Heavens, Agnes. You are too worried about small things. I am a happy king with a beautiful wife. There is nothing that needs to be forgiven." He smiled again.

"Now, tell me...Who is here, Agnes. Is it Drum? I didn't think he could get here this fast. I must send a note to Urhonordo immediately. They will certainly want to tell Drum their concerns. See if you can find me a messenger, would you please?"

Agnes was beside herself. This was not Drum who was visiting. It was the Royals of Urhonordo "who should have been expected and prepared for," added Cagnes snidely.

"The visitors downstairs are Queen Luna and family, Your Highness. I believe Horace is greeting them and taking them to the conference room. By your leave, I will fetch Prince June." Agnes hadn't bowed to Solem in a long time, but she bowed now. She felt she had failed in her duty.

Solem raised his eyebrows and nodded, as Agnes scurried down the hall to June's room. *The Urhonordorians are here before Drum? And without a note from me as to time?* he mused. *How strange. June was with them yesterday, but he came home early. Hmm.* Quickly he turned to the stairway and descended to meet his guests.

Agnes cracked open the door to June's room. As she expected, he was sprawled across his bed, deep in slumber. For a moment she recalled him as a small boy in the same child-like position. But he was not a little boy anymore. He was a man with responsibilities.

"We treated him as a child, Agnes," admitted Cagnes. "We let him rest when he had a duty to perform. That is the problem with us old Consciences. We tend to think of things the way they *were* instead of the way they *are.*"

Cagnes was right. Agnes hurried to June's side. "June...June...Wake up. I let you sleep. I shouldn't have done that. The Urhonordorians are already here and your father has gone to meet them. Wake up, dear."

June was in a fog. What day was it? Where was he? In a moment, however, Agness's words registered. Angrily he rolled over and yanked on his trousers.

Pulling his shirt over his head, however, he caught sight of Agnes. He stopped and stared at her worried face. She had cared for him when he was small, and she still did. But she was upset. Without hesitation, his anger dropped away and he reached down to her with a quick hug. "Thank goodness you let me snooze a bit. I needed a good sleep," he called back as he ran toward the stairway.

Queen Luna and Prince Armando were already seated in the conference room when King Solem rushed in and a map that included Drumsworth was on the table.. Horace had fetched tea from the kitchen and Manda was dusting off cups she had found on a display shelf.

Solem didn't feel like a king as he entered. He felt like a child that was late for tutoring lessons. He sat at his usual spot at the head of the table, however, and gratefully accepted a cup of tea for himself as June breathlessly entered.

"Try simple honesty here," Cjune advised.

"Please excuse our tardy and unprepared appearance," June began. "It is my fault. I fell asleep before I could tell my father you were coming." He grinned self-consciously at Manda and sat down on his father's right.

Manda and Queen Luna couldn't help but smile. Honesty was so appealing.

But serious matters were at hand. Armando was not going to be deterred by excuses, tea, or smiles. He felt the guilt of a Lowlander leader who had not yet rescued his own countrymen... whether there was a reason for his lack of action or not. As far as he was concerned, action was now required and it was his responsibility to get it into place.

Armando stood and addressed Solem directly. "King Solem, we are here this morning to inform you of a grave issue that is taking place on River Kingdom land. There are *enslaved Lowlanders* being held and used as *slaves* right now at Drumsworth!" He remained standing. He stared at Solem. He clenched his fists. He was furious.

"At ease, Armando," counseled Carmando. "This information is new to Solem. *He* is not the one who has been using slaves. You know this man."

Armando ran his hands through his unruly curls and stepped back. Carmando was right, but Armando continued. "I am sure that this is something *you* knew nothing about, Your Highness, but I am also sure that this is something that you need to know *right now*. That is why *we* are here. Not only are there slaves at Drumsworth but they are *Lowlander* slaves. I must admit that the fact that they are Lowlanders makes it even worse for me." He sat down and stared at Solem.

Queen Luna patted her husband's arm and continued. "The fact that there are Lowlander slaves is the first issue of concern that we bring to you today. In addition to this humiliating and gross use of our Lowlander friends, there is *the matter of the work* they are doing. They are enlarging Drumsworth Lake which now nearly *surrounds* Drumsworth Castle."

"And the lake has become a mote...a mote, Father," June interrupted. "This is the other horrible thing we are all here to tell you...*all* of us. We were there yesterday. We all saw these things. There is no doubt about what we were seeing. Lowlander slaves were working hard on a secret mote that goes all the way around Drumsworth...and it's out of sight. You couldn't see any of it from the front of Drumsworth Castle. The whole thing is being done on the sly!"

Solem sat very still as did everyone in the hall. Even Horace froze in one spot with his teapot. There was too much to take in... too much to absorb. The morning sun struggled to penetrate the darkness that now seemed to settle around the table.

Slowly, the table full of Royalty sipped their tea. Their unbelievable problems had been stated. What were they going to do about them?

This was the scene that greeted Willa as she breathlessly entered alone, smiling. Agnes had made certain that she was aware of guests and had helped her find "a quick something suitable" to wear. Her sure sunny smile was still visible as she offered her apologies for being late, but the smile quickly faded. Observing the looks on everyone's faces, there was no way to ignore that this room was full of dark thoughts and worry.

"Welcome, my dear," Solem uttered solemnly. "Come here and sit by me. I need your understanding of two things that I have just learned. Our friends and June have just informed me that we have slavery, *actual slavery,* taking place down at Drumsworth Lake. And, in addition, those slaves are being used to dig a mote around Drumsworth Castle. They have seen it, as of yesterday". Those gathered around the table nodded with the same morose look that Solem was wearing.

Willa sat down beside Solem and slid her hand under the table to pat him fondly. Slowly she moved her gaze from one face to another. Frustration, anger and concern were all evident on every face as each tried to focus on an answer to the problems they were facing. And then Willa focused on that awful word...slavery...

Surprisingly, however, Willa's memory was suddenly jarred by that word. She was remembering something she had heard a long time ago. But where? Before she was married? Something left over from her solitary, lonely life at Thorndike Manor?

With amazing clarity, Willa began to recall the sinister looks she had seen on the nobles faces as she passed the doors to her father's conference room on one of those days. Anger, plotting determination, sneering...

She had not really paid attention then but now that one word, 'slavery', began to unscramble bits and pieces she had seen and heard before in a life she had wished to forget. Why now? Was her injury unclouding her memory? It didn't matter. The memory was there.

Sneering determination had been on the faces of noblemen gathered with her father, but for what reason? It was that awful word...'slavery'. The word was the key...Now it became clear. The reason for the nobles' determination had not been *against slavery but for it! That* was what Drum and the nobles had been discussing!

"You are right!" Cwilla whispered with excitement. "*We* are remembering what we heard when no one knew we were there. Now you really understand why you disliked Drum so much. You must say what you remember. Your husband...and your friends need to hear it."

Willa stood slowly. "I see that this map shows Drumsworth Lake circling Drumsworth Castle like a mote...and you all say that slaves are building it, right?" she asked as a way to begin.

"That's exactly what we are saying, Willa", Manda answered. "We didn't even get to talk about Drumsworth problems on our little day trip," she added sadly.

"No, we didn't, but I have just now remembered something that I heard way back, even before I met all of you...something I haven't even thought about since I left home. My memory is a bit scrambled, I think or maybe becoming un-scrambled. I wasn't really paying attention when I heard it."

June stared at Willa with a mixture of affection and tolerance. *What does she think she knows?* he thought condescendingly. *She looks pretty but this is serious business.*

Willa closed her eyes, trying to organize the scattered memories that were now returning to her, unbidden. "You say that slavery has been discovered down at Drumsworth?"

All nodded.

"Well," Willa continued, "If I remember correctly, that slavery and its location will not be a surprise to the nobles who are so in favor of The Royal River Road. Slavery was part of their reason for wanting that road. It was to control Lowlanders."

The Royals around the table now refocused and began to listen.

Willa's eyes were still closed as she thought some more. Her memories continued to resurface. "The nobles said that a lot of slaves would still be needed so that they should build something at Drumsworth to control them. Then, the new road was what Drum said would be best for getting the militias down south easily and keeping the slaves in their place...I think that's what they said."

Willa stopped talking and looked around. Everyone was staring at her. She sat down and folded her hands in her lap as she always did when she was uncertain. *Did I remember something stupid?* she wondered. *I only said what I thought I remembered when I wasn't paying attention. Maybe it was crazy. Maybe it didn't make any sense.*

"You said what you heard," Cwilla said reassuringly. "You *had* to say it to your husband now, even if it might get your father in trouble."

Oh my. My father... I forgot. I heard all this in my father's house... She now added to her concerns. *Oh, dear...*

Queen Luna studied Willa. She didn't really know this young woman yet, but her sense was that she was honest. The Queen spoke carefully. "You believe you heard the nobles planning to use slaves to build something?" she asked slowly.

"Yes, Your Majesty," Willa replied. "It was mostly Lord Drum that was suggesting slavery, but all the nobles seemed to think it was a good idea. My father was excited about being able to move his militia south easily and use fresh troops to keep 'Lowlanders in place for The Plan'. You know how he is about his huge militia."

June was paying close attention now. "Militia? You father has a huge militia, Willa?"

Manda shook her head and snickered. "You really ought to read more, June."

Prince Armando frowned at his daughter. This was not the time. "Lord Thorndike has always taken pride in a large militia," he added, returning his attention to Willa. "He is very proud of it, I hear. Am I not right, Queen Willa?"

Willa nodded. "Indeed," she responded eagerly. She was sure about this, at least.

Solem returned to his seat beside his wife. "Your father and Lord Drum met often?" he asked.

Willa rolled her eyes. "Very often...more often than his female servants and I wished," she answered bluntly.

Solem smiled. "I saw that he was not your favorite person the other day, Willa. I now understand why. But why do you suppose he came to see your father so often? You mentioned 'The Plan'. What was The Plan?"

Willa thought back. What else had she heard. Sadly, she shook her head. "I really have no idea what the whole plan was", she answered. "I tried my best to stay out of their way. Drum was there more than the other nobles, though, so I assumed that the plan had to do with increasing the size of his lake and making it easier to get there. Flo used to tell me that the nobles' part of The Plan was to 'control the Lowlanders,' but I never actually heard that part".

"Flo? Who is Flo?"

"Flo was my favorite servant at Thorndike Manor. Her mother took care of me when my mother died but Flo was the same age as me. Flo was my best friend... very quiet and shy, but nice."

"And Flo told you that the nobles wanted to 'control Lowlanders'?" Armando's concern was increasing.

Willa nodded confidently. "That's what Flo told me, Your Highness. She didn't know much about The Plan either, but she said that that was part of it. Flo hated that idea. She didn't like Drum any more than I did."

Manda had not commented, so far. Her Conscience was very active, however. She was reviewing her discussion with Willa that had focused on her father's militia. She had also read about the Thorndike Militia of the past. That militia had been huge then, like the current one, and it had been aggressive. It had actually stormed Urhonordo. Maybe that history had nothing to do with today, but... *It's Drum and Thorndike together again,* she mused. *What Willa says she heard sounds like history repeating itself, doesn't it. Should we be worried about that?*

Queen Luna broke into Manda's thoughts. She had obviously been thinking the same thing, but she now offered a thought of restraint. "It is easy to jump to the most terrible thoughts that we all have now", she said calmly, "but I am concerned that we *still* do not know *exactly* what is happening down at Drumsworth or *why.*"

"We saw what we saw," growled Prince Armando. He was still on his feet and pacing. "My fellow countrymen were enslaved by Thorndike henchmen and were building a mote. Now we hear that Drum and Thorndike have some sort of plan to control Lowlanders. What more do we need to know?"

"We need to find out if what you saw was really what you saw and why they are doing whatever they are doing." announced Solem.

June stared at his father. "Are you saying that because you feel bad since you already gave Drum and Thorndike permission to build the new road, Father?" June said flatly. "*We three* know what we saw, and Willa even heard them planning it. What more do we need to find out?" June had ridden all night to say this. He was ready to challenge his father and do something about it

June looked around him. Willa and Manda were staring at him, their eyes wide with concern. *They agree with me,* June said to himself with assurance.

"Have you forgotten who King is here?" warned June's Conscience. "You are reacting with your anger and forgetting to think. Action should usually wait for thought, don't you agree?"

Queen Luna spoke with her usual calm. "I understand how you feel, June. I agree that there is a desire to do something immediate here, but your father speaks for me as well as himself. Willa has provided us with needed insight, but we need more. We must speak to Drum and Thorndike directly."

King Solem was looking at his son who was again on his feet and pacing. June heard his Conscience and was prepared to ignore him, but Queen Luna's Royal bearing and attitude joined his Conscience to help him find his composure. This was not what June wanted right now but he would listen.

Armando continued to pace, as well. His wife and King Solem were not keeping up with his intensity or June's. June and Armando did *not* want to wait for further information while there were slaves (and they were sure they saw slaves) being mistreated at Drumsworth.

DRUM'S TRIP

Down at Drumsworth, Balynn's father was still talking to her as though she was a person of worth. She knew about The Plan now (or thought she did.) Nobles had arrived again this morning as usual, and she had been allowed to attend their discussions. The talk was something about Lowland property lines versus those of the nobles again and had dragged on in a boring fashion.

Balynn couldn't stay focused. Her mind was actively anticipating the arrival of the twins she had met the other day, the 'Thorndike boys' as her father called them. They should be back again today she thought, finishing their work out back.

During a break in the inside discussions, Balynn donned her shawl and slipped outside. She had arrived at night and thought that she would catch up on the progress of 'the mote'. The air was chillier today so she gathered her shawl around her shoulders and head and then looked around.

Balynn was shocked. The River, although flowing into the lake at the quiet level that was common in the autumn, was already passing through the lake, gurgling past the east side of Drumsworth Castle and circling around to meet the rest of the mote in the back. The tangle of scrub brush that she had fought through to reach her vantage point only a week ago was gone and, in its place, water was lapping at a new wall that kept the flow from entering Drumsworth's eastern lawn. Access to and from Urhonordo was completely cut off.

Drumsworth was now almost an island. The only access point remaining was on the western corner of Drumsworth property and there were men digging there. This had all happened so fast. She couldn't believe it. First the new road was approved. Now a mote she had scarcely known about was almost finished. What was next and how did she feel about it?

She also wondered about the Lowlanders that were still working out back. Were they slaves? When she had asked her father about that, he had become immediately angry. She had thought for a moment that she was going to lose her newly acquired acceptance into his world.

The anger had quickly passed, however, when he explained that the Lowlanders were only the same young local workmen that he had transferred from working in his fields to work on the "lake expansion" that currently *did resemble* a mote, which, it was not...actually.   Balynn was satisfied with what seemed to be a reasonable explanation *and* she had thankfully remained on her father's 'good side'.

Balynn decided to dismiss her many questions for now. The day was brisk and perfect for a meeting with Jorge. He should be arriving with his brother any minute now. She strained to look upriver for their little boat.

And there they were...but there was only one and he was not in a boat. He was picking his way along The River's edge on a horse.

Only one of the boys is coming down here today? she thought with concern. *Which one? If it's Jorge, I don't care how he gets here. But if it's Jackson...* Balynn ran to the lake wall and strained to make out the details of the rider's face.

The result of her new vantage point was sadder than she had expected. She could now see clearly that neither of the twins was on the approaching horse. It was a stranger, and a stranger that was demonstrating a formal bearing. The rider carried a River Kingdom's Royal Crest on his tunic and his riding gear. He was unmistakably a special messenger from The King. No parties or Gatherings were anticipated. This was probably not good news.

Quickly Balynn hustled inside to deliver the news. Drum had just reseated himself with the nobles after their break and looked up with irritation as she entered. She was not deterred, however. She was part of the 'in crowd' now. The nobles all turned in her direction.

"A royal messenger is approaching on horseback, from the north," she announced with business-like brevity.

For some reason, the noblemen looked around guiltily, as though they had been caught doing something illegal. Balynn noted the looks darting between them but forgot that immediately as her father exploded with instant fury.

"Who told them we were meeting down here today? Who is the traitor among you? Speak up and confess!"

The man closest to Drum recovered quickly with a plausible explanation. "Relax, Drum. Didn't you say that the King asked you to return to The Castle to work out the details and sign paperwork and such? That's probably what this messenger is all about. Relax."

Drum blinked. The man on his left was probably right. He was getting too jumpy. He sat back in his chair and smiled at his newly transformed, helpful, intelligent daughter. "Sorry Balynn. I don't know what I was thinking. Would you do us all a favor and step out to receive the message for me? As you can see, we are very busy in here."

Balynn began to breathe normally as she now realized that she had stopped breathing, altogether. "Of course, Father," she responded formally and gratefully hurried back to the western corner of the front lawn.

The messenger had already arrived and dismounted. "I have a message from his Majesty, the King," he intoned formally.

Balynn smiled at the man and curtsied. "Good morning, honored Sir" she said politely. "Welcome to Drumsworth Castle. Would you like me to deliver your message to Lord Drum for you? After your long ride, I am sure that a bit to eat is something you might enjoy. I will ask the servants to escort you to a place to eat and refresh."

The messenger did not look at Balynn but stood at attention, eyes straight forward. "Thank you for your consideration, mistress, but I must give my message to Lord Drumsworth, himself, by order of The King. I will wait for him here to deliver it."

Balynn's new confidence slipped a bit. She couldn't bring the message in, herself? The message must be very important. What could it be? And how would her father react to this?

But Cbalynn chuckled. "Now you are on 'the inside of The Plan' and your father's affection. This is what you wanted. This strange situation is part of it. Go tell your father what the messenger said. This is what you signed up for."

Balynn swallowed hard and marched inside with determination. Drum looked up from his paperwork and scowled, holding out his hand. "Hand it over," he ordered bluntly.

Balynn held her ground and said plainly, "The messenger is under orders to give the message only to you, Father. He is waiting for you outside."

"What? A mere messenger is expecting *me* to interrupt my day and go outside to receive a message when I already know what it says? Hog wash! I will do no such thing."

"Here, here," grumbled the noblemen."

"You see?" Drum continued, leaning forward toward the cohort before him. "This is what we have been talking about, right? The King is so full of himself that he even puts protocol on his messages. He thinks that everything he does is more important than what *we* are doing. Royals have no regard for us nobility. That is why things must change."

"Here, here," The nobles responded.

"Shall I tell him that he has to come inside to deliver his message then?" asked Balynn, sure that she understood what was needed.

"No!" her father and the noblemen shouted in unison.

Balynn shrunk back out the door. *What did I say that was wrong? I don't understand. I am obviously in trouble again...*she thought miserably.

Drum was already on his feet. With an almost magical return to command of the situation, he held up his hands to the men before him. "Do not worry, my friends. This royal idiot will not dare to step inside Drumsworth Manor and report what we are discussing. I will see to this messenger myself."

With freshly donned confidence and swagger, Drum headed out of the room and down the hall toward the front door, beckoning his trembling daughter to follow him as he strode by.

Out on the lawn, the Royal Messenger still stood at attention. Drum approached him. *No arms presented...no accompanying militia...only one rider...no immediate threat.*

"So, young gentleman...You say you have a message that you will only give to me. Is that true?"

"Yessir. I have a message from The King for Lord Doran Drumsworth only, worthy sir."

"Well, I am Lord Doran Drumsworth. Give me the message."

The messenger bowed and handed the scroll to Drum, returning to his position at attention.

Drum laughed at the formality taking place on his lawn. "You have delivered your message," he said, smiling. "I know what it says. You can go now...unless you would like something to eat before you go."

"Please read the message, sir," said the messenger, still staring straight ahead.

Drum sighed. *This King is such a nut for protocol...among other things.* Carefully he broke the seal on the scroll and unrolled it.

The Call of the Auras

"To The Honorable Lord Doran Drumsworth.

Please be advised that an error in protocol took place recently in the meeting with Lord Drumsworth and Lord Thorndike in which approval for building The Royal River Road was granted by King Solem.

This approval was granted in error as this project will affect the flow of The River for both The Lowlands and Urhonordo Kingdom as well as River Kingdom. For this reason, all parties must be in attendance when approval for the project is granted.

Therefore, the afore-mentioned verbal approval of the Royal River Road Project is hereby officially rescinded.

As the timing for this project is affected by weather changes, King Solem herby commands that you, Lord Drumsworth return with Lord Thorndike to The Castle immediately for discussion of this project with all affected parties.

My messenger will return to The Castle with both Drumsworth and Thorndike today. If you do not comply, the project is permanently canceled.

Signed, King Solem

Drum dropped his hands to his side. He was in disbelief. This was all happening so quickly. A few minutes ago, he and the southern noblemen had been happily discussing property lines that would probably be changed with the construction of the new road and the finish of the mote.

Now all that was worthless...and now he would have to explain this ridiculous interruption to the noblemen.

The noblemen were his partners. They were all Birthers. Their side of the changes to come had been discussed here for months and were finally close to being realized. Now there was this stupid message. It was humiliating to say the least.

But perceptions were important, and Drum was the only one that knew what the message had said...he and his daughter. Drum groaned. Deception was so much easier to keep in place when only one person (him) knew about it.

With agility, Drum turned to Balynn and purposefully folded her into his next move. "Go directly inside and pack clothes for an overnight. I will dismiss the noblemen. The King wants my assistance immediately and you must come with me."

Without hesitation, Balynn, the new 'insider,' rushed into the castle to do as she was told.

But the messenger had something else to say. " You are requested to come alone, sir."

Drum shrugged. *I will leave her at Thorndike's,"* Drum said to himself as he hurried inside to confront the nobles.

 Entering the conference room, Drum regained the swagger he had worn when he left the space and offered an acceptable message. "I guess I have been requested to *consult* in a very important meeting at The Castle regarding materials for the new road. Our inept King doesn't think we have enough rock to fill in all the ditches. Obviously, he knows nothing about building a road." He rolled his eyes and the other nobles smiled knowingly, making moves to leave the room and mumbling their frustration with their King.

"Things will be different when you're in charge, Drum," one of them called back as he climbed into his carriage. He was one of the few that knew The Whole Plan. Drum looked around nervously, but no one else was listening.

In Balynn's upstairs bedroom, petticoats and bloomers were being stuffed into the largest satchel she could find. One of her Lowlander servants held out her apron full of combs and brushes, powders and soaps.

"You will need help with all of this up north where there are more noblewomen with fancy things and maids to keep them beautiful," she said with a wink and a giggle. "Ask the new Queen to borrow one of her Royal Dressers. You will be in The Castle overnight. If you look your best, Prince June will be at your mercy."

Balynn laughed nervously as she folded a dress into a trunk. "He will probably be in the 'meeting' with the other Royalty. And I'll tell you a secret, Pauline. I am more interested in one of the Thorndike militia men than I am in Prince June."

Pauline stopped short. She and Balynn had become friends in recent years but this confidence from the wealthy young girl was something she hadn't expected.

"You would rather have a militia man than a Prince? You are crazy. I am the one that likes militia men, not you."

Lynn began sorting her creams. "Not *any* militia man, Pauline... only one."

"Which one?"

"Jorge...I think his name is Jorge. I've only talked to him once."

Pauline stared at Lynn in amazement. "You fancy Jorge? I don't believe you. I just met his twin brother, Jackson." Both the young girls hugged and laughed as friends do. There in Lynn's bedroom, the fact that they were very different disappeared and they were very alike.

Into the giggles and masses of female essentials an angry voice now boomed. "If you are going to go with me, *Balynn*, get down here. My carriage is leaving, and I will leave you behind," Drum yelled up the stairs.

Quickly, Pauline and Balynn buckled all the fasteners and hauled the luggage down the stairs.

The noblemen departed quickly and Balynn's luggage was loaded and strapped to the top of the carriage. The Royal Messenger started upriver ahead of them. Balynn was excited as she climbed aboard and adjusted her skirts.

Drum, after waving his friends off, climbed into the carriage, dropped into an angry mindset, and stopped talking at all. His thoughts were crowded with competing imaginary scenarios and possible reasons that his victorious approval of the Royal River Road was now no longer a reality. *What had happened?*

The most probable voice of opposition was probably Prince-Consort Armando, the unofficial head of The Lowlanders *who is one of the main targets of The Plan,* Drum conceded silently. Drum had had dealings with this man through the years.

His disgusting, dark, Lowlander presence gives him a say over two 'kingdoms,' Drum groused to himself. *Without his interference, neither Urhonordo nor the Lowlanders would have had input on anything all these years...and that's the way it should have been, of course, and the way it soon will be.*

Drum settled back in his seat and a smug smile returned to his face. He seldom allowed himself the luxury of indulging in his ultimate goal...control and rule of everything near and around The Great River. Few knew of this plan completely, but today the thought of his final victory gave him strength...and inspiration to overcome the challenges that might await him. He closed his eyes and dreamed of Royalty...his.

The carriage lumbered along its well-worn route, revealing the splendor of fields of fruits and vegetables that decorated its path. "Who is taking care of our orchards and land crops, today, Father?" Balynn asked in an offhand way. "It looks like the trees are heavy with fruit."

Drum opened one eye and peered over his daughter to the orchard beyond. "Damn those Lowlanders," he growled. "They never do what they're told unless someone's on top of them every minute. I'll have to get the Thorndike boys to round up some more Lowlanders for me."

Balynn shrugged. "Probably these Lowlanders are just busy digging the mote, Daddy," she contributed.

Drum sat up straight. "Mote? What do you know about a mote? Who told you there was a mote?" His face got red and came so close that Balynn could feel his breath.

Balynn shrunk back. Of course, she knew about the mote. She had just seen the piece of it that was between Urhonordo and Drumsworth...she had even watched as it was being dug...but her father didn't know that, did he? Her mind raced. She was in danger again. *Quick! Quick! Think of something. Tell him a lie.*

Cbalynn came to the rescue instantly. "No need to think up a lie," she offered. "You learned about the mote in the map room at Thorndike's. That's the truth."

"You told me about the mote at Thorndike's," Balynn squeaked into Drum's red face.

Drum thought for a moment and relaxed. "Ah yes. I remember now. I had forgotten that." He smiled and patted her hand.

"And he *also* forgot to tell you that the Lowlanders that are digging that mote are slaves, didn't he?" hissed Cbalynn.

Drum was now awake and looking at his fields. Balynn had called something to his attention. He had forced the Thorndike militia to drag his farming Lowlanders to work on the mote build, but he had forgotten about his own fields.

This was simply a logistics problem, Drum concluded. He would have to send Thorndike's militia down deeper into the Lowlands for more slaves. He sat back in his seat again to enjoy the ride.

After hours of bumping along the rough river road, Thorndike Manor came into view. Balynn sat up straight, hoping to get a glimpse of Jorge. The daylight was beginning to fade, but the amazing view that now greeted her eyes was not obscured. The field that they were passing was crowded for some reason with workers of all shapes and sizes. And these workers were not harvesting or tending to the fields. They were hauling rocks, pushing them into place, shoveling dirt... filling in levees.

With a frightening thump, several large boulders were rolled into a ditch in front of the carriage, causing the horses to rear and the carriage to tip precariously in response.

"Sorry Master," hollered the huge farmer that had rolled the rock. "Didn't see ya comin' round the bend there 'til them rocks was out'a my control." He bowed fearfully toward the carriage, not knowing what noble was inside but fearing punishment from whoever it was. His dark skin revealed that he was a Lowlander farmhand.

Drum leaned out of his window in anger...but also in surprise. He knew what he was seeing. These people...farmers and their field hands were obviously filling in holes for the future road, a road they didn't have permission to build. Drum told his driver to stop the carriage and slowly, stiffly, he got out to inspect what he knew he was seeing.

"Been a long day on the road, Drum?" came a deep voice from behind him.

The startled Drum whirled around and came face to face with Thorndike and his horse, both staring down at him with a very definite air of superiority.

"What's going on here, Thorny?" he growled back, quickly regaining his composure. "Didn't you get the message from the King? Didn't you hear that the two-faced moron reversed himself? What *is* all this?"

Thorndike and his horse backed up but continued their stance of authority. Additional militia men pulled up impressively behind him. "We are on *my* land," Thorndike said calmly. "I sent my militia out with word to order my landholders to start on the road build as we said we would do. Then I got that stupid message from Royalty.

"When I got it, I didn't see any reason to stop doing what I was doing *on my land*. I figured you would be on your way up here today, so I decided to meet you here and show you how my part of the build would be done." Thorndike grinned and looked around proudly. "What do you think?"

Drum did not like to be bested but he had to admit that what he saw and what Thorndike said impressed him. He turned slowly, smiled at the workers, and began to clap, much as he had when he entered The Gathering. This approval was his way of re-establishing *his own* leadership in a situation which looked like it was forging ahead at this point without *his direction*.

"This is great, Thorndike...just what I hoped you would do! We are showing the King 'who does what and where'. That's just the way it is going to be from now on. We Birthers are in charge of our own land and what happens on it!" he shouted, raising a triumphant fist in the air.

"Here, here!" shouted the militia, responding to Drum.

They were acknowledging Drum. Thorndike's pride took a hit. Drum was taking over *his* bold leadership action...which was exactly what Drum's move was designed to do, of course.

Thorndike's *control* over *his own militia* was being challenged in front of *his* men. Drum snickered. He was willing to share *some* leadership with a partner in *his plan,* but not much.

"Listen up, men!" Thorndike bellowed over the voices of his militia. "We Birthers stand up for what we believe in, right? And I am going to do so right *now!* Tonight, I will show the King that the strength of a Birther militia cannot be ignored. We received his permission to build a couple of days ago and we will keep it!"

"Well said," coached Thorndike's Auras from their perch on his shoulder. "Don't let this Drum guy steal your thunder. You are the main man here!"

With the brashness of his Auras' encouragement, Thorndike spurred his horse to the front of Drum's carriage. He was inspired. "Follow me, Drum!" he shouted. "My militia and I are off to The Castle with a show of force to tell the King what we want and when we want it, tonight! He won't dare to say no! Let's go, men!"

With that, Thorndike masterfully re-took his command position and rode off toward The Castle, a phalanx of hearty, invigorated militia men in his wake.

Drum was furious. This was not what he had planned at all. But he had been left with no alternative. To regain his authority over this now Thorndike-led situation, he had to be at The Castle tonight as well.

He would, of course, take the lead back where it belonged when he got there. Dusting the clouds of militia-generated dust from his clothes, the irritated Drum marched back toward his carriage. Stunned and bewildered farmers stood back silently and watched.

Drum smiled again, however, and waved at the gathered farmers as though they were part of *his* team and should understand *his* true leadership. "I am sure that Thorndike expects you to work hard today, but I suggest that you take the rest of the day off," he announced with a snide grin. "If he says anything, he will have to deal with *me*." He smiled his most engaging smile and waved magnanimously.

The crowd of exhausted farmworkers cheered and clapped as Drum climbed into the carriage. They were on his side...at least for tonight.

On the other side of the carriage, however, a small but important incident took place. "I will follow behind," said a deep voice near Balynn's open window.

A brief glimpse of a smile was all she caught out of the corner of her eye, but it was Jorge. The carriage driver clicked his horses into motion, and they were off. Jorge followed at a safe distance.

There will be no stopover at Thorndike Manor tonight, thought Balynn happily, *but this is even better. Jorge is here.* She settled back contentedly.

THOUGHTS FROM THE SITTING ROOM

Solem stopped pacing and dropped into a chair across from his wife. "I have two serious problems to face, my dear. The first problem is of my own making. I simply didn't think before saying yes to the new road. This is the problem that Drum and Thorndike expect to discuss when they get here tomorrow.

"And I will be standing up against slavery, of course," Solem said quickly "but this will be slavery that *I* have not seen myself. Can I accuse a man of evil if *I* haven't seen it?"

"We know what Juna and Manda and Armando saw," countered Willa.

"We know what they *think* they saw," said her husband. "If Drum denies it, that clearly places June and Urhonordo on one side and Drum, Thorndike, and the nobility on the other. I will be in the middle between them."

"True."

"These are the ingredients of... of a war, my love. If the nobles and Thorndike's militia take their side, things could get dangerous. Most of the nobles are Birthers. I am thinking that I might ask June to take you away to someplace safe until I find a peaceful solution."

Willa stood up and glared at Solem. "Who do you think you are talking to, Solem? A stupid child? A mindless female? How dare you think so little of me? Who do you think you are?"

Solem winced but looked calmly at Willa. "I am the King of River Kingdom," he said slowly and sadly. "Don't be angry, my love. I must do what I think is right."

"Then do it!" Willa retaliated. "You have two *huge* situations on your hands, and you tell me that you are looking for *'one solution'*. Then you think of dismissing the *one person* who will be at the meeting that you *know* to be on *your* side. That is so wrong! Isn't your Conscience paying attention?" Willa stood over her dejected husband in fury.

Csolem spoke quietly and firmly in Solem's ear. "She's right you know. I have not been strong enough yet to advise *sure* solutions to these issues. And *I* am the one that didn't guard you against the Auras in the first place. Listen to your wife, Solem. Her Conscience is strong. Your bride and her Conscience will help us."

Solem stood and put his arms around Willa's staunchly positioned little body and felt it melt slowly into him. They and their combined Consciences would face the meeting tomorrow together, he now knew.

The Urhonordorians were asleep in Castle guest rooms. Willa had quietly taken inventory of the situation earlier. She had spoken with Queen Luna and Manda that afternoon and the women had decided together that journeying back and forth to Urhonordo while waiting for Drum and Thorndike the next day was impractical. Considering the size of The Castle, it was completely unnecessary. The Urhonordo Royal family would spend the night.

Willa now felt secure that half of the major players in this complicated scenario were comfortably housed in the guest rooms of The Castle. Luna and Armando had a room to themselves. Manda had been eagerly taken in by Mimi who, being completely unaware of the current difficulties, had happily offered to share her bedroom for an enjoyable 'sleep-over.'

Deftly, Willa and Agnes had martialed appropriate members of The Castle staff to ready the rooms and provide the occupants with private meals in their rooms. All were settled in for the night.

Solem gazed at Willa with admiration. "Well done," he said honestly. "Now, if I can just figure out an answer to these two problems that confront us the way you have done with housing our friends, everything will be fine." He shook his head sadly.

"You will," offered Solem's Conscience. "Your father and grandfather delt with these same characters successfully. You will, too."

Solem accepted Csolem's assurance, but he was not convinced.

"I am worried first about the slavery that June and Amanda saw down at Drumsworth", mused Solem aloud. "Roads and irrigation ditches are important, and I take responsibility for my blunder with these things but slavery is worse, don't you think?"

Willa nodded vigorously. "Indeed. I think that slavery is much worse than a road. And the Drum family really looks down on Lowlanders. One hundred years ago, Drum's grandfather even put Lowlanders in prison for no reason. I read about it. Manda says that the prisons he used are still there."

"They *are* still there," came a voice from outside the sitting room door. A knock followed. It was June.

"It's not polite to eavesdrop, son. Did you never learn anything about polite protocol?"

"I'm afraid not," June countered as he opened the sitting room door a crack. "Can we come in? Manda is out here with me."

Willa and Solem's eyes opened wide in surprise at the two young people standing and anxiously awaiting permission to enter their very private space.

Willa looked at her two friends with relief. *June and Manda have been talking together and appear close. They are a good couple,* Willa thought positively. *Maybe June fancies her. They are both smart. And all of us really should be talking together.* Cwilla agreed.

"Of course," answered Willa immediately. "Come on in."

June and Manda each grabbed a small side chair and pulled it up near the fireplace, seemingly unaware that they might be intruding.

"It's warmer in here than out in the hall," Manda explained with a shiver.

"So," began June as he relaxed and leaned casually toward the Royal couple. "Have you two figured out what to do when our visitors arrive tomorrow? I'm in favor of locking them both up for slavery, but Manda says I'm crazy. What do the two of you think?"

King Solem leaned back in his chair and shook his head. His son was so young and inexperienced.

Solem's Conscience began arguing with June's impetuous Conscience. "You can't be encouraging him to lock people up before you talk to them," he warned.

"I'm not," responded Cjune. "He's just all fired up. That's why Cmanda and I had Manda bring him up here. We hoped you and Cwilla would slow him down."

Solem's Conscience nodded. "The boy needs to calm himself and think," he advised Solem. "That's what fathers are for."

"I hear you, son," offered Solem quietly, "but there are many questions that need to be answered before we do anything rash like locking people up."

"Although that would be a fun thing to do," laughed Willa. "I have been wanting to lock Drum up for years. Flo and I used to dream about that back at Thorndike Manor when Drum would visit."

"Who is Flo again?" asked Manda.

"Probably my only friend before you and Agnes," Willa answered thoughtfully. "And that's the truth. The two of us could see even then that Drum was on some sort of evil mission. He and my father were always talking about 'The Plan'. We never knew exactly what that meant, but it always seemed to have something to do with 'controlling Lowlanders'. Now it looks like they think the road will get them closer to doing that somehow."

The Call of the Auras

"'Control' sounds like slavery to me", groused June.

"Me, too," added Manda, pulling her chair up closer for emphasis. "And after what June and my father and I saw down at Drumsworth, we now have a theory that fits with that idea perfectly.

"June and I think that Drum and Thorndike want to capture all Lowlanders, send them back to the Lowlands and take control of their whole country. That's why we came up here. We think Drum and Thorndike have decided to do that together. The Royal River Road is only the first part of that plan."

King Solem could not remain seated. "You two seriously think that Drum and Thorndike want to send the Lowlanders back to their homeland? Why would anyone want to do that? Lowlanders are part of both northern kingdoms, now. Why would anyone want to send them away? That is a ridiculous idea."

"Yes, it is," contributed Manda evenly. "But they wouldn't do that completely. They would keep 'the useful ones that obey'. That's where slavery would come in. The Lowlanders who didn't obey or were in the way, would be returned to their own area to be imprisoned there...or worse. Those two know that the Birthers would love that."

Willa shuddered. "I think she's right, Solem. That fits exactly with what I used to hear when I wandered by the open doors of my father's meetings with the nobles. Those there were all Birthers. What they beleve is that they are better than all Lowlanders by birth. Lowlanders have no rights because they don't deserve them."

"My God," Solem moaned. He sat down again, carefully, as though his thoughts could be changed by a quick movement. "So, the three of you actually believe that what you saw at Drumsworth is only a start, don't you? The Lowlanders down at the so-called 'mote' are only the first of the slavery that will eventually control all Lowlanders. Is that what you are saying?"

Only June spoke. "Yes, Father. That is what we are now thinking." Manda nodded.

"I see," Solem said solemnly. "If you are right, then why are Thorndike and Drum intent on building their road. What has their road got to do with enslaving the Lowlanders?"

The four stopped to ponder. "I think I know the reason for that," Willa answered slowly. "I think the reason is... my father's Militia." Eyebrows went up and heads turned to the young, inexperienced Queen, but Willa was ready with the answer.

"The Birthers have decided that the Lowlanders should be returned to their homeland, right? Who is going to capture them? The Militia.

Who is going to keep them in their place? The Militia.

Who is going to guard the border between The Lowlands and River Kingdom down south? The Militia.

And how will the Militia and their equipment travel to and from Thorndike Manor and the border on a regular basis? The Royal River Road."

Willa took a breath. "My father's militia has been gearing up for this plan all my life without me thinking about it. But now I see it. That is what he has been living for. With this road created for him and Drum and the nobles, his militia can take control of The Lowlanders at the border with River Kingdom and my father and Drum will have everything they have always dreamed of."

Silence settled over the sitting room like a shroud. The cheerful crackling of the fire in the hearth seemed out of place. The answer was too perfect, the rationale was too complete. The four would-be problem solvers were stunned at their sudden understanding. An immediate discussion of what to do about this revelation was needed now.

But it was too late. Into this new, wondrous understanding that Willa had just introduced came the unmistakable noise of arriving horses from outside. One of their summoned visitors had apparently already arrived.

Solem turned to Willa. "Would you be so kind as to accompany me to the colonnade? It would appear that we have a guest that we need to greet." The two of them would go together. He would not stand on his protocol again.

Drum and Thorndike were not scheduled to arrive until tomorrow but, undoubtedly, this noise meant that one of them was here.

"Indeed, my dear," responded Willa, taking Solem's arm. Together they walked down the grand staircase where Horace was already holding the door open for them.

But Solem was not prepared for the thundering horde that now appeared before him. The Castle driveway held Drum's coach and a set of winded horses that had obviously been driven hard. That was not as surprising, however, as a complete phalanx of militia men dressed to impress in full militia gear. Their leader rode rudely through the gate in front of them and onto the colonnade itself.

"Your Majesty," Thorndike called out from atop his horse. "We are here to receive formal, written approval of The Royal River Road as requested! Where would Your Highness like my men to stand? They have come with me tonight to witness the approval of Royal River Road!

"This militia doesn't hop and skip over ditches and logs. Militias need a real road... The Royal River Road, to be exact, and they need it *now*, as agreed to by *you!*"

Willa cringed at her father's absurd demand. What did he think he was doing? Was he trying to impress Solem? Did he think this was the way to do it?

Solem looked at the rowdy tangle of horses and men cluttering his usually peaceful colonnade. Thorndike was obviously making a display of his power. And that power was hard to ignore. The concern and anxiety that the King had been feeling earlier, returned.

With the noisy militia continuing to circulate its intimidating numbers on the colonnade, Drum now clambered out of his carriage and hustled past Thorndike and the confused horses with calculated calm.

Determined not to be bested by Thorndike's show of force, Drum approached the King with as much contrasting dignity as he could muster and bowed. "Lord Doran Drumsworth the Third, present as you requested, Your Majesty", he intoned respectfully. His eyes flickered snidely toward Willa for a second and then he lowered his eyes with a show of perfect decorum.

"It is a pleasure to see *you* again, Your Majesty," he offered smoothly to Willa. "I'm afraid that Thorndike is a little overly assertive tonight. Please allow me to apologize for him."

Csolem was not impressed. "These two grand entrance displays show more than a greeting", whispered Solem's Conscience. "I believe that they show clearly that these two 'partners' are now also rivals. They are *each* trying to show you that *they* are the most important tonight. Let them know who is really *most* important here."

Willa's hand clutched Solem's arm but, with a gentle touch, Solem covered it with his own and addressed the arrivals together. "I am pleased to see that you two have both arrived for our necessary discussions, Lord Drum, but discussions will not take place until tomorrow, of course.

"I asked that you both come alone, Lord Thorndike. These others will need to find accommodation other than here at The Castle as I will allow only the two that I invited to enter. I trust that I will see you both here in the morning." With that, he turned and escorted his Queen confidently back toward The Castle.

"Well done," whispered Csolem.

Thorndike could see that his impressive arrival had not achieved the result he had hoped for. "Trample the King and take control of The Castle tonight," coached

Thorndike's Auras. "Who does he think he is? Your militia could do away with him right now. They are excited and ready to take over. There is no Royal Guard in sight. Kill this stupid King right now. Your daughter is already here and she is already Queen. Daughters must do what their fathers tell them to do. This could be all over, tonight!"

Unaware of the Auras' coaching, Willa dropped Solem's arm and stepped forward into the moonlight. She appeared soft and quite beautiful, standing there, but she was determined. Thorndike had never really noticed how much she looked like her mother. It was unnerving. She was reaching out to him and smiling at him as though she was a trusting ghost... as her mother had reached out to him years ago. He shivered.

"Why did you come here so early, Father?" Willa asked calmly. "Your militia looks wonderful, and they are truly impressive, but they should really go home now. Then you can come back tomorrow to talk to The King...and to me...alone. I will be in the meeting with you, of course. I would love to see you for our discussions then."

Thorndike shivered again and halted. His mind was playing tricks on him. Who was this apparition, really? A daughter talking to him like a daughter or vision of his past? She must be his wife. An eerie fear ran down his spine. What was happening?

The whinnying, stomping, shuffling assemblage of people and creatures continued in a directionless jumble on the colonnade, but the Auras that had traveled with Thorndike throughout the afternoon recognized his reaction to the little Queen but misunderstood it. "Get on with it, Thorndike. You're losing focus. Your placement of your daughter here still appears to have no value. No need to push for her influence tonight."

The militia, weary from being on horseback all day began to grumble among themselves. Their commander had brought them here for what?...nothing, it would appear. What were they doing here? What was their mission?

Thorndike and his Auras turned hastily away from Willa but now confronted by the militia's questions. The trauma of seeing his wife here in front of him was still jabbing at him. He couldn't focus. His wife continued to stare at him. And there was Drum, standing right with her and smiling...no laughing at him. In confusion, Thorndike began to yell, his most dependable form of communication.

The Call of the Auras

"This King doesn't want us here!" Thorndike shouted, sneering at Drum and the Royals. "This King is not ready for what *we* have to tell them because Drum is standing in our way."

Drum's smile turned to a glare, but Drum's Auras understood their man. "Style over pushiness," they reminded him quietly.

Hearing Thorndike's snide remarks and his Aura's advice, Drum smiled again and looked accommodating. "Thorndike is a bit pushy, to say the least," he commented to Solem with a knowing smile. "I apologize for that." He could make this situation work for himself, he was sure.

"See there? He is already kissing up to the King," hissed Thorndike to his troops. "*We* Thorndikes are men of action. I can see that we aren't appreciated here, tonight. Tomorrow we will show the King who the real leader of this operation is."

Thorndike kicked his exhausted horse in the ribs and rode off into the darkness that led toward Thorndike Manor. His militia regrouped and obediently followed him.

The accompanying Thorndike Auras were not aware of Thorndike's night-time vision but accepted this unexpected denial of their forceful approach, rationalizing that an exhausted Thorndike at tomorrow's meeting would be even more ready for anger and turmoil. They would plan and counsel with him tonight. Tomorrow could be wild.

*****

Balynn had been talking to Jorge outside her carriage window since they arrived. Unconcerned, she stepped down happily with the help of Jorge and waved to her father who was now standing alone on the colonnade.

Drum did not wave back. He had fallen for Thorndike's exciting 'forceful approach' but obviously it hadn't worked. Drum's Auras were not deterred, however. They had seen Drum work. He could turn this minor miscalculation into a major victory if he played it right. And he had already started. Finesse over obvious power... this would be the Drum approach...the velvet glove...the deceptively well-placed smile...but wait...

Drum suddenly realized King Solem and his Queen were walking arm-in-arm toward The Castle without him. He couldn't believe that this was happening. He was being ignored. This sort of off-hand dismissal had never happened to him before.

I am important. I am always welcomed profusely. My presence is coveted, he thought in disbelief. *I must deal with this strange behavior immediately.*

"Daddy!" Balynn shouted to him with a smile. "Where shall I take my luggage and who should take it? Jorge was going to help me, but he had to return to Thorndike Manor with his militia. Why did everybody leave?"

Willa was just entering The Castle with Solem when she heard Balynn's shouts. Turning back, she spotted the source. Drum was still standing motionless on the colonnade, but his daughter was running happily toward him from the gate, fully expecting to be accepted into The Castle and assisted with her things.

Willa stopped." Did you know that Drum brought his daughter with him?" she asked.

"No." Solem shook his head. "What in the world did he do that for? I told both Drum and Thorndike to come alone."

"Probably to give her a chance to see June," snickered Cwilla.

Willa looked up at Solem. "But you aren't going to allow Balynn Drum to sleep outside in her carriage all night, are you?"

"Why not?" Solem retorted firmly. "She shouldn't be here. She wasn't invited. It's against protocol to expect welcome where you aren't invited."

Willa grinned. "Isn't it against protocol *even more* to allow a young woman of quality to sleep out in a cold carriage all night...and just outside *your* gate?"

"We have no room in The Castle," grumbled Solem. "The Urhonordorians are in the prepared guest rooms.

"My father already left in a huff" whispered Willa. "Drum's daughter can have his room."

Now Solem grinned back at his persistent wife. "I will tell both Drums to come into The Castle for the night, if you insist, dear. And I will send Horace out for their luggage."

A PLAN WITHIN A PLAN

Thorndike had ridden all day to demonstrate his power and might to the lame King. He had envisioned The King immediately granting the approval that he and Drum requested for their Royal River Road ...and more.

Then Drum got in the way.

That's the way he remembered things tonight. But Thorndike had lost something vital in this denial of his expected victory…He had lost his grip on logical thinking.

The power he had felt at the meeting with the nobles had started it. He had not needed Drum that night and they had loved him. Tonight, Drum had blocked his powerful demands of The King on the colonnade. Thorndike had now determined that Drum had become a problem. He, Thorndike, would now need to re-direct The Plan *his way.*

All day, Thorndike had believed that his new-found leadership and this move tonight would advance the preparations for his plans with the noblemen…and these preparations would also have brought him closer to his own *personal* plan.

Drum was a Birther and, like all Birthers, the return of River Kingdom to white rule with Drum as a strong king was a long-term goal. In this, Thorndike completely agreed.

But increasingly, Thorndike felt that Drum had no real interest in anything but himself. Thorndike was planning to help Drum with *his* plan…but now, only if it helped Thorndike with his own plan and Urhonordo was increasingly part of that plan.

Thorndike's dream twisted and turned in his mind. He now felt he was smarter than Drum. He had always known it, but Drum had a way about him, a way that drew people to him. For that reason, Thorndike now viewed Drum as 'useful'. The nobles loved him. They had contributed their private militias to him for that reason. Control of *all* River Kingdom militias into The Thorndike Militia with him as Commander was one of Thorndike's most constant dreams and always had been, but now there was so much more…

To Thorndike, Drum was now like the carved statue on the prow of a ship… useful as a motivating attraction but not able to steer the ship. That would be Thorndike's job. Drum could take his place in The Castle…maybe even *two* Castles if you counted Drumsworth, but *The Thorndike Militias* would be in control of everything…and they would *own* Urhonordo!

Tonight was an unexpected hitch in what was now Thorndike's plan within The Plan. Thorndike loved power and intimidation. He had envisioned tonight as a triumphant capitulation of The King and the signing of the rights for the road building that had been given already verbally.

The new road was essential to Thorndike's Urhonordo goal. It was the quiet, unsuspected road to his surprised control of every human being on *both sides of The River!* Signing papers for *his road* was merely a formality in his mind.

Of course, the lame, protocol-bound King would have required the co-signing of Urhonordo and The Lowlands for the road, but this would have been granted without difficulty. These kingdoms had nothing in the way of standing forces and The Thorndike Militia would eventually control both.

No wonder then that Thorndike was morose as he returned to Thorndike Manor tonight with nothing. He was in the grip of what had now developed into an Aura-assisted mania.

The Auras were impressed. Seldom had they had this kind of fervor on their side. Immediately, they began to strategize. How could they use this quirk to their advantage?

Thorndike's Militia was also dejected. They could see their leader's disappointment. They could feel it, so they also reflected it. He was pushing them every day, forcing them to prepare for...*something*. No one knew exactly what the 'something' was, but they knew it would be physical and military and require the skills they had trained for daily.

No, they didn't know who they would eventually be fighting, but they felt confident that they could defeat anyone who deserved it. Their Commander had made that clear.

The word had grown in the villages and towns was that the Thorndike Militia had the power to conquer anyone they wished. They had been bred and trained for that power all their lives and they were proud of their reputation. They were Birthers... although the militia was not quite sure what 'Birther' meant.

On this night, Flo had not known when they would return so their noisy arrival had the same surprise effect on her that it had had on the inhabitants of The Castle earlier.

The good news was that she *had* expected them at dinner time. She had heard that they would be supervising the filling of drainage ditches all day and she had expected them to need a hearty meal. To that end, she had concocted an enormous pot of lamb stew which was still sitting untouched on the hearth.

With the sound of horses and men in the driveway, Flo went immediately to the hearth and stoked the fire under the stew. With a deft hand, she began to stir the pot with the large paddle that her brothers had made especially for her and this kind of big job.

But why are they so late, she wondered. *Did something go wrong where they were working? Was either of my brothers injured?* She always worried about that.

In addition, what kind of mood was Thorndike in? His mood would have impact on who ate her food and how much they ate if he had had a bad day. On a hunch, she decided not to invite her brothers into the kitchen. This might not be the right time since the hour was so late.

Soon she heard the men milling about the militia's quarters behind the manor. She rang the bell. That was the signal that the militia was waiting for. Food cooked in large amounts inside the manor would need to be carried out to the men in large pots on the instances when Flo cooked for the whole group. Soon there was a cluster of hungry men gathered at the hearth, anxious to tote the pot. They were obviously hungry.

Flo followed the steaming kettle out to the barracks where long tables were located. *She* would serve the men to make sure they all got a fair share. Shockingly, Thorndike himself was seated at the far end of their enormous table tonight, however. That hardly ever happened.

With discipline developed by not noticing things she was not supposed to see, Flo proceeded to dole out equal portions on each plate coming to her, the first being handed to Thorndike by one of his sons.

When she was finished, Flo stood behind the pot which had shrunk to less than half full and waited for those who wanted seconds. She also waited to find out what the reason was for Thorndike joining his men for a very late meal, especially after such a long and difficult day of work.

She hadn't expected Thorndike to be friendly and appreciative of the work they had done, and she was right. He stuffed his heaping portion of stew into his mouth without speaking, his head down and his shoulders hunched. The rest of the men at the table ate much more quietly than usual and appeared to be waiting for something bad to happen.

Unobserved by anyone, the Auras followed the men in from their chilly midnight ride. The Auras knew that this was a time for their expertise. Everyone was tired. The effect that they had hoped to create at The Castle had not been appropriately received. The militia knew that Thorndike was furious, even if they were not sure why.

The Auras were there to ride the wave of discontent. "Tell them what they did wrong," the Auras whispered to Thorndike. "Tell them what they must do to make up for it. Tell them that their future depends on it."

They'll do what I say, Thorndike thought peevishly. *That is the power that I have here.*

Thorndike finished his bowl of stew and looked up at his men. "You failed me tonight," he growled as he rose suddenly before them. "You were at The Castle to present a show of force," he shouted. "Instead, you allowed a weak king and his queen to dismiss you like a gathering of children! You embarrassed me! You have been trained to be warriors! Tonight, you were gutless wimps!"

The militia stared at their plates.

It occurred to the Thorndike twins that they had not been ordered to 'show force' at The Castle. It occurred to them also that their father was blaming them for behavior that *he* had ordered and tonight was not their fault.

But the Auras crept up to their chins and nestled in their ears. "Your father knows best," they whispered. "He always has and he always will. You are his blood. You know this. This is your time to show what you are made of. Support your father!"

"Here, here!" shouted Jackson. "We failed you tonight, but we will not do it again, Sir!"

Jorge felt guilty without any coaching from the Auras. He had been flirting with Balynn when the rest of the militia was stomping around on the colonnade. "We will not fail you again, Sir," he proclaimed now, standing tall with his fist in the air. "What would you have us do to make up for it?"

The rest of the militia took their cue from the brothers and stood with a rousing "At your service, Sir!"

Thorndike and his Auras had what they wanted...support inspired by guilt.

Thorndike puffed out his chest and issued his next statement with new determination which was designed to overcome the humiliation of earlier tonight. It was full of Aura-inspired bravado, if not well thought out.

"Listen to me, men. Get some sleep tonight because you will get up at dawn tomorrow. I want men who are loyal to the road-project we started today to be at The River early. That work must be done before anyone tells us we can't do it. Make sure that *every* landholder that lives on *any* of my land is there with his family and farmhands to work hard. Together we will teach Royalty what it means to be Birthers on our own land! Do we need permission from Urhonordo? No! Do we need permission from Lowlanders? Hell, no! We are Birthers!"

"Huzzah!" shouted the militia in unison.

"I *will* do what I want on my own land and just see if anyone can stop me!" Thorndike shouted. "Let's see if anyone, including The King, has the guts to tell me that I can't fill drainage ditches on my own land!

"I am tired of being told what to do. There are steps I can take tomorrow that will ensure my right to finally achieve my own destiny...steps that no one will expect! That is the power of superior, white thinking! Be ready!"

Amid the cheers of his militia, he marched proudly back to his manor. *The road to my future will not wait,* he thought fearlessly to himself.

Flo watched quietly and did not exchange looks with her brothers. What did this new enthusiasm mean? What did Thorndike have in mind?

THAT NIGHT AT THE CASTLE

Back at The Castle, Drum's thoughts were very different from his partner's at Thorndike Manor. Drum was strolling the grand halls of The Castle itself, a castle where his grandfather had been King and where his dreams in The Crown Room had promised him, *he* would be soon. He felt like he had come home. The Auras of his famous grandfather were still here. He could feel them.

Blissfully, Drum meandered down the ornate hall to his prepared room. There were bedroom doors here and there. He could even hear Balynn and Princess Mimi giggling behind one of them. *My daughter is as comfortable here as I am,* he thought, as his imagination took ownership of his surroundings.

Drum's personal Auras had followed him into The Castle. They had been together at Drumsworth. The Auras had sat with him in "The Crown Room." They knew that *this Castle* was not only part of Drum's history, but his dream for the future. It was also the objective of *The Plan*. It had taken him years to get to this position and he was determined to take advantage of the momentum that he had developed... soon.

"You are not going to let the obvious snub that the King gave you tonight go without notice, are you?" snarked his Auras. "If it were not for the Queen's intervention, he would not have allowed your daughter to come inside tonight. He seemed pretty strong. And he didn't even greet you with any of his usual protocol. Are you going to stand for that?"

The fiasco at the colonnade was Thorndike's fault, Drum rationalized. *He thinks everything can be gained by force. Thorndike is too stupid to understand. Finesse, strategy, and manipulation are the tools of those with a higher mind set such as mine.* Drum's Auras agreed. They were partial to his way of thinking.

Thorndike is useful with regard to muscle, however. His Militia had his landholders already at work on the road that would be necessary for The Plan, Drum thought. Drum had seen them supervising workers this morning. He knew that Thorndike would put the militia back to work herding the local farmers to do his bidding. He was planning for that.

The Call of the Auras

The Royal River Road was key to the rest of Drum's plan. This road would allow militia troops to move efficiently to and from The Lowlands, the initial target of The Plan. Thorndike's militia, enhanced by the militias of the other nobles, would travel to The Lowlands and take control of the area. Drumsworth would become The Lowland's reigning headquarters for control of The Lowlands and Drum as monarch of River Kingdom would benefit from the Birther's demands 'controlled usage of Lowlanders.'

All wandering Lowlanders in River Kingdom would be rounded up and returned to their homelands, where they would be allowed to leave only for necessary work as defined by the nobles of River Kingdom. There would be no risk of intermarriage or unnecessary social interactions. The Lowlanders' role would be to serve Birthers under Drum rule and Thorndike Militia supervision.

This is what Thorndike and the other nobles expected, what they had been meeting about and working toward. Easy transportation up and down the length of The Great River was essential for troops and their equipment. Drum or his minions could initially direct the operation and rule the Lowlanders from the protection of a moted Drumsworth Castle. Like his grandfather, Drum was a born leader. This time, however, his rule would not be ended by a know-nothing King.

What the nobles did not yet expect, was Phase Two of Drum's plan. Phase Two called for the very same militia that established control of The Lowlands to invade The Castle, itself on his behalf!

This was Drum's primary dream. This had always been his dream, and the Auras knew it. Thorndike militia was huge. It would defeat the Royal Castle Guard and Drum would regain in his rightful position as King of River Kingdom, from north to south. Thorndike would force The Royal Guard to join his own militia as Drum knew he desired.

Drum's Auras were completely in favor of Drum's plan. Birthers would use their own militias to oversee Lowlander laborers, thus insuring progress in the Kingdom and Drum would be King of Birthers...all the nobles of River Kingdom.

It was perfect. Existing rules, laws and most of all, *protocol* would vanish at once. Greed, selfishness, prejudice and cruelty would rule immediately and be enforced happily by Drum's orders to Thorndike to keep everything in its place.

Urhonordo, of course, would not have the strength to resist and would no longer object to 'unauthorized use of their territory', such as very large pool that was shaped by the mote that already flowed between Drumsworth and Urhonordo farmlands.

Drum continued to enjoy his dream as he wandered back to the room that was his for the evening. *The mote is also key,* he thought as he reviewed his dream more fully. *From Drumsworth Castle, the militia can take turns watching and controlling the Lowlanders to see that they don't get out of hand...or simply squash them in case they don't behave properly.* He snickered at the thought.

The Auras around him settled down with Drum in a comfortable chair. A hearth crackled and sparked, as though it, too, was entertaining his dream.

The mote will be my own, personally controllable barrier...to Lowlanders, Urhonordorians and even wayward nobles...if they bother me. I will control the universe of The Great River, myself from the south as well as the north... as only a Drum can.

Fatigue was claiming Drum's conscious thought, but his fantasy continued... *Tomorrow, I will show everyone who has underestimated my strength among the nobles. Years of my distrust of Royalty and Birther hatred of Lowlanders will result in this smug King's capitulation. This will display the importance of my cunning versus Thorndike's dumb use of force.* He drifted off contentedly. *Tomorrow will be special.*

PRELUDE TO A MEETING

"Horace just told me that my father has already made the farmers start work on the River Road," Willa announced sadly to Solem. "Did he think he could do that because I am here? Did he think I had influence?"

Solem's 'protocol loving nature' took over. He stood back, every bit the Royal sceptic. "How did Horace find this out?" he asked with a squint.

Willa smiled. "Hector told him," she answered happily.

"Hector?"

"Hector, the farmer...Horace's brother," Willa explained. "They tell each other everything."

"You know Hector?"

"Yes, of course. I met him when I went exploring with Manda. His land is in the Center region of The Kingdom, right below the land of my father's landowners. "

Solem stared at Willa. His brow creased. "I know that, of course," he responded. "Why did Horace tell *you* all this? He should have repoted this directly to me as King."

"Because *I* was there in the conference room when he came in from talking to Hector, I guess. I was helping get the conference room ready," Willa stated matter of factly. "I was just coming up to tell you but..."

Willa now looked at her feet. Her old self-doubts rushed back. *What do I know? I thought what Horace heard was important but maybe not. Maybe Hector was wrong. Maybe I should have just referred Horace to Solem. What is my father really doing? I don't have any real knowledge. I was out of place, here. I was brought up by a servant and a father who is now acting like a traitor. I shouldn't even be here.*

"Oh my. "Aren't you the picture of self-doubt?" Cwilla snarked in Willa's ear. "I had no idea that you had this ability to put yourself down for no reason. Your upbringing was a little strange, but do you really think that Horace wasn't telling you the truth or that *you* are the reason your father acts badly?" She crawled into Willa's hair and waited. Honest questions deserved honest answers from her human. She would wait.

Seconds passed. *I don't know why I thought I could be a good queen,* Willa continued to think bleakly. *Look at me. I was all set to arrange seating downstairs. Now I don't even know if I deserve to be down there, myself...or if my husband will trust me down there.*

Cwilla rejoined Willa's thinking. "You're thinking like a loser," she chided..."You just brought your husband valuable information," Willa's Conscience continued practically. "For that reason, I suggest you stop feeling sorry for yourself and do what you know you need to do *now* to help Solem. *This* is going to be a tough day. You have a job to do. Do it."

Solem was watching Willa stare at the floor. Now Solem's Conscience spoke up to *his* human, a protocol-bound man with a new young wife. "This woman just brought you some valuable information, Solem. She may be the only person in the meeting today who is completely on your side. She is trying to help you. Despite having a difficult father, she is doing her best. Quit asking stupid questions and appreciate her."

Slowly Solem looked at Willa's face. He reached out and pulled her close again. "I'm sorry, Willa...I ask too many questions," he sighed. "My protocol is a force of habit. I should just have said thank you. "You gained Horace's confidence, and he has shared important information with you. I know he wouldn't have done that if he didn't trust you to tell me. I trust you, too. I'm nervous about today, but I have no other excuse."

For a moment, the two stood together, re-confirming their trust in each other. They would need that trust today in a room where little trust would exist...and perhaps little would be deserved.

*****

Watching this from the far end of The Castle's visitors' wing, Drum and his daughter perceived the royal couple hugging in very different ways. Drum thought their affection a disgusting display, but he also took note of the King's softness.

This softness was one of the important reasons for Drum's dislike of King Solem. The King had no backbone. He was unnecessarily kind and thoughtful when he should be assertive and strong.

His 'food for the poor program' was a perfect example of that warped decision making. Some of the recipients were Lowlanders. All that program did was discourage their hard work by giving food to lowlifes who hadn't earned it.

This King seemed to have no idea of the greatness that natural, *white* drive could have on a kingdom with the right leadership, but his softness made him vulnerable. This was a good thing..

Drum's grandfather had had the right kind of leadership and that was what today's meeting was all about in Drum's thinking. Today would show this inadequate king what real leadership was all about.

Balynn Drum viewed the embrace of the Royals very differently. She was jealous. *Look at the love the King and Queen share,* she thought enviously. *I wonder how one gets that. Willa is just a new Queen, so they just learned to love each other recently. How did they do that?*

"I need to tell you that their obvious love is based on deserved trust and basic caring, to start out with," Cbalynn contributed sadly. "Unfortunately, you have had no examples of that in your life. Your mother was never very good at it and your father does not acknowledge love as a worthwhile emotion. But I do think *you* could learn to care for others if you tried. "

How do I learn to care or trust, thought Balynn skeptically. *I don't know anybody who cares or trusts somebody that much.*

"Well, I guess you could start by watching other good people. That would be a start. Then you could practice what you see those other people doing...That's just a thought."

Balynn thought back on her life for a minute. Some of her servants were nice to her...even kind. I might try watching them, she thought slowly.

Just then, Princess Mimi popped out of the bedroom that she had shared with Manda the night before. Both had mouths full of the muffins. They still held a few in their hands.

"Oh, my goodness!" Mimi bubbled. "It's The Drums! Hello Lord Drum. Hello Balynn. You both look rested and ready for the day. Have you had anything to eat this morning? I got these muffins myself down in the kitchen. Want one?" She held out a warm muffin with a smile.

Drum stepped back and scowled. This was the lack of Castle Discipline that offended him. He didn't know this girl... although she *was* pretty. What was she doing out here in the hall? He decided that he would not answer. He was on his way to an important meeting, and she was in his way. He scowled again and brushed by.

Balynn, on the other hand, smiled back. "Muffins?" she asked shyly. "What kind?"

"Blueberry," crooned Mimi with a knowing grin. "They are the very best." Quickly, the young princess linked arms with Balynn and handed her a warm morsel. "Take a taste right away," she advised. "They're best when they're warm."

Balynn was charmed. Maybe this girl could be one of the people who know how to care and trust, one that her Conscience told her to watch. She would try it. She smiled and shyly took a muffin.

Manda hadn't said anything and Cmanda noticed. "You don't have any reason to be snooty to Balynn," counseled Cmanda. "She hasn't done anything to you. She just has a father you don't like. You should be nice to Balynn like Mimi."

Manda shrugged and smiled. "Try one of my muffins," she said to Balynn, holding out one of her muffins with a friendly snicker. "Mine already have butter on them, so mine are better."

"Thank you so much," answered a grinning Balynn. "Maybe I should just eat them both." Balynn now took bites of both muffins. The three girls laughed and proceeded down the hall together.

"See?" whispered Cbalynn. "Here are two girls who are generous and kind. You were nice...and funny in response. Maybe you can make friends with them and learn some of their ways. I know you can be friendly. You just need a little practice."

Balynn grinned and bit another one of the muffins...the blueberry one. She and 'these two who could be friends' followed Drum toward the stairway. The King was already descending the stairs and Drum was following... but Queen Willa remained at the top of the stairs.

The Queen is waiting for us, Balynn said to herself. *It looks like she could be a friend, too, someday.* Happily, she approached the young Queen with Manda and Mimi.

Willa smiled. She had meant to explain this to the girls earlier but... "I forgot to explain to you ladies about who King Solem wishes to attend the meeting down in the conference room, didn't I?" she asked in a friendly fashion.

The three young women stopped, still licking the muffin crumbs from their fingers. "We are all heading down there together right now, Willa," Mimi answered happily for them. "Just as soon as we finish our muffins."

Willa sighed. *I feel like an old grump,* she thought.

"Today you must act with responsibility more than the three friends that you are facing. It's just this 'queen position' today that makes you feel bossy," hinted Cwilla.

Today is going to be difficult...It already is. I know what Solem wants but how do I say it? Willa thought.

Manda looked at Willa's hesitancy and read her thoughts. "I don't think the King wants us all to attend," offered Manda helpfully. "They will let Willa be there because she is the Queen. They can't keep her out."

The young women giggled knowingly.

"They will let my mother be there because she is a Queen, too. They all know that. The only reason they will let me in is because I have firsthand information about slavery that I saw. This meeting is really a 'men's decision-making meeting', as usual, however, by order of the protocol-loving King Solem. You know how Solem is about women and protocol."

"Slavery?" The word resonated down the hall.

 Two other women heard the word and froze in one spot.

Manda looked around and blinked. *Did neither of them know about the slavery at Drumsworth? Did no one know that this was going to be what the main fight was about today?* she suddenly thought. *Have I spoken out of turn?*

"Indeed," answered Cmanda. "You need to explain yourself...carefully. And remember...Drum's daughter is here."

Manda swallowed hard. "I am not sure, but I think slavery will be mentioned today, "she answered quickly. " We are talking today about The Royal River Road, right? *Someone* is going to build it."

"But not slaves, right?" blurted the soft-hearted young Mimi.

"Hopefully, not," answered Manda, in a way she hoped would sound conclusive. Apparently, the Royals had not taken Mimi into their confidence about this yet and Balynn only looked unaware.

Mimi breathed a sigh of relief. "In that case, I will gladly *not* attend another boring Water Rights discussion." She linked arms again with Balynn. "Come with me, my new friend. We can be up in my room or out in the Royal Gardens together. We'll have a much nicer time than those who are locked in with my father and his protocol all day."

With that, Balynn smiled at the young women and the friendliness she had experienced so seldom in her life. She and Mimi turned together to make their way back to Mimi's room. Balynn was thrilled with her new (first) girlfriend. The young girls seemed content. Manda and Willa breathed a sigh and headed downstairs.

PROTOCOL

Solem was first down the stairway and into the conference room. Drum, however, had passed the conference room and headed out to the colonnade. Solem noticed Drum's detour with surprise.

"Where did Drum go?" Willa asked Solem as she and Manda entered behind him. "I thought he would be more than ready to get started on the discussion of his new road right away."

"So did I," Solem said with a scowl. "And, actually, I would like to talk to Drum alone. He needs to discuss the possible slavery issue before we even begin to talk about the road."

Horace escorted Willa to her chair and took that opportunity to offer a thoughtful comment. "I bet Drum's gone out to wait for Thorndike, Your Majesty," he whispered confidently. "Those two work together all the time usually, but Thorndike may not have told Drum that he was going to start the road without the King's permission. If that fact is news to Drum, this could make this meeting *very* interesting. Drum thinks that he is the boss." He rolled his eyes and glanced from side to side. No one else seemed to be listening.

Willa nodded. That was probably what Drum was doing, she agreed, but Solem would not be pleased or willing to wait for anyone who was not here at the meeting on time. And would her father even show up at all? According to what Hector had told Horace, Thorndike was probably out building the road right now.

Queen Luna and Armando were entering the conference hall. It was almost time to get started. With warm smiles to her fellow Royals, a nervous Willa was determined to behave as a confident but quiet queen. She smiled and motioned them toward the chairs they had used in their previous meetings. "Good morning, friends," she said with what she hoped was confidence. "Please make yourselves comfortable in your usual chairs. Horace has put blankets on them in case you are chilly this morning as an early storm appears to be approaching from over the northern peaks. Agnes has fresh-brewed tea in the pots before you."

Queen Luna smiled in return and reached down to hug her daughter who had already taken her seat. She also smiled and nodded down the table to Solem as she sat down, herself.

Luna was intuitive. *There is an undercurrent of some kind happening here*, she observed silently. *I don't know what the problem is yet, but I can feel it.*

She feels the Auras that are already here, thought Cluna proudly. *She is wise and sensitive.*

Armando went straight to Solem to greet him, man-to-man. The two had become friends over the years and today was going to be a difficult day. "Looks like we are almost all here, Solem," he began, setting a casual tone. "Hopefully we'll be able to change the evil that appears to be happening at Drumsworth peacefully, once everyone gets here. We *must* and we *can*... if *I* can stay calm." He smiled ruefully and reached out to shake Solem's hand.

Solem relaxed and shook Armando's hand with both hands. "I am so glad *you* are here, Armando. You and Manda and June have seen what I haven't seen. I hope you know that slavery is even bigger in importance to me than the new road. You and I are together on this," he stated. "We will bring a halt to anything that is remotely like slavery from Thorndike and Drum, or *nothing* will happen on the road. Sanctions and punishment will also occur if slavery is currently taking place. Slavery will not happen in River Kingdom...ever!"

"Drum *will* come to understand that or else," Armando answered with assurance. "I'm here to make sure of it.By the way, where is the rest of our group? We seem to be missing three very important people. Do you think they're close?"

"Drum is out on the colonnade for some reason," Solem admitted. "Horace, would you ask Drum to come into the meeting immediately, please." Horace nodded and proceeded on his errand.

"And I don't know where my son is," Solem continued. "He knows how I feel about punctuality. If we need to, we may begin the meeting before everyone gets here. If they do not wish to come to a meeting on time, they will not be heard, and we will make our decisions without them."

Willa, Queen Luna and Manda rolled their eyes and looked at each other. When Solem was not happy, his need for protocol always took charge.

Suddenly June rushed into the room. "I'm sorry I'm late, Father," he gasped, "but you should see what's happening out on the colonnade. Noblemen are arriving... *lots* of noblemen! And they are acting very strangely for nobility...no decorum at all. They are chanting and saying that they are all here to tell you that they 'demand the new road!'"

The meeting room occupants immediately rose and rushed outside. Lots of noblemen? Noblemen making demands of their King? This was not the meeting anyone had anticipated. It was totally against all protocol.

But June's report was true. Carriages at the gate were currently off-loading portly, well-dressed occupants, one by one. These wealthy individuals were not used to running but they were puffing forward, doing their best to display behavior which someone had suggested would represent enthusiasm. They were all clapping their hands in the same manner they had used at The Gathering and there, waiting on the colonnade to greet them, was a smiling, clapping Drum.

Drum was strolling back and forth, clapping his hands, smiling and confidently and pointing to first one nobleman and then another, calling out their names in recognition and acknowledgement of their presence.

Willa stood beside her husband as her familiar feelings of dislike for Drum rose to the surface. "This display is the same as occurred at The Gathering," she whispered to Solem. "This is nothing more than a Drum fan club. I recognize these men. They used to meet with my father and Drum at Thorndike Manor. What does Drum think they are doing *here* now?"

"Royal River Road! Royal River Road!" the growing group chanted now, raising fists to punctuate their enthusiasm. "We want our road, and we want it now!"

Solem stood firm and stared straight ahead. "Drum brought the nobles here today to do what your father had his militia doing last night," Solem whispered to Willa. "They were brought to show noble support for what Drum wants. He hopes this will sway me to do whatever he wishes."

June slid quietly up behind Willa and his father. "But it won't move you an inch, will it, Father," he stated fiercely. "They are actually trying to bully you. You should call the Royal Castle Guard and have them all arrested for trespassing illegally on Castle Grounds!"

I would truly like to do that, Solem acknowledged to himself. *Drum thinks he can push me to do what he wants with this stunt. He breaks the rules of our meeting today as though my rules do not exist. I should just call for The Castle Guard and have them thrown off the colonnade.*

"Drum is using the nobles to get to Solem as he did at The Gathering," thought Willa at Solem's side. *" He thinks that their presence will force Solem to give him his way."*

Willa's Conscience listened to Willa's thoughts and passed on the thought to Csolem. From his vantage point by Solem's ear, Csolem could plainly see violent Auras swirling around the noblemen and Drum, almost knee deep. Csolem knew this was a dangerous, Aura-inspired situation.

The Call of the Auras

As Conscience for a King, Csolem knew that this mob of determined nobles was designed to threaten the King's control of today's meeting. Loss of control here could have a catastrophic result. Csolem needed to help Solem focus. "These noblemen are the basis of your Kingdom's wealth and *your* success, are they not?"

Yes, indeed, Solem confirmed mentally without speaking. *They own most of the northern land of the Kingdom other than my own.*

"It would be dangerous then for *all of them* to be angry with you right now if you call in the Castle Guard?" Csolem continued.

This is true, Solem suddenly realized. *It would mean anger. It could mean eventual overthrow. Most of them don't like me already...and they all have private militias like Thorndike's that could cause my Castle Guard real problems.*

Solem rubbed his chin while time seemed to halt as he waited for inspiration and a solution.

One evil Aura that was near Solem slithered up *his* robe in an attempt to influence him toward a violent reaction. (Auras can, of course, try their tricks on either side of an argument.) "These stupid noblemen can't invite themselves into a meeting like this," the Aura whispered in Solem's ear. "Show them who is King! Lock them all up for busting into your meeting without an invitation. Use your famous protocol as an excuse."

Solem's Conscience heard the Auras' suggestion and quickly offered a counterargument. "Protocol, yes,' Csolem whispered to Solem. "But using protocol to create more anger in the nobles?... No.

"You *are* famous for protocol, indeed, Solem. "Use your protocol for peace, this time," *offered Csolem. "Protocol organizes activities into predictable, expected practices. Slowing anger here with protocol might be helpful...and peaceful. Try it."*

June was getting impatient, but today, King Solem did not agree with his young son. Instead, he stood quietly, looked out over the nobles, and raised his hand. All these noblemen had been raised to honor this royal request for silence. It was only a familiar gesture, but they responded to their King with almost immediate quiet. It was protocol everywhere.

Drum was not facing the King. When he turned and saw Solem's hand raised he, too, stopped clapping in response. It was automatic.

The shocked Auras dropped to the ground temporarily. Was this a new strategy? This kingly gesture and its *immediate* response were obviously not what they expected. Neither was the smile that Solem wore as he approached the nobles with confidence.

"Welcome, my friends," he called out calmly over the crowd. "I am so surprised to see you. I wish you had told me, Drum, that all our fine noblemen would be here today. I would have prepared for them more adequately. They *all* must be seated, of course, so that they can have their say." Solem smiled again at the crowd.

"The King isn't mad at us at all," murmured an influential nobleman. Further surprised murmuring of nearby noblemen followed.

"Drum said that the weak King would fold, if we all burst in on him unannounced."

"Drum said that strength in numbers would get us our way with the Royal River Road."

"Drum said that the King wouldn't have the nerve to turn us *all* down and we would just get immediate approval to build the road."

Solem turned to June who was simply standing beside Willa with his mouth open at his father's response. "Ask Horace if he will help you bring out some chairs, Son", Solem called out loudly. "These good nobles will have to sit out here today. We are not set up for this many surprise visitors inside. We will simply conduct our meeting out here, so that all of them can be heard."

June blinked and turned to do what he was told. Then he stopped and stood still, waiting to be sure. "Really, father?"

"Do hurry, June," the King continued. "You know how I feel about proper protocol. *All* these fine gentlemen wish to be heard. *That* is proper protocol. Standing about in bad weather, however, waiting for their turn to speak *inside* is *not* proper for these fine gentlemen. In addition, I believe a fall storm is gathering to our north and approaching more quickly than anticipated. It will be raining soon. Would you like to help us arrange chairs for your fellow noblemen quickly, Lord Drum? We should really hurry."

The noblemen froze in a motionless quandary on the colonnade. They looked at Drum. They were not having the group pressure impact that Drum had promised their numbers would generate.

A breeze from the north began whistling around the nobles, reinforcing the King's prediction of bad weather moving in. The formerly enthusiastic Auras at their feet seemed confused.

Drum no longer wore the smile that he had worn only a few minutes before. Instead, he quickly donned a look of disdain. "You think you can head off every challenge you meet with your lame protocol, Solem?" he sneered loudly.

"Not every challenge," Solem countered calmly. "Just this one. If you and your *uninvited* friends wish to wait to speak your minds out here in the rain, fine. If not, I feel that appropriate protocol now would be *your* apology to them for having encouraged them to arrive here unannounced…again. Or, if they wish, they can appoint you as their spokesman as I requested before."

A long moment followed. Drum had suddenly lost his power. He didn't know exactly how it had happened, but he knew that this odd, inappropriate use of outmoded protocol had thwarted him…difficult to believe, but true.

Regrouping with pride seemed to be Drum's only remaining option as big rain drops began landing on the faces of the gathered nobility.

But Drum was resilient. He spread his arms wide with all-encompassing confidence and announced his miscalculation to the noblemen with such skill that it sounded like a win.

"The King is not *prepared* to allow this many of his countrymen to speak their minds individually as you expected to do today, but *your numbers did* speak loudly *for you*," Drum shouted to the crowd. "Nobility is too important to be left to the rudeness of the coming storm. With your permission, therefore, *I, your loyal spokesperson,* will carry your message to the King *for you*, thus allowing you to return safely to your homes."

Without hesitation, the dampening nobles shouted out something that sounded like "huzzah!" and dashed off through the raindrops to their waiting carriages.

"They got turned away, Father, but they still think they just won a victory of some kind," marveled June as he, his father and Willa waved goodbye. "How did Drum do that?"

"I'm not sure," whispered Willa, "but he has been doing things like that to my father for years."

-

SLAVERY

The King and Willa followed the rest of their meeting participants back into The Castle conference room, each trying to take stock of what they had witnessed. Everyone in the group was aware that they had narrowly avoided an orchestrated push of some kind from the nobles and the unease that Queen Luna had sensed earlier continued. Cluna made note of the increased number of Auras that the departing nobles had left behind.

Drum marched back into The Castle with confidence, completely certain of the positive impact that the gathered nobility had been transported there to produce. He did not look left or right but focused ahead, reluctantly finding a chair at the center of the table, only because all the important chairs at the head and foot of the table were filled by Royalty. He pulled some papers from beneath his vest and laid them out, relaxing casually into his chair. "Good morning, everyone," he said with a smile and a nod. "Sorry for the delay. Shall we begin?""

No one answered. The royals were past pleasantries and important issues needed to be discussed. Everyone but Drum was now focused on one issue. Solem nodded to Prince Armando to begin. *Armando's* need to defend his fellow Lowlanders was painfully acute, although everyone in the room was leaning forward tensely.

Armando rose with the steady, purposeful demeanor that burning anger can produce. Carmando leapt to the ear of the Crown Prince and took a protective stance with only a few words. "Be calm and thoughtful, Armando. Don't let your anger lead."

Armando swallowed hard and began." Lord Drum, there is a matter that I feel must be discussed before we begin any discussions of the new road that the nobility was so obviously interested in promoting."

Drum stared ahead. He had had dealings with Armando before. Although Armando was the Crown Prince of Urhonordo, he was also a lowly Lowlander by birth. Drum was not about to be intimidated by this dark-faced lowlife. He rolled his eyes and came directly to the point. "What is your issue *this* time, Prince? Did one of my workmen get Drumsworth mud on your precious Urhonordo land?" he snarled sarcastically.

"As a matter of fact, that *does* happen on a regular basis, Lord Drum, but that is not my issue today," Armando answered in a business-like tone. "My issue is not land. My issue is people...and the slavery of people...Slavery!... Slavery! Do you hear what I am saying?"

Standing tall with anger that had been building for days, an infuriated Armando glared down on Drum with a withering stare.

Drum simply raised his eyebrows with a calm and questioning look. "What would you like to say about slavery, Prince Armando? I know nothing about the topic. Why do you want to talk to *me* about slavery?"

The heads of all that were seated around Drum swiveled in his direction. Armando stepped toward Drum as his fists clenched and cold perspiration rose on his brow.

"Steady, Armando," Carmando rumbled in his ear.

Queen Luna raised her hand for calm as King Sol had done earlier. The room became very still as all breathing also seemed to stop. "I feel that your response may not have been a completely true one, Lord Drum," Luna stated without emotion. "Slaves have been seen on your property recently, Sir...many *Lowlander* slaves. Lowlanders have been seen digging a large *unpermitted* mote in the back yard of your estate...which is partly on Lowlander land. Lowlanders have also been observed to be thrown in a prison behind your 'castle'. What do you have to say about these observations?"

For a moment, Drum seemed startled, and color drained noticeably from his face. But it was only for a moment. Almost imperceptibly his confidence resumed, and he swept the faces that were lined around the table with a haughty, belligerent stare.

"*Who* would tell you such a bald-faced lie as this? Just *who* has made such unbelievable accusations against me? There are no slaves on my land. There never have been slaves. Someone just wants to ruin my good reputation and slow down the progress on my beautiful new road with lies. The person who is telling you this is giving you fake news."

Drum's stare moved boldly from one face to the other around the table, coming to rest on his most likely target...Prince Armando, himself. "So, this is your new tac, heh, Armando? I never thought you would stoop this low just to make your '*two* kingdoms' seem better than they are. Are you really that jealous that my new road is going to be in River Kingdom and not yours?"

King Solem held up his hand again and again silence prevailed. "I will have no sniping between participants at this table, my friends," he stated firmly. "We seem to have two different opinions about the very serious issue of slavery. When we have an accusation this serious, we must be sure of what we *know*, not what we *think* we know."

"I agree with you completely," Queen Luna added.

"Mother...*You* know that June *and* Father *and* I all *saw* this slavery with our own eyes. Didn't you believe us?" gasped Manda. "All three of us saw *slavery*, pure and simple. Did you think we were wrong?" The three that had watched the slavery scene from the bluff above Drumsworth broke into noisy, handwaving confirmations of what they had seen.

King Solem raised his hand again, but Drum's Auras had already formed a defense for him to deliver. He rose with an air of righteous indignation and glared at those around the table. "*The three* of you accuse me? Just *when* and *where* were you three when you *thought* you saw this terrible 'slavery'?" he asked smoothly.

"From the Urhonordo bluff that overlooks your "mote", Sir, stated Manda firmly. "And we all saw Thorndike's militia men behaving like slave masters, standing over the Lowlanders, and yelling at them like they had to do what they were told immediately! We can all verify what we saw."

 Drum took note of the faces around him. He could easily see that the three accusers knew *exactly* what they had seen and wouldn't believe a word he said. *The King and Queen had not.*

Drum sat down calmly and smiled. "You only saw what you *think* you saw," he said smugly. "You were all quite a distance away. You made an assumption about what you were seeing ...but you were wrong.

"You really shouldn't accuse people of such terrible things until you make sure that they are true. There is no slavery going on at Drumsworth. I can tell you that, for certain. *I* live there."

Drum smiled broadly and relaxed again into his chair. "I forgive you three for making a mistake, but don't accuse me of this again. People might believe you."

June jumped from his seat. "We saw what we saw," he exclaimed. "We saw the men throw a slave into a prison. We heard them yell at the slaves and watched as slaves cowered in fright. And what about the mote? Are you going to try to tell us that we didn't really see that either?"

Drum turned slowly toward June with a studied nonchalance. "You really *are* an angry young man, Junior," Drum commented insultingly. "You are obviously disappointed that you and your lovely companion were wrong about the slavery you *thought* you saw. Now you are seeing imaginary motes, too?"

Drum turned toward King Solem and shook his head. "Isn't it too bad that the younger generation is so easily misled, Your Majesty? They come up with an idea that was obviously fed to them by a misguided older gentleman. *He* feels guilty that his countrymen are so unproductive in their homeland that they need to be supervised to build a pool *for themselves.*

 "I feel sorry for the youth of today. The Lowlanders that they saw were digging a pool for *themselves on their land.* If my accusers were not so intent on proving that I am a slave master, they would have seen that."

The Call of the Auras

King Solem looked around the table. June and Manda had been temporarily silenced by Drum's confident denials and were staring at him in disbelief at his artful response. Willa kept silent but felt that she knew June and Manda well enough to believe that what they had seen was what they said they had seen. Nevertheless, Drum was amazingly believable. For a moment, silence reigned.

Armando began pacing, clenching and unclenching his fists. King Solem and Queen Luna stared ahead silently. These two were experienced monarchs. This was not a time to overreact to incomplete, unproven information from either side. They also knew that neither side would be satisfied with that reaction.

Drum looked around with satisfaction. "You have the older monarchs doubting the slavery and mote charges," whispered his Auras, "Now is the perfect time to discuss the Royal River Road. That is a safer topic, and they won't want to argue about it as much after half of them were just put down as 'misinformed'. Go for it."

Drum heard the Auras' encouragement and leaned forward over his paperwork, stacking it and restacking it with a look of confidence. "You know that they were right, and you were lying, right?" whispered his usually always-present Conscience. (He had to *try* to have an effect, at least.) But the Auras were stronger. He held up his slim stack of papers again.

Drum shrugged. "I have taken the liberty of drawing up a simple contract for the proposed Royal River Road that we discussed last week, Your Majesty. I am sure it will please you as it is nothing more than what you and I discussed, and *you* already approved. Maybe we can get this paperwork done today. I am willing to forgive the accusations of the younger, more impressionable members at today's meeting. The Royal River Road is a separate and already approved issue."

He handed out a simple, one page copy of 'his contract' to Solem and Luna only, purposefully demonstrating that their signatures on the contract were all that was needed.

"We are not ready for your contract, yet, Drum," growled Prince Armando. "We have not yet concluded the discussion of slavery."

Drum turned and stared at Armando. " *I* have concluded the discussion, Your Majesty", Drum drawled slowly to Prince Armando, his voice reflecting the contempt that he was no longer trying to hide. "*I* will not sit here in this room with a liar who accuses me of slavery when there is no proof anywhere of such an outrageous accusation. I came here to finalize a contract with the King of River Kingdom, a contract that was already approved verbaly and needs only the signatures of The Water Rights Committee to make it official. That is my position on today's proceedings."

"But there can be no signing of anything by anyone until this outrageous accusation of slavery is put to rest," stated King Solem. "That is my position on the Royal River Road. If you propose to build a road...or enlarge a lake...or create a mote...or whatever you plan to build down south, understand this: I will visit this area soon and decide what is being built and what kind of labor is being used to build it. There will be no major building *of any kind* in my kingdom until I can make sure that there is *no slavery taking place.*"

THORNDIKE PROPOSAL #1

"I am here to sign The Royal River Road Contract," rumbled a deep voice from the conference room doorway. Thorndike did not notice the cold atmosphere in the room after the King's declaration. He knew what he was here for, and he wouldn't tolerate questions about why he was late. He had no interest in anything that had occurred before he got here, nor did he wish to discuss anything now. He was here for signatures, only...or else. (The Auras had remade him.)

Without looking left or right, Thorndike marched from the doorway and pulled up an empty chair. "Where is the contract? Where do I sign? I am a busy man. I have a plan and I don't have time for any chitchat", he stated loudly as he seated himself across from Drum.

"That is not the way things are done here," stated Solem. "We have not yet discussed water rights, irrigation issues...

King Solem was a man of protocol, but Thorndike appeared to be *without* an understanding of that. He stood again with clenched fists. "Just give me the papers," he growled.

Solem nodded slightly. From the conference room doorway, two very large members of the Castle Guard entered and yanked Thorndike unceremoniously from his seat.

Thorndike was caught off-guard. Although he was stranded in a humiliating position, his anger was not slowed, and he responded with a yell that echoed throughout The Castle. "Unhand me, you lawless henchmen! I am a nobleman! I know my rights!"

Thorndike's chair clattered to the floor and the women gasped aloud. Solem and Armando stood strong and waited. Drum was apparently as surprised as everyone else. Even Drum's Aura was surprised.

Drum was the first to recover. "Nice of you to drop by, Thorny," he declared sarcastically. "Your entrance was a little loud, but you arrived just before we actually got down to discussing the real reason for us to be here. I've passed out our contracts for the road. Have a seat and get ready to sign."

"Just give me the papers," Thorndike snarled again.

Willa wanted to disappear. Her father was being rude in *so* many ways. She stared at her hands. She was embarrassed. She was *related* to this creature who was late and making a scene in front of everyone.

Cwilla climbed down from her perch on Willa's crown. Reading Willa's thoughts, she had a few remarks to make. "First, let me just remind you that you are your father's daughter, not his mother. That means that you are not responsible for his behavior. Secondly, I agree that your father appears strangely out of control but try saying something soothing to settle him down before either your father or your husband do something they will regret."

Like what, for heaven's sake? Willa thought hopelessly.

"Try a simple greeting, for a start," whispered Cwilla.

"It is good that you finally arrived, Father" Willa said politely. "Lord Drum has just furnished us with a contract for the new road that he wishes to have approved, but no one has read it yet. If you will, please take the time to read it, then we can all discuss it."

With as much queenly dignity as she could muster, she looked into her father's eyes but what she saw there was strange. There were wild eye twitches coming and going rapidly. There was fury, surprise, determination...and more fury. It was frightening.

"Maybe we got him too excited", said the Auras to each other. "Calm down, Thorndike," they crooned softly. "Think about your proposal."

Suddenly, there was a change. The huge man shook his head and became unnaturally calm. He smiled crookedly. Willa wondered if he was well. He was not.

Thorndike's head lowered for a moment. This intervention from the Auras caused him to see his daughter in a completely different light than he had ever imagined. *My daughter has taken this position as Queen because I put her here to help me,* his reeling mind thought proudly. *She will be queen when Drum takes over as* *king. My daughter is now in a position to help me with **my** new proposal. I can do this.* For a moment he smiled and remained confused. Then he shook his head again and gathered his thoughts.

Now, Thorndike's head turned almost mechanically, and he faced forward, looking directly at Willa. His Auras had taken advantage of his mania of fury all night last night and they had now created a new person, a person who did not need Drum anymore, a person who was in control. This was the way he now must act.

"The Queen of this castle has called me on my behavior, Your Highness" he said calmly. His focus now went to Solem's stoic face. "I hereby apologize for my late arrival and disruptive behavior. I was working all morning and lost track of time. Please ask your men to release me. My behavior will improve."

King Solem was shocked at the change but nodded to the Castle Guard and sat down. Thorndike quickly gained his footing and his new confidence. Completely silent, he slowly picked up his toppled chair and lowered himself onto it.

Willa's knees were shaking but she resumed her seat as well. Her Conscience quietly voiced approval but both she and Willa wondered how this strange version of her father would behave next.

The remaining royals re-gathered around the conference table and exhaled as a group. The immediate crisis was apparently over.

But Hector's hand quivered as he poured another round of tea. No one had said anything yet about where Thorndike had been before he arrived or what he had been doing. He had been breaking the law. Chector sensed that there was much more disruption coming. He could feel the Auras. Was this what had been talked about in The Conscience Alert months ago?

"I suppose *everyone* would like a copy of the Royal River Road contract, now," Drum stated, looking sideways at Thorndike. Magnanimously he pulled out another sheaf of papers from his vest and passed more contracts out.

"You will all notice that there is really nothing special about this contract" he began. "It only states the obvious. The road will begin here by The Castle and simply run south along the bank of The River until it arrives at the Lowland Border near Drumsworth. It will be smooth and rut-free. It will be wide and safe and built by the men that will use it, namely, the farmers and landholders of River Kingdom."

(At the mention of the word 'safe', Drum looked directly at Solem for only a moment, but he had made his point, one more time.}

The Call of the Auras

The room was quiet as everyone read the simple, one-page contract. Without looking up, however, Manda made her observation clearly, one more time. "I see no mention of the fate of the numerous small levees that are currently the main source of irrigation for the crops of River Kingdom," she said quietly.

"Why is this absurd female in this meeting?" Thorndike hissed, his eyes fixed on the contract before him. He jumped again from his seat. "I was subjected to her ignorance previously! I do not intend to sit quietly now and allow her to interrupt proceedings again!"

"Amanda is a guest in my castle and a member of this Water Rights Gathering." King Solem declared bluntly. "You will not only listen to her concerns, but you will respond to them."

"Well said," whispered Csolem.

Drum's Aura reacted swiftly. "Your partner seems to be weirdly in and out of control this morning. Take care of him or he will ruin your future."

"I can take care of this issue," offered Drum hurriedly, unfolding a small map which was among his papers. "As you can see, there will be two larger rivers cut into the land for irrigation, one in the north and one in the south. These two rivers will irrigate the farmlands much more efficiently than the hundreds of current, small levees do."

Queen Luna asked that the map be passed to her for closer inspection. "I see the rivers that you mention, Lord Drum. It would appear that they are both on land belonging to Drum and Thorndike estates. If this is true, how will water irrigate the property of others...the Central farmlands, in particular? There is a large amount of land in *between* the Drum and Thorndike estates."

"An interesting question, Lord Drum," murmured Solem, scratching his chin as he thought the proposal through. "What is your answer to that?"

Thorndike stood again. "I will answer that easily." he said with bravado. He had a plan to make this question unimportant, part of a new proposal he and his Auras had crafted in his frenzy last night.

"I know that the central landholder, Hector, believes that the land between the two larger rivers will go dry and the Central landowners will be blamed for their lack of production from un-irrigated crops," contributed Willa. "He believes that the Central landowners will suffer if the new road is constructed without more thought."

Horace nearly dropped his pot of tea. With great effort, he attempted to conceal his pleasure with his Queen and her insight. He was thrilled. How would The King react?

Thorndike glared at Willa across the table. *My own kin speaks up against me,* he thought angrily. *How dare she? I didn't approve of her marrying into this disgusting Royal Family for betrayal such as this.*

Thorndike's Auras had been with him all night. Now was the time for him to deflect any irrigation questions with his 'new proposal', the one that he and Aura had developed using the creativity that only manic fury could generate.

Thorndike regrouped. *I will now play the first of my new and important cards,* Thorndike thought to himself. *Wait until this assembled group of know-nothings discovers how they have been outsmarted.*

With more restraint than he had ever exhibited before, the plotting Thorndike swallowed his anger. Drum was staring at him from the other side of the table, clearly afraid of his reaction to Willa's remarks.

But the Aura-reconstructed Thorndike was ready to unveil *his* idea. It was a beauty, and he, as 'Commander of the Militia', was the only one that knew of this plan...he and his co-conspiring Auras. Drum was no longer in charge.

Thorndike's driven mind raced. *The half breed princess is smiling her approval of my daughter's remarks, of course. The ignorant Prince Armando is focused on Drum's map. Now is the perfect time for me to put my proposal into action.*

Thorndike stood again, but this time he knew that he and his Auras were in control...in *complete* control. He smiled, a congenial look that few had ever seen on his face before. "I understand your concern, daughter," he offered smoothly. "I am impressed with your knowledge of this situation. You are smart, very much like your father, eh?" He smiled crookedly at his little attempt at humor.

The Auras surrounding Thorndike were impressed...so far, so good. Due to last night's mania, they were not completely confident that the idea they had created with him had been absorbed enough for him to activate it. But things were looking good. The Auras would stand by to assist when needed.

Thorndike caught the eye of King Solem and spoke directly to him now, man to man. "I know that you are concerned about all of your farmers and land holders as well you should be, King Solem. For that reason, I have arranged a field trip to the territory that we are discussing so that you can see for yourself the exact water system that Drum and I have proposed!"

He smiled engagingly at everyone else in the room. "Of course, I am inviting *all* of you to visit the area at the same time. Drum and I are sure that our irrigation system will work to your satisfaction, aren't we Lord Drum?" (He didn't wait for a response.) "And, of course, the best way to illustrate our plan is to show it to you in person. What do you all think of a visit to Thorndike Estates?" Thorndike sat again and gazed smugly around the room, waiting for reactions.

The room appeared to be in shock. No one had expected this creative offering. Drum, in particular, was staring at him in confusion. Drum had never thought of something like this. A field trip had never even been discussed. Where did Thorndike get such an idea?

Drum's own Aura was blind-sided by this suggestion. A quick and appropriate response was required here. What should they advise? Hurriedly, Drum's Auras settled on 'acceptance of the idea.'

"Thorndike is on your side," his Auras whispered. "Go with what he says. You will need him and his militia in the future. He will probably fill you in soon on the particulars of this."

"Great idea, Thorny," Drum offered as heartily as he could, under the circumstances. "Excellent. Why didn't I think of that?"

Thorndike nodded and winked at Drum. He had wondered what Drum would say to *his* idea, but it appeared that he was accepting it, at least for right now.

"Good move," whispered Thorndike's Auras.

Queen Luna watched the pair inquisitively. After her initial surprise, she sat back to analyze the situation. Instinctively. She recognized Drum's initial surprise. *He didn't know about this idea until Thorndike announced it,* she deduced silently. *Thorndike has assumed a new, more authoritative personality. Drum is as surprised by it as the rest of us. Who is the leader now? This is very interesting.*

"This is an unusual idea, Lord Thorndike," Luna offered without further comment. She glanced at Armando. What did he think? They had never had much contact with Thorndike in the past. They had no reason to doubt his motives...as they would if the suggestion came from Drum. Due to past dealings, Luna knew of Armando's distrust of Thorndike's partner, but not Thorndike, himself.

Armando looked Thorndike in the eye. Actually, Armando had thought of this idea himself earlier. The irrigation issue was important. Manda had talked to him about it several times since The Fall Gathering. Basically, he was in favor of the idea of observing the situation firsthand.

The young people are skeptical of my motives, Thorndike recognized immediately. *June and the half breed don't trust me. I can tell. And then there's my daughter. What side is she going to take with my idea? She knows nothing, of course. Like her mother, she lived in my manor house and was told only what she needed to know... At any rate, she makes no decisions here. She will do what her husband dictates...as she should. Solem is the decision maker here. Whatever he says is what will occur.*

Solem returned Thorndike's gaze quizzically. "This is a very thoughtful offer, Lord Thorndike ...very thoughtful, indeed," he stated slowly. "What brought you to this idea? I was under the impression that you and Drum wanted a decision to be made today so that you could begin work on your project right away."

"That *is* what I want, *exactly,* Your Highness," drawled Thorndike with a barely noticeable smirk.

"That must be why you started working on the road *yesterday*, right, Father?" asked Willa innocently. (*This is an almost acceptable reason for Father breaking the law and starting earlier,* she thought hopefully. *Maybe Father was just over-anxious to get started.*)

But Willa's innocent question hung over the table like a bird of prey, now swooping down on Thorndike who had been blissfully gaining control until that moment. He held his breath.

This was *his daughter again*, blatantly questioning his actions where she, as a woman, had no right to speak, let alone offer information he didn't want disclosed. How had she discovered that he had already started on the road? Who was her informant? Was she against him now?

" Calm yourself, Thorndike," the Auras cautioned.

Thorndike tried to shake his concern. He had this phony Water Rights Gathering moving in his direction so far. He was almost ready to follow through with the rest of his amazing new idea.

Drum, however, recognized that Willa's innocent question was not going to help his partner with his new, brilliant idea of the 'field trip'. He had seen the early work on the road from his carriage the day before, so he knew that Queen Willa was right. Thorndike *had* started early, but he also knew that Thorndike's amazing new field trip idea could advance their road approval.

"Your father was not starting on the road early, Queen Wilhelmina," he contributed smoothly. "I saw what he was doing yesterday as we drove by on the way here. He was simply filling the current road's potholes with dirt and boulders from his own land. He was just trying to make *the current road* safer. I'm sure Your Highness can appreciate that."

Drum smiled a sympathetic smile which might have been believable if Willa had not seen Drum's fake emotions before.

Solem had not learned to distrust Drum's smiles, however, and this explanation made sense. He, of all people, should be sensitive to the need to repair ruts in the road, ruts that had surely contributed to the accident that had killed his wife. He nodded sadly. This was a reasonable explanation for what he had assumed to be disloyal law breaking.

Solem smiled at Willa for informing him of the possible 'early build' but he was now relieved. "I understand very well how that kind of repair needs to be done regularly," he acknowledged to Thorndike. " Now, let us get back to your suggestion. I, for one, feel that a visit to the site of at least one of the smaller irrigation rivers is a wonderful idea. When will your men have it ready for viewing?"

Thorndike breathed a sigh of relief at Drum's unexpected intervention and The King's acceptance of it. He was again intent on his *new* idea, the part that 'useful Drum' knew nothing about.

"I suggest that we go there immediately," Thorndike responded abruptly. "The sooner the better."

Solem chuckled a bit at the suggestion of an immediate trip. Royalty didn't go anywhere without planning. Planning took several days. It was a matter of arranging protocol for Royal Trips.

"I understand that you are in a hurry to begin actual work on the new road," King Solem said politely, "But, of course there are preparations that must be made first. I will see to it that those preparations are made without delay."

Thorndike, a man still in the grip of the Auras' magic and last night's mania, was not ready for any kind of delay. His proposal today did not include any wasted time. His frenzy was for building *his militias' road*. For his total plan to work as he had contrived it, it had to begin immediately.

His usual temper began to rise unbidden, but his Auras were aware of his temperament. "Speak calmly, Thorndike. *You* are in control here. Only *you* know today's plan that you have *already* put in motion. Too much anger will reveal your trap too early. Be calm."

The local contingent of Consciences heard the Auras' instruction to Thorndike clearly, but only the ones who had been in attendance at The Conscience Alert months before understood its implied danger.

The Auras had mentioned a 'trap' and a plan that Thorndike had 'already put in motion'. It appeared that Drum didn't know the particulars of this plan. Cwilla, Cjune and even Chorace came to alert. Was this the beginning of what Cwilla had seen coming so long ago?

Thorndike appeared to have regained control of his temper, but not his sense of urgency. "Yours must be a delicate explanation", counseled the Auras. "This explanation must emphasize the importance of time. Timing is important to today's unique idea."

Thorndike assumed a diplomatic pose and launched into a logical reason for the Royals to hurry. His explanation was Aura-dictated genius, but no one could have suspected that.

"Your Majesties, the need for work to begin on the Royal River Road is immediate. Today's short cloudburst was only an early indicator of the weather problems that River Kingdom...and Urhonordo... will face if we don't construct the road now, in the Autumn. Winter storms may make construction impossible.

"The Great River is low now at its best place to assist us with building the large levees needed for my militia... I mean, to establish the safety of the new road. We need to take advantage of our mild weather right now. This is why I have arranged for my household to make ready for your visit immediately."

King Solem raised his eyebrows but nodded. Thorndike's thinking was admirable.

Willa, Manda and June sat motionless and watched Thorndike deliver his prepared speech. It was too organized and unlike his usual way of speaking to be anything other than a prepared speech. Cwilla, Cmanda and Cjune were now convinced that this was the work of Auras.

Drum had always thought ahead and calculated outcomes. He took pride in the preparation of each strand of The Plan to create *his* goals. He had, until now, carefully groomed the 'less able' Thorndike with the promise of the position of 'Commander of the Royal Militia' which everyone knew the noble coveted.

What Drum was witnessing in Thorndike now, however, did not fit what Drum thought he had developed in Thorndike. He was too focused, too determined, too articulate. This Thorndike was different.

Drum's Auras were equally perplexed. They had not predicted this move by Thorndike either. What did this 'leadership behavior' mean to their man, Drum? Was Thorndike trying to take over?

Strategically, they decided to advise Drum to show caution rather than complete surprise or disapproval. That would give Drum a position from which his usual calculating prowess could re-establish control. "Stay calm," the Auras whispered, "while you assess this strange, new behavior in your 'second man'.

Drum leaned forward in his chair. "This idea is quite amazing, Lord Thorndike. "I had no idea you were so thoughtful, but your reasoning is indeed correct. Perhaps we can send a runner to the other noblemen so that they can meet us at the proposed site when we get there." Drum's support calculations were already in gear.

"You mean big batches of nobles like the ones that you encouraged to come here to The Castle *early* this morning?" asked June with a sly smile. Manda snickered quietly to herself.

"Good question," whispered Cjune. "Thorndike and Drum are now making separate moves without telling the other. This 'partnership' is getting very interesting."

Thorndike stared at Drum. "Big batches' of noblemen were *here* this morning?" Thorndike snapped again. "Why were *they* here? Why didn't you tell *me* they were coming? Were you and your nobles going to leave me out?"

Drum blinked oddly. The *secretly arranged* demonstration that *he* had designed to impress the King with *his popularity alone* was now on record with Thorndike. He had not planned that.

June smiled inwardly and continued. "They were here to support Lord Drum, of course. They chanted "Drum, Drum, Drum," like they did at The Gathering. There were quite a few of them here, too many to fit inside, so they had to leave because of the rain. That was a pity."

That's what I thought last night! I knew Drum wanted to take control of everything, himself and leave me out! Thorndike's mania fumed jealously. That suspicion had been his motivation for today. Thorndike's Auras came to his rescue one more time with advice of calm.

But Thorndike's focus was wavering. He had been up all night and exhaustion was setting in.

The Auras could tell they were losing control of their man. "But today, you and your excellent planning have the upper hand," the Auras whispered quickly. "When your proposal is in place, Drum will either follow your lead or be eliminated."

THORNDIKE PROPOSAL #2

Thorndike had created this proposal in the gloom of last night's royal rejection and the retreat of his wonderful militia. This proposal was his cure for that humiliation. His actions today would display *his* leadership.

"So, your majesty, due to the importance of making a decision on the road-build quickly, we must proceed immediately to Thorndike Manor. My staff has made ready for you".

Again, Solem smiled at this man's lack of insight into the complexity of such quick movement for one royal contingent, let alone two of them if you included the Urhonordorians. "Traveling to your manor today is quite out of the question, Lord Thorndike, but I do recognize your need for haste. I am sure that we can be ready to travel tomorrow morning. Will that suit you?

"Thank you for that, Your Majesty," he responded. "But I am afraid that will not be quick enough." Thorndike grinned weirdly. He had one more card to play.

"You see, the young ladies of your family, King Solem, and yours, Lord Drum are *already* on their way to Thorndike Manor as we speak."

"What?"

"I took the liberty of ordering a few of my most trusted militia to escort them personally. They will be arriving there shortly. When you leave with me now as I request, we will all meet at my manor for a lovely luncheon, followed by a day trip to the site of the small irrigation river on my property."

No one moved. In truth, no one was sure that they had actually heard what Thorndike said. Their daughters were already on their way? This must be a joke.

"I know this *un-* preapproved move of your daughters is a bit unusual, but I needed to ensure the fact that all of you would visit my manor *today,* as I am requesting," Thorndike explained as though this reason made perfect sense. He tried to look confident and assured that his actions would meet with the desired approval.

"You what?' shouted King Solem and Drum in one strangled duet. All the assemble gasped in shock and the men at the table stood en masse, staring at Thorndike with disbelief. Thorndike smiled at all those present. He looked almost gentlemanly. He appeared self-confident to the others... but he was no longer himself.

Thorndike nodded smoothly but his actions camouflaged a sudden knot of panic that began boiling in his stomach. "Stay calm," advised the Auras again quickly.

"You expected this kind of reaction. Remember...*you* have the upper hand. *You have the girls so you have the power.*"

Thorndike's wavering self-control was bolstered by the Auras' statement of fact. *I am the boss, here,* he reminded himself. *Only I know where the girls are. That puts me in a position of strength.*

But Thorndike's creative fervor was wearing off. After plotting all night, he was exhausted. He had not braced himself enough for the fury of the response that now descended on him.

"How dare you do such a thing, idiot!" Drum snarled. He was beyond the excuses that he had just offered to keep his unpredictable partner in an acceptable position with the Royals. The fact that *his daughter* had been moved to Thorndike Manor without his permission was beyond excusing.

His anger was not based on concern for his daughter, however. This was all about a loss of power. This was a threat to *his* mania. The main issue for Drum was the fact that Thorndike had done something *very* significant (and questionable) without *his* instruction or even knowledge. Thorndike had acted *on his own to accomplish something big.* He was *taking charge.* This could not be allowed!

Solem's response was completely personal. He was in Thorndike's face with such speed that no one actually saw him do it. He grabbed Thorndike's collar and lifted him to his feet. "Who do you think you are to move my daughter anywhere without my permission? Have you no understanding of protocol? Where is my daughter at this moment? Return her to me immediately or find yourself in the dungeon!"

The Castle Guard sprang into action and took over, yanking Thorndike completely off the ground...again. Thorndike, a novice in the area of personal interactions, had not anticipated this level of negative reaction to his new plan.

"Gentlemen, relax. Why are you so excited?" Thorndike croaked from his mid-air position. "Your daughters have simply left early on a very enjoyable day trip. You act as though they have been kidnapped. They are being transported and well cared for by my militia. Unhand me and let me explain."

Solem released his hold with Thorndike's collar but not his glare. He stepped back and stared at the face before him with an odd combination of fury and fear. And, he had just discovered a truth. *I am a King, but this man knows where my daughter is, and I don't.* With careful control, Solem waited for the explanation.

The Castle Guard allowed the nobleman's feet to find the floor, but they did not release their grip on his arms. Drum elbowed his way between the two and shook his fist in the nobleman's face.

"What in the hell do you think you're doing, Thorndike? I had a good thing going here before you arrived *late* and decided on this 'day trip' idea. And what does *my daughter* have to do with *our plan*, Thorndike? Any part she might play in my plan will happen *later...much later, when she...*"

Drum stopped short. He blinked into the face of the man that he had been grooming as only his helper for years. What he, himself, had just said about his daughter was a reference to a greater plan that he was not ready yet to reveal. In his anger, Drum had almost said too much.

Drum stepped back and stared at Thorndike. With sudden clarity, he realized that years of planning could have been unraveled here and lost forever. With as much dignity as he could summon, he returned to his chair.

The surrounding Auras recognized the danger as well. This was a time to unite the separate Auras of these two conspirators into one force, at least for the time being. The Auras were self-preserving. They would now adjust.

Thorndike had been accustomed to following Drum's lead for years. Only in the wake of his humiliation on the colonnade the night before...and the ghost that had suddenly appeared out of nowhere...only then had his fury evolved into this scheme that was putting *him* in the lead.

But today, after thrashing out his idea all night, this new idea was showing itself to have serious flaws. Slowly, Thorndike lowered his head in this realization. "My mistake, Drum," he mumbled, humbly. "I should have let you know what my idea was before I put it into action. I'm sorry."

Drum recognized that Thorndike had just relinquished his new-found power. Drum would need more groveling than this to atone but his own near revelation of his ultimate plan, caused him to accept Thorndike's apology. This was no time to quibble over small issues like the movement of daughters. He would gather the details of Thorndike's idea later. Right now, he was probably close to a goal that he had been working toward for years. He needed to be smart.

Solem and Csolem were frightened for the first time in years. "This man has *your* daughter as well as Drum's," Csolem whispered in Solem's ear. "This move is without attention to protocol, but right now, protocol is not our concern. Mimi is your concern. Where is she? Is she alright? You must find Mimi and make sure that she is safe. *Then* you can find out why this was done."

Willa reached out to Solem. Her Conscience was frightened, too. "Solem needs to regain his control of this situation", Cwilla told Willa flatly. "In truth, he and your fellow Royals no longer have control. They are at your father's mercy. Whether or not his movement of Mimi and Drum's daughter was just a breach of protocol or something more evil, Thorndike now has the upper hand. *He* has Mimi. Both Mimi *and* the Kingdom are in danger."

And my father is the main threat to both, Willa thought with shame.

ON THE WAY TO THORNDIKE MANOR

The decision was immediately made to transfer the Water Rights Meeting to Thorndike Manor. The King knew, much to his chagrin, that he had no choice. His daughter had been taken from The Castle without his knowledge. He was consumed by anger and fear.

Drum did not have the same concerns about Balynn. His concern was, instead, about the new, assertive attitude of Thorndike, a man that he thought of as an underling...with a very large militia. He knew that he didn't need Thorndike himself, for his plan to succeed, but he *did* need the Thorndike Militia.

After a few minutes, it was decided that Willa, due to her recent accident, would ride in the Urhonordo carriage with Queen Luna. Most of the men and Manda chose their own individual horses.

Drum, however, had arrived yesterday in his own carriage, complete with an entourage of Drumsmen. His elegant method of travel was not about to change. He would travel in the comfort of his carriage by himself, but he would not be happy about it.

The idea of viewing a small river on Thorndike land was a good idea, Drum admitted. Requiring everyone to travel to Thorndike Manor *today* was even better. But these ideas had not come from him. How had the slow-witted Thorndike managed to cause this trip to occur *now*?

As Drum bumped along the rutted road between The Castle and Thorndike Manor, another troubling question surfaced. Who had made slavery a part of the discussion earlier today? Slavery...the mote... the fact that he had been conducting this movement secretly...all of this was now out in the open. Had Thorndike disclosed it all out of spite?

Drum knew that his use of Lowland slaves was not going to bother most of the nobles. Many of their serfs and farm hands were treated like slaves, themselves. After Drum's plan to take over The Lowlands matured, slavery wouldn't be a problem at all, of course. It would be accepted.

But King Solem had always been a 'pro-Lowlander' monarch who thought of the Lowlanders as equal to white men. He didn't approve of taking advantage of Lowlanders...and, of course there was Urhonordo whose female monarch had actually married one. Ridiculous.

In order to get my road built, this slavery issue must be dealt with soon, Drum decided to himself. *The new road is key to the movement of Thorndike's militia, my militia actually, and key to my whole plan.*

The Thorndike Militia is the only necessary thing that this ridiculous Thorndike is good for. I now know that I must watch Thorndike and stay on his side, but I must also regain my control... and the use of his militia.

Thorndike, however, was back in the middle of his own creative operation. He was jubilant. His earlier exhaustion had disappeared in the light of his current success. With joyful anticipation, he conjured up the sight that would greet them all when the Royals entered his domain. He had ordered the suits of armor that usually filled the entry hall removed. He had ordered the floor polished and the dining table decorated. A wonderful meal was being prepared.

Flo is good at these things, he thought to himself. *She is as handy as her mother was. I'm glad that I didn't give her away as I first intended when she was born. She wasn't the boy that I wanted, but she became useful after her mother passed.*

"Does it ever occur to you that women are wonderful creatures as well as useful... even as wonderful as men?"

Thorndike looked around. He was ahead of the entering procession. Who was talking to him?

"You probably forgot that I even existed," the voice came again.

Thorndike looked from left to right with caution. "Where are you?" he snarled under his breath. "Come out and say what you have to say like a man."

"I *am* out but you can't see me," the voice continued. "I am your Conscience. You can't see me, but I am right here beside you. I figured that you might be too tired to fight me off like usual, so I decided that this was a good time to speak up and keep you from doing anything you'll be sorry for. You are successful right now. Don't mess it up."

The Call of the Auras

Thorndike shook his head. *I'm not hearing anything*, he thought to himself. *I'm just tired*.

"You can ignore me if you want to", Cthorndike whispered, "but I've decided. I am not going to sit by silently anymore. I gave up on you years ago, but I just heard you appreciate your daughter. To me, that was a sign that you may learn to listen to your Conscience after all these years."

"Balderdash!" Thorndike shouted as he climbed down from his horse.

"What was thar?" asked King Solem as he climbed off his horse nearby.

"Oh, nothing." mumbled Thorndike. "I just stubbed my toe."

Riding now in Thorndike's ear, Cthorndike chuckled.

Everyone was stopped in front of the impressive entry to Thorndike Manor. The manor's steps were swept, and the windows were polished as Thorndike had ordered, but the sight that greeted the guests was enhanced by something unexpected. Three beautiful young women stood at the front door, smiling at the arriving group as though they were the welcoming contingent for a grand ball.

Willa jumped out of her carriage and ran up the front steps, straight into the arms of Flo. "Oh, Flo! You look so beautiful. That dress looks beautiful on you! And look at the rest of you. You all look so beautiful! You make me feel all rumpled and dowdy. Can we go upstairs and find a clean dress for me?"

Manda scrambled from her mother's carriage and joined the giggling gaggle.

All of the girls exchanged hugs, but then stopped short. This was not a party. They were here on serious business. Bunched together like gayly colored flowers, they turned as a group and looked to Thorndike.

Cthorndike asked Thorndike the question that was on the girls' faces. 'What was going to happen now?' Thorndike simply stood in front of them, motionless. This had not been part of his Aura-sponsored proposal. They had never gotten this far.

"Smile, idiot," hissed Cthorndike. Thorndike managed to do so in a clumsy, completely uncharacteristic way. He decided to just stand back.

The men gathered, one by one on the elegant stairs of the Manor and lowered their hats with respect. The view of the three...then four, now five beauties laughing and hugging together was an unexpected surprise in a day that had already been full of surprises.

Solem and Armando stood back and surveyed the scene with curiosity. Even though they were now relieved, they too, wanted to know what was going to happen next.

Balynn and Mimi nudged Florence forward. These young women now felt that they knew each other since they had just endured a surprise 'daytrip' together. "You are now the woman of this house," whispered Balynn. "Do what your mother would have done if she was here."

Flo stepped forward timidly, glancing at Thorndike for approval. Thorndike was still smiling oddly and did not move to say anything. "Welcome, everyone," Flo began in a manner that she had seen her mother do. "I am so glad you are all here. Please follow me to the parlor. Luncheon and tea are almost ready. Jackson will take your wraps." With that, she turned and walked into the front hall. The males of the groups followed obediently into the parlor.

The luncheon was held in what Balynn recognized as the former map room. It had undergone an amazing transformation. The dusty curtains had been taken out and shaken until the tassels fairly gleamed. The maps were gone from the walls and lovely paintings of imposing military men appeared on the walls.

Flo, still in a dress-up dress but now covered by an apron, bustled about giving directions to recently acquired servants as well as doing much of it herself.

"This table looks beautiful, and the food is great," whispered Balynn as Flo passed by. Flo smiled gratefully in return.

Willa added to the compliment. "I didn't know what a talented cook you would become when we were growing up, Flo. Your mother really taught you well. I remember her wonderful cooking...although I never gave her the credit she deserved. I miss her. I'm sure you do, too. We should talk later and bring each other up to date."

Flo nodded. Willa was now a queen, but her words showed that she remembered their childhood together. Tears sprang to Flo's eyes, but she wiped them away, undetected. *This really is a perfect day so far,* she thought as she continued her chores.

"This would not be a bad time for you to tell Flo that she did a good job, too," announced a voice in Thorndike's ear. Thorndike's Conscience was at it again.

"Shut up, whoever you are," he hissed.

"What was that?" asked Solem.

A swallow of the ale that Thorndike had just ingested threatened to rise up in his throat. "Nothing, Your Highness. Nothing at all. I was just thinking, though, that now would be a good time for us to take our ride out to the fields to view my irrigation river. If we go now, we can return in time to discuss what you see and maybe sign some papers."

Making sure that he was still important, Drum echoed the notion. "Indeed," he offered emphatically. "I was just going to suggest that we go there right away. That's what we all came here to do. I know that *I'm* ready. I'm *past* ready. Let's do it right now so we can get down to business."

King Solem frowned at Drum and Thorndike. *First he steals my daughter away without permission.* Now we have barely *eaten and he getting pushy right away,* he observed. *No understanding of protocol, but very focused. I didn't know Thorndike had this in him.*

He has 'Aura focus' inside him right now, Csolem thought. *Without it, he would never have pulled this off.* "I hope you know that your signature is the *only* reason that Thorndike invited you here," warned Csolem. "Thorndike and his Auras have the whole Water Rights Meeting here for just one thing. He wants the Royal Approval signed *today*. That's it."

You're right, of course, Solem agreed silently, *even though seeing the necessary irrigation in person was a good idea.* He motioned to Armando. "Are you ready to take a ride, Armando?"

Armando swallowed his last sip of ale and nodded enthusiastically. "Perfect. I would like to get our carriage home before dark. That road's ruts can get treacherous at night."

"You are so right. We all know that for sure," Thorndike agreed, looking with meaning at Solem. Solem again nodded.

Willa patted Solem's leg under the table. "I have a thought," she said quietly. "Why don't you men ride out to the irrigation river while we women stay back here at this manor and chat. It looks like it has been cleaned up. Maybe I can take all the ladies on a tour of my former home. How does that sound?"

"I, for one, am not going to stay behind and go on a tour," Manda stated flatly. "I want to see this so-called 'irrigation river' for myself. I want to see how much land it can irrigate other than Thorndike land."

"I thank you for your kind offer, Queen Willa," Luna said with a smile, but I would have to agree with my daughter. I want to see the irrigation river that is on this property. If it is not strong enough to flow south to the center, we may have to add another irrigation river somewhere. But Armando and I would like to see it for ourselves."

Mimi just grinned happily. "A tour? Really? That would be so much fun, Willa. I always wanted to see where you grew up. I hardly remember any of Thorndike Manor. My mother used to take me here, but I was really small when the parties stopped."

"Me too, and that was a long time ago," added Balynn. "I only went to one party, I think. I hardly remember."

Suddenly, Balynn stopped short and looked around for Flo. Balynn had been on 'the tour' with Flo a few weeks ago. Not all of the manor might be open to 'a tour'. Flo was busy in the kitchen. What would *she* say about this idea?

THE TOUR

But there was no time right now to find the answer to that question. The men arose from the table immediately and went out back for their horses. Queen Luna and Manda followed close behind. Drum was somewhat at a loss for a horse as all of his were still in harness for his carriage.

Thorndike smiled magnanimously at his partner and offered the loan of a gentle mare as he knew Drum was not much of a horseman. Drum was chagrined but he knew he had to do this. This small river might be the key to everyone's acceptance of the new road, even if it was on Thorndike property.

Drum had not seen the small river in question, and he had no idea what the irrigation river on his own property looked like. He wasn't even sure that there was one. He only hoped that Thorndike's 'small river' looked like it would do the job that Thorndike had advertised. Approval for the road was all that mattered,

Drum accepted Thorndike's offer with little more than a curt, "Thank you, Thorny," while Queen Luna gratefully accepted her loaned ride. Within a few minutes, an inquisitive group on horseback was ready and headed for the important little river in question.

But they had left behind a group of ladies that was hardly going to miss them. Mimi eagerly rushed to Willa's side. She loved an adventure and exploring the big, old manor house where Willa had grown up was exciting. To Mimi, Willa was still somewhat of a mystery. What did she like to do as a child? How did she spend her days as a little girl? Did she have any boyfriends before she met Solem?

Mimi was full of questions, chattering and laughing and trying to make up silly answers of her own. Balynn was more reserved...and she was concerned. There were secrets hidden within the walls of this manor. Hopefully, Flo would soon return from the kitchen and volunteer to 'help with the tour'.

Willa grinned at Mimi. "You want to know everything, don't you?" she laughed. "What about you Balynn? Don't you want to know everything about my 'sordid past.'?

It was time for Balynn to join in with the fun. "Of course, I do," she giggled. "I was stuck mostly alone in my father's house all my life, just like you. I want to know what *you* did when nobody else was around. Did you do crazy things, sometimes? I used to dress up in my mother's old dresses and pretend I was queen. What about you?"

"You did that, too? Willa chortled with a grin "I used to parade around the halls upstairs and put my dolls on chairs to watch and adore me. I thought I was very special. Who knew I might actually get to be a queen someday."

"You make a very good queen, Willa," commented Mimi with a low, playful bow.

Balynn now joined in the fun of her new friends. "I acted like a queen, too, but so far I haven't become one," she giggled. "I guess my time was spent in reading and exploring the rooms that I could get into in our manor. My problem was that a lot of the rooms upstairs were locked off and never used. I thought that was a waste. How dull can that be?"

"I know what you mean," Willa responded. "My father had some rooms locked off too. But some of his books were pretty interesting...especially the history books about the old River Kingdom. There were some wild things that went on back in the day. My father has practically memorized all those old history books."

"I hate history books," pouted Mimi. "I pretty much hate books of all kinds. Books are boring. So why are we just sitting around here talking about them. Exploring is much more fun. Come on, Willa. Let's start the tour!" Mimi danced down the hall.

The tour was on. Like the children they had been before, the young women scurried from wing to wing, peeking into abandoned guest suites and storage rooms, sifting through trunks of out-of-date clothes, and trying them on with impromptu fashion shows, laughing and mocking their elders in the safety of the empty halls.

The truth was that all of them had had tragedies in their young lives. All of them had spent lonely days. Gradually, they discovered that they had a bond because of this. Little by little, the laughter died away and a serious tone slipped up on the trio as they wandered down the last empty hall.

"I think we've seen all there is to see," Willa sighed. "I think that it's about time to go in search for lemonade. "Let's see if we can find Flo. She will know where there is some, for sure."

"We haven't seen everything, Willa," announced Mimi with a grin. "There's one more room we haven't poked into down at the end of this dark spooky hall. I bet there is some scary stuff in that room."

Balynn kept walking. She knew where they were. They had avoided this room so far. "I only want some lemonade and Willa said she would find some for us," she teased.

But Willa had stopped. She was looking toward the room at the end of the hall with fascination. "You know," she mused quietly," I actually have no idea what is in that room, and I have always wondered."

"Why didn't you just go in and find out?" Mimi asked with a practical tone. "You should have just gone in and looked. Why didn't you?"

"It's always been locked. I asked for the key once or twice, but I got in trouble with Flo's mother. She was very strict about it."

"Hmm... A locked door that no one has been in for years...very spooky...very *interesting*. Let's go!' shouted the newly energized Mimi, running down the dark hall.

Balynn stopped. She did not look back. "You are on the edge of something that could change Willa's life, forever," warned Cbalynn. "Willa is a good person. If this room contains what you think it does, it will crush her. Save her from that if you can."

What can I do? I have no right to tell her or not tell her. What should I do? Balynn thought in panic.

Cbalynn reverted to her standard advice. "Tell the truth but don't lie. Perhaps time will be on your side."

"I am thirsty, too, Mimi. Do you think you have to see *everything?* Come on. We've seen enough. Let's get some lemonade."

"Oh, alright. I'm thirsty, too," Mimi acknowledged and slowly marched back to her friends. "We'll probably be missing out on a lot of dust that we haven't seen yet, but I guess we can pass this one by."

Willa laughed and the girls resumed their chatter and continued down the hall toward the kitchen wing. Balynn sighed in relief.

<u>DISCUSSIONS AND DECISIONS</u>

Drum and the Royals followed Thorndike to the top of the rise that overlooked a valley and The Great River flowing lazily beyond it. The River was low in its banks revealing river rocks that had been submerged a month before. It was now late fall causing thirsty deer and other wildlife to creep over the banks of The River for a drink.

The Water Rights Committee halted. The grass on the hillsides was dry and golden, crunching softly under the feet of their horses. But a brisk breeze whisked around them in a pleasant way and all were determined to feel positive. None among them enjoyed the conflict they had been experiencing and the proposed road had threatened to have them taking opposite sides on the issue.

Queen Luna, in particular, was willing to look for solutions that both sides could accept. She had hope that this little journey would solve part of their concerns. Perhaps Solem would approve the road if the little irrigation river proved up to the job. That would still leave them with the concern about possible slavery, however. Luna sighed.

"You can solve one problem at a time, you know," Cluna whispered helpfully. "The two may not even be connected."

Very true, thought the patient Queen.

Everyone but Manda hoped that Thorndike's 'Irrigation River' would prove to have the beginnings of a positive solution for both sides. Manda was sure that it wouldn't. As she strained forward from her horse's position in the group, she saw nothing that changed her mind.

"There it is," announced Thorndike proudly pointing across the hilltop.

"Where, exactly?" asked Solem, straining in the direction of the point.

"Right there, Your Majesty," Thorndike responded, inching his horse toward a pile of rocks and a few cat tails wafting in the breeze.

The group followed Thorndike's lead and worked their way slowly toward the rocks. And there it was. Gurgling gently from under the rocks was a stream of water splashing happily into a small puddle at its base and trickling into a small stream below it.

Thorndike leaned back in his saddle with pride. "See that? I told you that there was water up here, and here it is," he announced. "There's a spring up here."

Manda dismounted to the grassy slope and walked closer. She walked around the spring and stuck her fingers in it. She didn't like Thorndike, but she was determined to be appropriate. "I see that this is a lovely little spring," she said as she rose, "but where is the rest of your 'Irrigation River?' "

Thorndike didn't like girls in general and Manda, in particular. He had just eaten, drunk some good ale and now he was on his own land. In his opinion, he didn't need to be appropriate any longer. "I have put up with you and your know-it-all questions for the last time, missy" he snarled. "You are on *my* land. You are on *my* horse and now you are questioning *my* river! You know nothing and you have no right to be interfering with something you know nothing about! "

Thorndike's horse reared and stomped in response to his anger and the other horses in the group jostled and pranced. Drum hung on to his horse and prayed that he wouldn't fall off.

Manda swung onto her horse and rode silently to the back of the group. She had asked what she needed to ask. Her parents must make the decisions that counted.

Luna and Armando turned to Thorndike with disgust. They were not worried about Manda's feelings. They knew she could take care of herself. But she had asked a logical question and Thorndike had only been rude. Solem, too, stared questioningly at Thorndike while his horse settled itself in response to Thorndike's rude lack of answer.

"You are so stupid," barked Thorndike's Auras. "You worked so hard to get this far with folks thinking positive and then you blow it with one of your usual fits of temper. You may have lost this whole opportunity to get Royal Approval for your road! You had better get humble and ask for forgiveness or give up and forget your whole plan."

Thorndike climbed quickly off his horse and got down on a knee. "I humbly apologize to all your majesties," he mumbled almost convincingly. "My temper is my downfall. I have been up since before dawn, and I lost control. (He looked slyly from side to side and continued.) Please forgive my outburst, Your Majesties. Please ask your daughter to forgive me, as well. I swear I will definitely attempt to get better control of my temper."

"Good job," cheered Thorndike's Auras.

"That apology was very well done," whispered Cthorndike. "I am proud of you."

(This apology *was* truly *amazing.* It was a rare time when both a man's Conscience and his Auras had suggested the same behavior.)

"Take care of your partner," whispered Drum's Auras.

Drum chimed in immediately. "Timing is crucial here." His Auras continued whispering and Drum followed, word for word.

"There is nothing much to see here right now," Drum explained carefully. "But in the spring of the year, this spring begins to flow at a more rapid rate and can be diverted to irrigate a lot of farmland on its way down hill."

"Hmm...I can see that this will probably serve much of your land, Lord Thorndike," King Solem commented thoughtfully, "but how much water will arrive at the farmlands that are not on Thorndike property?"

"That's *their* problem," answered Drum with a wave of his hand. "I have a spring on my property that can do just as this one does. The rest of the nobles will have to solve their own irrigation problems for themselves as I do."

"According to my daughter, the central farmers have accomplished their irrigation forever with only their levees that bring in water from The River," commented Luna. "They coax water inland in the spring and hold it in low ponds to water with during the summer. The Central farmers are very worried about the new road stopping that."

"Exactly," agreed Manda with a worried sigh from her less visible position. *My parents and the King must work this out for Hector and his fellow landholders. I wish Willa was here to back me up.*

"Do the nobles who own land in the Central valleys of the kingdom realize that the new road may cut off The River water to their land?" added Prince Armando.

Thank you, Father, Manda thought gratefully. *Father knows.*

"Of course, they do," lied Drum with convincing bravado. "Why do you think they all support *me* and *our plan* for the new road."

"The nobles certainly know and are all for *our road*," added Thorndike, sitting tall in his horse. "They know where the water comes from and where it goes. They also know where *their protection* comes from when they need it. They know where their militias need to go when their land needs protection as much as they need water."

Drum was not finished. "A smooth road will enable safe travel for both the nobles *and you,* Your Majesty, as well as the Royal Guard and the militias that may someday be needed to keep everyone safe. That's why the nobility of River Kingdom is so much in favor of the new road. You saw them gather on your colonnade this morning, Sire. That is what they came there to say."

Their arguments were convincing. King Solem *did* remember the impressive group of nobles from this morning as well as the nagging sadness of his first wife's disaster. There was no doubt that the river road needed repair and these two men were ready to get it accomplished.

"I truly see no reason to deny approval for the building of this road, my friends. It will be a great benefit to the safety and protection of all who travel."

Armando and Manda were not as positive as Solem and were certain that many of the River Kingdom nobles had no idea about how the new road might interfere with the irrigation of their land. Nobles had always left these details to their farmers and land holders. But the nobles *were* Drum fans.

The new road would be constructed on River Kingdom land. The ultimate decision about the road was for King Solem to make for *his* kingdom, Luna and Armando concluded. They should not object if he felt strongly that the new road should be built.

But that was the issue of 'The Road', alone. Armando was *not* going to lend his approval to a road that was championed by a man who used slaves. "What about your slaves?" he asked bluntly there on the hilltop. "Will slaves be used to build this beautiful, *necessary* new road? We still have not dealt with the issue of the slaves that I, myself, have seen building a mote for *you*, Lord Drum. I, for one, cannot approve of a new road which will end at *your* Drumsworth Lake...which is now being enlarged by Lowlander slaves."

"Well said," echoed the Consciences who were listening intently from within the other humans. "Slavery is a more important issue than a road," they chattered to their humans...

"Who will build this road? More slaves like the ones we already saw?"

"You can't build a road to a place that is built by slaves."

"It is against all that is good and right among humans."

"That would be a road built on evil!"

The disgusting thought of slavery had slipped out of sight earlier with the forcefulness of Thorndike's unorthodox proposal and their travel to Thorndike Manor.

How could something this important have been tossed to the side? In truth, it had been dropped out of discussion after merely a simple denial from Drum. The Consciences of the Royals were ashamed of themselves. They should have stayed focused. Thank goodness for Armando.

The Auras that sponsored the evil practice of slavery were ready for this discussion. Drum's Auras were particularly ready as, of course, they needed to be. Drum *was* a slaveholder. Defense of the indefensible was a necessary skill for Drum's Auras.

"Just deny slavery's existence", suggested Drum's Auras, confidently. "That always works. "Noone can prove slavery if they don't see it or hear it. If someone is working hard, you might hear them complain, "I am a slave." How can anyone prove that this person is or isn't a slave? Use your head, Drum. Deny, deny, deny. Your accusers can't *prove* they are right. It's just that easy."

Drum smiled at Armando. "I agree with you," he said nonchalantly. "I don't approve of slavery, either but what you thought was slavery was not. It was just hard-working Lowlanders building a pool."

"They were slaves building a mote for you on *their* land," Armando countered.

"The Prince is right," contributed June. "I saw it with my own eyes. They were slaves building a mote on Lowlander land!"

"They were Lowlanders building a pool on *Lowlander land,*" said Drum calmly. "*You* were watching from a bluff on Urhonordo territory. You couldn't hear or see well from up there. You jumped to a conclusion, both of you...Manda, too, if I recall. That was a mistake.

"You must only report on what you have actually seen and heard up close. I forgave you for your mistake earlier today and I will do it once more now...but I will not do it again. Your lies hurt my feelings."

"Awesome response," murmured Drum's Auras.

"Drum's response was reasonable," sighed Carmando. "I almost believed it myself."

"'You made a mistake' was a good response," whispered Cjune. "But it was too smooth. You *know* what you saw but you *only saw* it from a distance. Will Drum be willing to show the pool and the workers to everyone...up close?"

"You are right, Lord Drum," said June politely. "I *did* only see what I thought I saw from a distance. I truly apologize if I got it wrong. For that reason, I will be riding down to Drumsworth within the next few days just to prove to myself that 'I was wrong'."

"Bravo!" whispered all the nearby Consciences.

King Solem smiled at June proudly. "That's a good idea, son. Maybe you and Manda could go down south together", he suggested, catching a glimpse of Manda who had been shaking her head at what she thought Solem was deciding about the road. She was not smiling.

"I will go there with you and my daughter, as well, if you don't mind," added Prince Armando "That's the best way to find out the truth of this issue, once and for all."

DISCUSSIONS AND DISCLOSURES

While decisions and deceptions were being contemplated on the grassy slopes of Thorndike's hill, new and old friends were sipping lemonade in Flo's kitchen. Balynn, Mimi and Willa had come for lemonade but stayed because Flo needed to finish the cleanup from the luncheon which she had magically prepared for this important occasion.

It had been a success. Flo was exhausted but happy. Willa sat down at the kitchen table and slipped back into the childhood friendship she and Flo had always shared. As little girls, they had been close but had drifted apart. Still, they were glad to see each other again no matter how different their lives had become. It was like old times.

"You are amazing, Flo" sighed Willa. "I can't believe how much you got organized and still managed to come fetch the girls from the garden for my Father. How did you do all that? And why?"

"I asked some of my town friends to mind the ovens while I was gone," Flo answered with a grin. "And I think your father got the idea that he could get everyone here to approve his "Irrigation River" idea if he got both girls over here for bait." Flo answered honestly. "That's what I finally figured out."

"Pretty clever," laughed Balynn and Mimi.

"Pretty sneaky," sighed Willa. She had already figured that this was true. This impromptu invitation to luncheon and now, 'a trip to the 'Irrigation River', had just been too sudden. There was more to this she was sure. She would keep her eyes open.

"Well, his plan worked, you've got to admit", Balynn offered. "Here we are and all 'the important people' are up looking at a little river on a hill. But how did Thorndike choose *you* to come and get us, Flo? I thought that he would have just sent his two "sweet" militia men."

"My brother told him that I knew you from when you were here before, Balynn. He thought you would come with me if I asked you."

You have a brother, Flo? thought Willa suddenly. *Why didn't I know that before?*

Balynn just grinned. "Thorndike didn't know that I would have gone with Jorge if he had come by himself and asked me to follow him home. I would have gone along shamelessly without a question."

"You fancy my brother?" Flo gasped. "He is only a lowly militia man. You come from money, Balynn. You could do a lot better than Jorge. I thought you would go for somebody higher than that... like Prince June. He is so handsome."

Mimi snapped to attention. This was a new piece of good gossip information. "Wait a minute...Jorge and Jackson are your brothers, Flo? Really? I didn't know that."

Willa didn't move. She stared at Flo in amazement. "Jorge and Jackson are both your brothers, Flo?" she repeated. "Why didn't you ever tell me?"

Flo blinked. What had she just said? There was no lie that would serve her here. Balynn already knew the truth.

"Indeed," Flo said quietly. "I'm sorry, Flo. The boys were always supposed to be a secret, even from you."

"Why?"

"It's a long story, Willa. Father just wanted us to keep all three of us a secret."

"Father?" gasped Willa.

Now Flo *was* in trouble. Her guard had come down. She had told Willa the unmentionable truth...and after all these years of hiding it. She was too tired to be as sorry as she should be. She couldn't think. Slowly she bowed her head to the tabletop where four young women had just been enjoying mindless chitchat. But now four minds were churning.

Willa couldn't move. *Flo is my sister.*

Memories and circumstances from her past life swarmed around her. When she was little, she had been a motherless, lonely child, playing with a servant's child. They had been friends. Were they now sisters, as well? Did she have brothers, too? Where and who were the mothers of all of them?"

Mimi was not a deep thinker. Worrying and thinking carefully about things had never been required of a beautiful young princess. But quickly she realized that Flo was upset. She slid into a chair next to Flo, patting her softly and saying, "There, there, Flo," the way she remembered her own mother doing when she scraped her knee.

Balynn felt instantly sorry for Flo. She knew how close Flo had kept the shame of this secret for years and how rigid Thorndike was about keeping it that way.

Balynn could also clearly see that this information was as new to Willa as it had been to her. That was extremely strange. Now she felt sorry for the young Queen Willa, as well. Balynn knew almost nothing about her own family but neither did Willa. How strange it was that they were so much alike in this way. Their fathers had so many secrets...

"What you just heard is secret, Mimi," Balynn whispered quickly. "Flo will get in trouble if her father finds out that she has told us these things."

Mimi jumped up happily. "All that stuff is a secret? Wonderful! Secrets are great! I can keep it a secret. I won't tell a soul." She ran to the kitchen door and looked around. "There is no one here but us girls either."

Flo looked up. She could tell that Mimi didn't understand the importance that 'the secrets' had just had for Willa. How could she? It was such a bizarre story.

Shyly, Flo took Willa's hand and waited. *What must she think of me?* she thought. *I have always been the daughter of her servant. That is the way I still see myself. Now Willa is a Queen. How can we be sisters...and will I ever have the nerve to tell her the rest of what I have been hiding?* she wondered sadly.

Mimi was intuitive. She looked at Willa and Flo sitting hand in hand as Balynn looked on sadly. She wondered what it was that they knew that she didn't. But she was used to this. As the youngest in a royal family, she had always been the last in line when important information was being shared. "She's too young" they had always said.

MIMI'S DISCOVERY

Mimi had developed the patience that went with her youthful position. *When I need to know what they aren't telling me, they will tell me,* she thought stoically. *But today, having patience isn't going to be easy.*

She could see that the older girls had things to discuss, but that was okay. There was more exploring to do in this creaky old manor. She might even find some exciting secrets of her own.

Like many old mansions, there were huge portions that were closed off and never used. She and the others had only seen a few earlier. There was a whole wing that was off to one side and obviously not part of the everyday activities of the manor. She was determined to investigate that wing, for sure.

First, however, she would start with a room that was just down from the kitchen. It had been open a crack earlier.

"I saw you wandering out this way," said a young militia man with a smile. "You didn't go to the hill with my brother and the Royals. Why not?"

Mimi blushed profusely. She had noticed him when she first saw him in the garden of The Castle. He was one of 'the brothers' of Flo that had brought her and Balynn here...and he was very handsome. What excuse should she use for the fact that she was "snooping."

"Try the truth, as usual", counseled her Conscience.

Mimi shrugged and agreed. "I was just snooping, trying to kill time while other people talk about serious things," she announced with a grin. "That is what I usually do when people don't want to tell me what they are talking about."

"Sounds like you are used to that," Jackson said with a shrug. "I understand. That happens to me with my father all the time, too." Jackson looked more closely at Mimi.

When he had fetched her from The Castle's Garden, he had been 'on duty'. His father had taught him to never mix duty with pleasure. Now, however, he was on home territory. This young lady was truly beautiful. She also seemed to have a jolly personality. What other attributes did she have? He decided to find out.

"Why were you snooping? Were you looking for something special?"

"Not really. I just thought I would wander and try to stay out of trouble." Mimi smiled up and Jackson. Now it was his turn to blush. *She really is beautiful...and I think she likes me,* he thought.

"I am a guard with nothing to guard, today," Jackson offered. "Thorndike just wanted to make a peaceful impression on all of you royals."

"Why do you call him Thorndike and not father?" Mimi asked sweetly.

"You know that Thorndike is my father?" Jackson gasped, looking cautiously from side to side.

"Indeed," answered Mimi confidently. "And I also know that that's a secret...but not from you, of course. I won't tell anyone else."

Jackson stood back and looked again at Mimi. "You must be very special if Flo told you our secret. Nobody is supposed to know. It hasn't been shared with anyone for years."

"I don't think that I am that special," Mimi said honestly. "I think that it just popped out when Flo was talking, and I heard it. That is probably what Willa and Flo are talking about now. I guess Balynn has known for a longer time than Willa. That is weird, her being a queen and all. All families have secrets though, I guess."

"I don't have any secrets," said Jackson proudly. "I am an open book. Ask me anything." Jackson was liking little Mimi more and more with each passing moment.

Mimi was not a young lady with experience. Since her mother had died, social life had been slow. Being this close to a young man alone was exciting. And she could tell that he felt the same way. But they were there *alone together*...in an empty hall...without interesting things to say to begin a conversation.

Mimi decided to try an idea. Maybe they could walk around in the manor together. That might be a nice type of 'get to know the other person better' thing to do.

She took a few steps down the hall. Jackson followed. "You have lived around this manor all your life, haven't you?" she began conversationally. "I bet you know what's in every room, don't you?"

"I guess so," Jackson agreed, "although I don't get to come any farther than the kitchen and dining room, usually. Flo sneaks us in whenever she can and Thorndike doesn't seem to mind as long as the rest of the militia doesn't see us. He says he doesn't want them to think that he treats us special."

"I can see that," replied Mimi as their stroll continued. "You must be very proud, though, knowing that *your father* is the leader of such a huge, awesome militia. I've heard that it is even bigger and stronger that The Royal Castle Guard."

"I think it might be," Jackson agreed proudly. Then he stopped short. Bragging that his father's militia was bigger than the King's was probably not a smart thing to do. He would change the subject.

The pair had reached the edge of a long, dark hall. It was different than the other halls, barren, with only a few small lamps lighting its corridor. "I wonder what father keeps down there," he mused as a good subject-changing statement.

"I wondered that, too," Mimi agreed readily. "We were going to see what was behind that door at the end of the hall when we were here earlier, but we went to get lemonade, instead. Can we go look down there now?"

"I don't see why not," Jackson answered, squaring his shoulders. "I *am* part of the militia. I *am* the guard of the room where all the armor is kept. Because of that, *I* have the keys that unlock every room."

"Really? Every room?"

"Indeed. Every room...Would you like to see what is lurking behind that mysterious door down there? You and I can check it out and no one will ever know."

Jackson's Conscience was seldom heard by Jackson. He always followed the rules laid out for him by his father---militia rules---rules of loyalty. His Conscience was clear on these issues.

Now, however, Cjackson had a question. "Do you think you should unlock a door that you don't know anything about? You were trusted with all the manor's keys because you are trustworthy, right? What would your father say about opening this door without permission?"

"I would love to see in that room!" Mimi declared happily. "It's so dark and mysterious down there. Maybe there's a ghost...or a spooky criminal stashed away in a cage. Wouldn't that be great?"

"You have a very active imagination, Mimi," Jackson laughed nervously.

She took his hand and smiled. "Please...?"

The smile made the decision for Jackson. Cjackson had said what he thought he should say. But Jackson was a young man who needed to show his bravery. He *would* open the mysterious door.

As the pair neared the door, they slowed, exchanging excited looks. For both, it would be a discovery of something they didn't know anything about. How fun would that be?

Jackson dropped Mimi's hand and reached for the large brass ring of keys that dangled from his side. Squinting at the opening that gaped darkly below the doorknob, he fingered the larger keys.

"This one should do it," he announced proudly. He tried it. Nothing.

"Try another one, Jackson," Mimi whispered, "but do it quietly. We don't want to wake up the boogieman inside." She giggled.

"Don't worry. I'll beat him to a pulp if he jumps out," snickered Jackson.

But there were many large keys on the ring. Time after time, Jackson's huge keys clunked to a stop in the uncooperative lock.

"That's alright, Jackson", Mimi sighed. "The boogieman is probably asleep in a pile of dust, anyway. No need to wake him up."

But then it happened. The huge key stopped in the lock and then clicked. Jackson turned it some more. The lock grudgingly gave way and the door edged open.

Mimi and Jackson grabbed each other's hand and slowly pulled the door open wide.

Now they could see inside. The couple didn't move, but simply stood in amazed silence. A welcoming fire crackled in a fireplace. A small oil lamp glowed cheerfully on a table beside a large chair.

"Flo, is that you? I was wondering when you would come today. You're a bit late. I'm sorry I take up so much of your time but I'm glad you are here. Did you bring Willa with you?"

Jackson and Mimi stared at each other in silence. There was a person in the large chair that was snuggled up to the fireplace. How could that be? *Who* could that be?

Jackson was a militia man, born and bred. He was trained to face danger without flinching. There was no doubt about it now, however. This sight filled him with fear. He had lived in and around Thorndike manor all his life. How could a real person exist at the end of this dark hall, a person that he had never heard of...a person who, until this moment, had been locked in.

Mimi had the opposite reaction. She was thrilled. "Look, Jackson' she whispered. "Someone lives in this room, relaxing in front of the fireplace, snug as a 'bug in a rug'. It is so cozy there. Who is it?"

Jackson couldn't answer. He had no idea. Against his Conscience's advice, he had opened a locked room without permission, he now remembered.

Cjackson was not one to say, 'I told you so', but now that he had Jackson's attention, he felt he needed to speak up. "You have to wonder why the door was locked on this scene," Cjackson whispered. "Was it to keep people out or keep people in... or both?

Mimi was conditioned by royal protocol to certain behaviors, even in unique situations such as this. They had barged in without an invitation and the person inside was expecting Flo, a servant. "Pardon us, milady," Mimi said sweetly. "Flo is very busy today, so we, Jackson and I, are here, instead, if it please you."

The crackly voice they had heard before responded. "I don't know anyone by the name of Jackson, but you sound very nice. Please come forward into the light so that I can see you."

Jackson held his ground in the darkness near the door, but Mimi had received an invitation. Without caution but with appropriate decorum, the young princess stepped forward and curtsied gracefully in the direction of the person in the chair.

"Charming," crackled the voice softly. "Look up so that I can see your face, child. I see clearly that you are well mannered. You couldn't possibly be of royal blood, could you? It has been so long since I entertained royalty. Come closer and let me look at you."

Responding to the request, Mimi stood straight, ready to meet an elderly occupant of the chair she had only seen from the back. And there they were...two women, on young and one old, meeting each other for the first time...or was that true?

"You are truly even more charming than your voice, my dear," exclaimed the elderly woman. "Let me look at you closer. You look very familiar. I do believe that I have seen you before. Could that be true? What is our name, child?"

(Curiosity began to invade Jackson's military thinking and he eased forward protectively, still out of sight behind the chair.)

"I am called Mimi, Milady, but my formal name is Princess Milinda."

"Milinda? Really? Of course, you're Queen Linda. I thought I recognized you. But it has been such a long time, Linda. Where have you been? How is Solem? Is he here with you?"

Mimi stepped back in shock. *She knew my mother. She knew my father. Who is this woman?* she thought, her mind racing.

Mimi's Conscience was confused as well but managed to whisper some quick advice. "She is old and confused, but not hostile. She believes that you are your mother. Be kind."

Mimi smiled and said what she hoped would do, under the circumstances. "Did you know my mother?" she asked gently.

"Heavens no, my dear, but I know *you*. It has been such a long time since you were here at Thorndike Manor, Linda. I heard that you were 'with child'. Is that true? That would be so lovely. You can bring both your children over to play with Willa. Is that little scamp, June, here with you today?"

"She has gone back in time, Mimi," whispered Cmimi.

Timidly, Mimi joined the discussion of the fragile woman in the big chair. "No, June isn't here with me today, Milady."

"Oh, that's a shame. Thorny loves little boys. He wants another child, you know, but he only wants boys." The lady sank suddenly into her chair with a frown. "He insists on it...insists!...But you said something about a Jackson person with you? Were you just joshing me? Did you really bring June with you?" The woman brightened and looked around. "Where is he? Is he hiding? He *is* a little scamp."

Mimi caught Jackson's eye as he stood behind in the shadows. Smiling a sympathetic smile, she motioned him forward. Reluctantly, he obeyed the gesture. This role playing did not fit his idea of a manly thing to do but Mimi's smile was irresistible. He sidled up next to Mimi, his scowl reflecting his discomfort.

Neither were prepared for the old lady's reaction.

As though she had been struck, the shaking woman immediately recoiled into a ball and shrieked into her hands with fear. "No! No! Who let you in here? You mustn't beat me anymore, Thorny. I see the Auras. They have come with you to hurt me again. Flo promised she wouldn't let you and the Auras hurt me anymore! She promised! Why did she break her promise?" The woman began to sob uncontrollably,

Jackson didn't know what he had done wrong, but instinctively he knew that *he* had caused this horribly fearful reaction. Quietly he backed toward the door. Mimi's instincts directed her to the frail, quivering bundle in the big chair.

"There, there," she repeated as she had with Flo only minutes before. "There is no one here to be afraid of, Milady. Look around you. There are no Auras. There is no one here to hurt you. You are safe. I won't let anyone hurt you."

Slowly, under Mimi's calming touch, the woman lowered her hands from her face and hugged the kneeling Mimi. "Is he gone?" she asked fearfully. "I'm so glad you were here to protect me, Linda. Thank goodness. But that was very wrong of Flo to let Thorny and his Auras sneak in like that. I must scold her. Could you please send Flo in when you leave?"

Again, she looked around but soon her eyelids began to flutter. "I will take a little nap now, if you don't mind," she murmured. "Just be sure to lock the door on the way out. I don't want my horrible husband and his Auras to sneak in on me." In a second, she was asleep.

Quickly Mimi rose from her comforting position at the woman's feet and tiptoed out the door to where Jackson was waiting. He was slouched against a wall, his head down and his shoulders slumped. "She thought I was my father," he said quietly. "She actually thought I was him." The young man shuddered. "The old dogs of the militia told me and Jorge that Thorndike was mean to his wife back in the day. I guess they were right. We thought they were joking."

"You should be proud. The old men of the militia must have suspected that you two were his sons already and trusted you. Isn't that weird? That old lady saw the resemblance right away. Who do you think she was, anyway...and what is an Aura?"

"I *know* who she is," announced Jackson, stepping away from the wall and shaking his head. Slowly he put his hands on Mimi's shoulders. His eyes turned dark and serious. His words were intense. "You must not say anything about this to anyone, Mimi. This door was locked for a reason. The person you saw is dead. She is only a ghost. She has been dead for twenty years...and I have no idea about Auras."

Mimi was dumbfounded. Was Jackson joking? The old woman was very real. She was only asleep now. Whoever she was, she was now content. " Don't be silly, Jackson. She only yelled at you because she didn't know who you were. Don't be upset."

"You're right. She didn't know who I am, but she *did* know who she thought I was, and she was scared."

A ruckus was erupting in the hall. Jackson and Mimi came to alert and hustled back up the dark hallway and into a more traveled part of the manor. Trying to look casual, the two hurried back to the kitchen, just in time to meet Jorge bounding through the back door.

DECISIONS AND DILEMMAS

"What's happening, brother?" Jackson asked quickly. "How did the big trip to the irrigation river turn out?"

Willa and Flo and Balynn were still deep in discussion but looked up with interest. They all knew now that Flo and Willa were sisters, which gave them an unexpected feeling of solidarity, each gradually perceiving the current situation and their past from each other's point of view.

As part of the recent sharing of information, Balynn almost felt like another sister, but they were anxious to hear the answer to Jackson's question and leave their current discussion for another day.

Jorge stopped for a moment. He had expected to see Flo and now he was being greeted by five people, his brother and four young women. He had news but what should he share with whom?

Flo waved away his concerns. "Say what you have to say, Jorge", she said, shrugging encouragingly. "Everyone here has been sharing secrets. Now we all want to know what went on up on the hill."

Jorge looked around and smiled, his warmest smile landing on the face of Balynn. (The intuitive females around her nodded knowingly to each other.)

Now Jorge began to enjoy being the first to deliver his news. "I was up there on the hill with the Royals the whole time," he announced with pride. "It looks like Father's plan worked. Most of the Royals headed for the dining hall just now to sign papers for the building of The Royal River Road just as he wanted.

"They sent me to find you, Queen Willa, so that you could sign too... and here you are, Your Majesty. Mission Accomplished."

Willa rose immediately and then stopped and looked around. She was looking at her family, some of whom she had never known as her family until today. It was overwhelming.

"I love you all," she whispered quickly, passing her gaze from face to face. "I want to stay and talk, but I will come back. We must catch up on our history. I have so much to learn. I must go sign papers now, but I will be back soon."

Jorge scratched his head at her remarks, but everyone else knew and understood. Balynn gave Flo a quick hug and followed Willa down the hall.

The room had been cleared of luncheon leftovers and made ready. Willa and Balynn arrived breathless but assumed a more formal demeanor as they saw all the others sitting formally at the table. The King's protocol was again in place as it had been at The Castle, but the young women quickly realized that some of the group were missing...June, Manda and Armando.

Willa settled herself next to her husband and reached to her right for a friendly pat on his knee. "I see that you have all gathered back in a formal way, Your Majesties and esteemed Lords," she began formally. "It would appear that this formal seating indicates that some decision has been made about the road. Would one of you care to tell me the result of your trip to the hilltop?" She smiled around the table and received smiles in return.

"Have *all* issues been resolved?" whispered Willa's Conscience. "You really need to find out about the slavery issue before you sign anything."

Drum spoke up instantly. "We have solved the issue of 'the need' for The Royal River Road. I believe that this need has never been in question."

Willa nodded. "Agreed," she responded amiably.

"Yes, indeed," continued Drum patiently. "The issue of irrigation has also been met satisfactorily as well. I believe Lord Thorndike was able to display the existence of a spring which will water the crops around it in the spring."

"And the crops of the nobles whose land is not near the spring?" Willa asked cheerfully.

"The nobles who own those lands know that they must find springs for irrigation on *their* land," barked Thorndike. "We will, of course, not begin filling in the smaller irrigation levees that depend on The River until all of those nobles take care of their obligations, daughter. We spoke of this on the hill. You should have come with us."

"You are a liar," murmured a wispy voice.

Everyone looked around. Willa had heard what sounded like a voice. Solem's hand found hers on his knee. He had heard the voice, too.

Luna glanced about the room. Even the Auras were perplexed but they were not deterred. "That must have been a stray breeze...or maybe even a trick," they whispered to Drum. "Don't pay any attention. Move on."

Thorndike ignored the voice, shuffled his papers and shook his head. The exhaustion of the last twenty-four hours began to tug at his patience.

He began again. "As I was saying, we have solved the irrigation issues, and the issues of labor and materials will be provided happily by each of the nobles on each of their portions of the road as the road crosses their land. My militia will direct them."

"With cruelty and force for the Lowlanders," the mysterious voice contributed again softly but clearly.

Thorndike jumped from his seat in a rage. His temper was now shredded. "That's it!" he shouted. "I will not be mocked in my own house. How dare you whisper and insult me here...*here* where I have taken pains to see that everything is perfect for the approval that our road deserves? How dare you?"

...I dare because I know you," the voice murmured again.

Strangely, Willa felt sorry for her father even though she felt that the strange voice was speaking the truth. She knew how hard he had planned and worked for this moment. Now he was being plagued by a mysterious voice. Was it the wind from down the hall? Was it the creaking of the rafters in the old hall? Willa had no idea.

"Don't worry, Father," Willa said. "I hear what sounds like a voice, too. It must be the wind or someone's idea of a joke." She reached over and patted Thorndike's arm.

Her Conscience said nothing.

The huge man looked down at his young daughter suspiciously. He almost recognized the voice. Was it her? Slowly he sat down with a slight look of confusion.

"This is all very strange, Lord Thorndike, King Solem said politely. "You must have a wayward servant in your home who thinks they are a trickster. You should investigate that.

"In the meantime, however, the time is getting late." He turned to Willa gratefully. "Thank you for your calm, my dear. Let me just conclude our summary. I have decided that Lord Thorndike and Lord Drum have sufficiently complied with my concerns about the new road, and, with your acceptance, the necessary approvals will be signed here and now by those who wish to sign.

"These signatures will allow work on the road to begin soon, due to concerns with the weather. Lord Drum has assured us that we are mistaken about the issue of slavery. Prince Armando will add his signature later, after he and Manda and June assure themselves that there is no slavery taking place down at Drumsworth. They have already left to do so."

Willa received the approval sheets from Thorndike and read them quickly. The agreement was short. She could see that it relied heavily on her father and his militia for supervising and assisting with the labor.

Drum wouldn't be using a militia. He would be using Lowlanders that lived in his area. The agreement indicated that these Lowlanders were already working on his fields. (Were these the ones that Manda and June had thought were slaves? She could see that this might have been their easy mistake.)

Happily, Willa turned to Solem." If you think The Royal River Road is a good idea, I do, too," she announced, dipping a waiting pen into the inkwell before her.

Solem and the others followed. The Royal Agreement for the Royal River Road was finally complete and officially signed.

Thorndike, himself, could hardly speak. He had accomplished his goal and he reveled in his achievement. After the night before had left him with such feelings of inferiority, he had risen above it all. And it was all *him,* he believed...*him and his ingenuity and drive.*

"...and the drive of the Auras" ...the tiny voice whispered, but no one seemed to notice.

Slowly the signed copies passed from one to another. As each person added their signature, they gathered their personal copies together and stood, passing the final signature page to king Solem to keep and file in The Castle Archives. Then they stood and worked their way toward the door. Everyone was tired. It had been a long, eventful day.

THE RIDES HOME

Luna exited first. She had the longest way to go and she was going to be driving her own carriage. Drum was in a very good mood and offered to allow one of his entourage to do the driving for the beautiful Queen. She smiled upon him, surprised at his generosity but declined.

Drum could afford to be generous today. Although Thorndike had maneuvered today's accomplishment to its successful conclusion, Drum considered this success to be his, alone. *He* had put all the players in today's position with his months—years--of planning. *His* gathering of the noblemen in the early morning hours today had sealed the King's approval of the plan. Thorndike had been helpful today, but Drum and his Auras concluded together that this accomplishment was *Drum's* and he was still in charge.

It had been long day. Quietly Luna drove her carriage through Thorndike Gate. She was not altogether satisfied with today's final decision, but at least the approval had been signed by all but Armando and peace seemed to be restored.

After only a few miles, however, Luna saw plainly that work had, indeed, already begun on Thorndike's part of the road and was still going on now, even in the dark. What she was seeing were huge boulders being hauled out of The Riverbed and rammed snugly into position where small levee openings had once been...mostly by Lowlander workmen.

Already, there were mammoth rocks jammed along the sides of The Riverbank with still more rocks and dirt topping them off to form a knee-high platform on top. The new levee was tall and solid and already stretched far ahead before her carriage wheels. This work had obviously begun days ago.

The little voice that had called Thorndike a liar and mentioned Lowlanders had obviously been right. He had been lying about this. What else had he been lying about, Luna wondered.

"Why do you think they were they so determined to build this road so fast?" Luna wondered as her carriage rolled by on the new surface.

"There must be another motive beside smooth travel, for Thorndike to be pushing his farmers this hard. They look like slaves." Cluna mused.

Luna would talk to Armando about what she was seeing when he returned from his slavery fact-finding mission. She now prayed that he had been mistaken about the slavery he thought he had seen. Maybe they had been just hard workers like these farmers.

But Armando was protective of his people. And he was inciteful. She had always trusted his judgement. If he was right and Drum was lying as well as Thorndike, The Royal River Road Agreement might be a big mistake. Was that what the strange voice she had heard was warning them about?

*****

The Drum carriage exited south from the Thorndike Gates with satisfaction...and concern. Balynn was asleep before Drum's driver had cleared the main gate, but Drum was wide awake. He was watching the workers on both sides of his carriage with pleasure. He had seen the work yesterday but now, with his precious note of Royal Approval tucked firmly in his waistcoat, he had something else to worry about.

Down at Drumsworth, he had had his Lowlander slaves locked in their prison after work each day while he was gone, and he had been gone now for two days. Four of Thorndike's men were guarding the Lowlanders, but if anyone found them in the prisons...

Prince Armando, June and the irritating Manda were already on their way south. Something must be done right away. Drum decided immediately to send Drumsmen from his entourage ahead to Drumsworth to implement the "no slavery here" declaration he had made to the doubting Armando.

These Drumsmen would have very specific orders. They were to drive most of the slaves that were locked in the prison west to work on Drumsworth fields...which had been their former jobs before they were building the mote. That part of the plan should be easy.

A few of thr most cooperative slaves from the prison were to be washed in Drumsworth Lake and redressed to look like Lowland construction workers on the mote that Drum had recently labeled 'Lowlander Lake' (since it *was* on Lowland territory).

Drum sleepily munched on a roll he had stuffed in a trouser leg at Thorndike's luncheon. The issue was covered.

*****

The day had been eventful for Willa to say the least. A beloved playmate had become a sister. Two young men that she had never really known before were now her brothers. How could that be so? Why had she never even suspected that these revelations were true?

Solem had fastened his horse to her carriage so that he could ride inside with her and Mimi. Mimi drifted off into a mixture of romantic dreams of Jackson and the swirling face of a tiny, frail woman silhouetted alone in front of a cozy fire. Every few minutes she would issue small questions and smile in her sleep..." Who is she?'..." Why?"

Solem smiled at his daughter. "At least Mimi had an enjoyable day today," he observed as he gathered Willa's hand tightly into his own. "I think she fancies the young militia man that 'escorted her' to Thorndike's Manor this morning...and there were two of them that looked alike. Did you notice that, Willa? They were twins. That was very interesting, I thought. Did you remember them from when you were growing up?"

"Not really," she answered. Willa was uncomfortable. Should she tell Solem that these militia men were her brothers? What would he think if she did?" Would he think she was strange because she had siblings that she had never told him about?

"It's the *situation* that is strange, not you," reflected Cwilla. "Your husband already knew that your family was a bit odd, and it didn't seem to matter to him. My guess is that these new family members won't matter to him, either."

"I was never allowed to go near the barracks of the Militia when I was growing up," Willa began.

Solem grinned a sideways grin. "Smart move of your father's part. I wouldn't have let you near a building full of men, either."

Willa smiled. "Flo and I had a nice talk while you all went up to see the irrigation river," Willa began slowly. "And guess what she told me that I didn't know."

'That she thought you were beautiful," said Solem with a squeeze of her hand. "I agree with her."

"Not exactly, Solem. She told me something that I never knew, even though the two of us played a lot when we were young. She told me that...she was my sister."

Solem turned around slowly and looked at Willa closely. "You mean 'sister' as in 'close friend' or..."

"I mean sister as in *a real, blood sister*, Solem. Flo is my real half-sister, and I never knew it. I am not sure that *she* even knew it until we were older and by then, she just accepted it.

Her mother, 'Mother Alana' really acted as mother to both of us...when she had the time. Mother Alana was a hard worker and *my* mother...my mother died so early. I guess Flo and I just accepted life as we found it."

Solem continued to look at Willa through the dusk that had now settled into the carriage. The evening had grown chilly. Solem put his arm around Willa's shoulder and drew her close. He could feel that she was shivering but probably not because of the cold. Perhaps she was thinking what he was thinking. "Your father has had many secrets in his life, hasn't he Willa?" he asked as though thinking out loud.

"And that isn't all that I learned today." Willa sat up straight and looked straight into Solem's concerned face. "I also learned that I have two brothers! They are the twins that came to fetch Mimi and Balynn this morning! How could I have a sister and two brothers and not even know it?"

Willa began to sniffle into Solem's chest. "You have married into a family that is so dishonest and weird that they don't even admit their own children. How terrible is that?" Solem's answer was to pull her close, again.

Mimi sat up on the other side of the carriage. "What is all the noise about?" she yawned sleepily. "I was just having the nicest dream. Are we home yet?"

*****

 Armando, June and Manda had left directly from the hill with a mission. They would check the so-called slavery at Drumsworth and report back to The King.

They too, had noticed the intense labor that was taking place on the bank of The River that ran along Thorndike property. The building of the new road had obviously not waited for royal approval.

"That man is such a liar," observed Armando. "I saw a bit of this on the way up but he insisted that he was just repairing the old road. I doubted him at the time. I should have doubted him more."

"What I see is a whole bunch of lies that Drum and Thorndike have been telling from the start," contributed Manda. "I know for a fact that the farmers in the center of River Kingdom have no idea how they are going to irrigate their crops when the new road blocks their river water. They don't all have a 'hilltop spring' like Thorndike."

"I hadn't thought of that," said June thoughtfully, "but you're right. And I don't think Drum has a spring either. It's dry down there. Why is he so eager to build a road when he won't have enough water for his own crops?

"And why is he so excited about the enlarged pool that we all know he is digging for himself," Armando wondered aloud, "no matter what he is telling your father, June. And Thorndike's new drive to get the new road done quick doesn't make any sense to me, either"

They rode on in silence. "This levee system is pretty impressive," June said they passed over a bridge that spanned a particularly large ditch. "This is what you showed Willa, right Manda? She saw what I now see. The new road will wipe out most of this existing system. It's not like we don't need a good road from north to south, but someone should have figured the irrigation issue out first."

Suddenly, the group heard hoofbeats behind them. Knowing that they were riding slowly, they eased themselves to the side of the path to let the approaching riders pass by.

In a cloud of dust and debris, the riders plowed past them, whipping their horses for more speed. With no acknowledgement, the riders hurtled past and disappeared around the bend ahead.

"Did you see who that was?" Manda coughed into the lingering dust cloud.

"I couldn't see a thing," sneezed Armando, "as fast and rude as they were. Do you think someone is chasing them?"

"I think not," grumbled June. "And I did see the markings on the horse that nearly trampled me. It had Drum markings. I saw that very plain."

"Drumsmen on horses going by in a hurry? Drumsmen too much in a rush to be civil?" mused Armando as he resumed his ride. "You don't suppose they were running from some sort of trouble back at Thorndike's, do you?"

"They would have stopped if they needed help, Daddy, don't you think?" Manda started forward again. "But they were certainly in a hurry about something."

June's horse had not moved. "They weren't running *from* anything, Your Highness," June declared. "They were running *to* something. My guess is that we will not see slaves when we reach Drumsworth Castle. We will only see 'workmen'. The slaves will be somewhere else. The Drumsmen that just passed us will "fix" the slavery situation so that slavery no longer exists for us to see."

Prince Consort Armando was a proud man. June's announcement struck him in the heart. With one movement, he pulled his horse around and drew up to June.

"You are a thoughtful man, June! Why didn't I see this? What kind of a blind leader am I? You are so right!"

Manda turned back as well. She didn't say anything, but she also knew that June was right. "June is so smart. Why didn't I think of that?" she thought. The trio looked at each other in anger and despair.

"We need to arrive before the Drumsmen do," Armando said quietly, as though speaking to himself.

Manda looked around. "Think, Manda," whispered Cmanda. "You've been here before. I recognize where we are and so do you."

Suddenly Manda recognized her surroundings. She knew exactly where she was. As a young recruit would speak to an honored leader, Manda addressed her father directly but with respect. "I know where we are, Father," she said. "There is a narrow, rocky shoal up ahead where we can cross to Urhonordo side easily. From there I know a shortcut to Drumsworth. We can beat Drum's team to where the slaves are if we hurry."

June stared in disbelief at the father and daughter. Manda is so smart. Why didn't I know something like that?"

Armando nodded in approval. Manda directed her horse forward. The shoal appeared ahead as advertised and they crossed. With Manda in the lead the three climbed Urhonordo's bank and started off at a gallop. No one spoke. They all knew where they were going and why.

THE DRUMSWORTH RESCUE

Paulina was a Lowlander. She had been a house slave at Drumsworth Castle since she was a child. There were other servants at Drumsworth that were not slaves, but they were men...and white.

Drum seemed to like his workers inside to be Lowlander women who had been brought to Drumsworth at one time or another and not allowed to leave. Pauline had just recently learned a new term for this. It was called, "Slavery".

But today was a new day. Paulina, the head of the inside slaves, oversaw getting the manor ready for Drum's return.

The current men who had been recently guarding the workers on the mote were much more cruel than the twins that were usually there. They had been working them longer hours, beating them often and feeding them almost nothing.

Paulina didn't like to watch the work taking place on the mote when the current bosses were in charge. They were true 'slave masters'. She had learned that term from one of the twins who seemed to be unhappy with his job. Lately.

They were the ones that called themselves 'slave masters'. They were the ones that taught her the new word 'slave.' People who were forced to work and not paid...and held captive were "slaves." They had told her that they were being 'slave masters' and that she was actually a slave, too. They seemed to feel very bad about it.

Pauline didn't think her life or that of those who were working on the mote were good lifes, but they had all accepted it. They didn't have a central government. The only time someone told them to do something in The Lowlands, it was for their own good. When they were told to work in River Kingdom, their bosses had told them that they *must*. It was for their own good. They just didn't think they could say no.

And it was true. If they didn't do what they were told, they were beaten. Working *was* for their own good.

Since the early days when Lowlanders' migratory civilization was discovered by explorers from River Kingdom, the citizens of the Kingdom had told Lowlanders what to do and they had obeyed. In these days, however, many good people asked for services rather than demanding them...paid Lowlanders instead of offering nothing in return for their labors.

Pauline now understood the finer points of this slavery idea and she knew what she was seeing at the mote. It was slavery.

Whatever it and the miserable conditions that her fellow Lowlanders were enduring were called, the situation was inhumane. Every evening lately, she had secretly talked to her friends who were always locked in the prison out back. Paulina had spent time out there, herself, when Drum had punished her for what she had done wrong inside. She knew what a miserable place the prison was, and she tried to help now by bringing the prisoners food whenever she could manage to sneak it out to them. Soon, however, she felt that she should be doing more. Her Conscience was bothering her. These were her people. But what could she do?

The mote's drawbridge had finally arrived yesterday, hauled down from forges that were usually making harnesses and wagon wheels up north in the town. Drum would be thrilled with this, Paulina knew.

But the drawbridge had not been put in place without a human cost. The pieces that needed to be joined together to make the contraption were heavy, constructed of forged metal pieces and huge, oil-soaked timbers. Each piece needed to be joined to the piece before it and the whole thing was propped over a span of logs that were unevenly covering a slimy ooze of mud. All as wet. All was slippery. All was dangerous.

The Lowlander workers had suffered broken bones and gashes in their flesh as the slave masters had shouted, cursed and (yes) whipped them into lifting and carrying loads that were too heavy for them. But the drawbridge was at last in place and complete.

The last section of the mote itself had just been completed between the mainland on the west and Drumsworth Castle. The ground beneath the new drawbridge was still mostly a gully full of mud and rock but the mote, the lake and The River were now joined. The River water had begun to flow into the mote. Drumsworth Castle was now an island, surrounded by water.

The final piece of the drawbridge had been labored into place at midnight. The exhausted Lowlanders had been locked away each night with only a pot of gruel to share after their day's exhausting labor. Paulina scurried through the bushes on the east side of Drumsworth Manor from the secret door that only she and a few others of her fellow workers knew about.

Just inside was the stash of food, water and clean clothes that she had been gathering to take out to the prison. *They will need everything that I have tonight for sure,* she thought. *I should be doing more. How can I let my people be over-worked, abused and not even be fed enough to keep them alive without doing something?*

Paulina's Conscience perked up immediately. "I thought you would never wake up from your own worries enough to think of your brothers and their slavery."

This is slavery, isn't it? Paulina thought. *Just like the twins said, Lowlanders are slaves here at Drumsworth and that's wrong. I guess that slavery has gone on for so long that my people and I got used to it.*

But the slavery of the men working on the mote is worse than my slavery. I have not been beaten or starved. I have not been ordered to do work that is as cruel and mean as my brothers have had to endure. I need to do what I can. I must tell The Great Leader. I must go tonight. Drum is not here. No one will know that I am gone. That's it! I will go tonight...after I take my stash to the prisoners.

Paulina snuck across the front lawn to the new draw bridge. This was the only route to the prison out back, now. Cautiously she started across, keeping low and out of sight. The 'slaves' and 'slave masters' alike were tucked away for the night, the slaves in their dank prison, their masters probably taking advantage of unrestricted food and drink inside Drumsworth Castle.

Paulina struggled over the bridge. Her arms were full. Her basket was heavy and the dry clothes she was bringing were draped over her head so that she could barely see.

For this reason, it was not a surprise that she had no idea who grabbed her and pulled her to the ground under a bush on the other side of the bridge!

She tried to scream, but someone was sitting on her and a large, heavy hand was over her mouth. She twisted and kicked. Being a hard worker, she was strong. She rolled and bucked, determined not to be taken without a fight.

"Be quiet woman and don't wiggle. If you promise not to scream, I'll tell my daughter to get off your stomach and let you stand."

Slowly, Armando removed his hand from Paulina's face and waited. She didn't scream. She knew the voice.

"Great Leader? Is that you?"

Manda pulled the cloth from her face so that Paulina could see and Manda got up from her spot on Paulina's middle. Quickly she put her fingers to her lips whispering, "Shh. Don't say anything. My father and June and I are not here to hurt you. Shh... Follow me."

Without hesitation, Paulina followed the three shadowy figures and disappeared with them further into the brush. When they were far enough away, the three turned and faced their captive, still hanging onto her basket of foodstuffs and other supplies. They knew she would be confused.

'Great Leader," she whispered. "Why are you here in the dark of night? Am I in trouble?"

"Not at all, young woman, but we all may be in trouble shortly. Drum's men will be arriving soon. Do you know the way to the prison where Drum's workers are sleeping?"

"You mean 'Lord Drum's slaves', don't you?" asked Paulina using her new knowledge proudly.

Manda and her partners smiled together at once.

"Indeed," answered Armando. "Are your brother Lowlanders being paid for the work they are doing on Drum's Lake?"

Paulina wondered if these three were here to give her the help she had just decided to ask for. If they were, how did they know that she needed help? It was almost like magic.

"No Lowlanders are being paid, Great Leader" she answered. "Our brothers are all locked up. That means that they are truly slaves, doesn't it? That's wrong, isn't it? They are slaves, right? I think I'm a slave, too but that doesn't matter. These brothers are *abused slaves* and that is *very bad*, don't you think? That's where I was going just now."

With innocence, Paulina's chatter ended in a new realization. *She* had obviously been forced to work for Drum without fair treatment and she was just learning that she this was slavery, too. She had accepted it. She had traded in her life as a free nomad of The Lowlands for control she had thought she must accept. Feeling like a 'lesser person', she had not questioned it.

Armando was stunned by what had been happening here at Drumsworth, right under his nose. He was ashamed of his lack of understanding.

"All Lowlander slavery needs to be eliminated," Armando's Conscience acknowledged, "but don't get lost in guilt about your past faults now. You need to focus on *tonight.* Drum's men will be arriving shortly. Do you want to scold them for what they have done...or fix the current situation that you wish you had fixed days ago?"

"We need to hurry. Father" Manda whispered. "What do you feel we should do right now?"

June spoke to Paulina directly. "I see that you were taking supplies to the prison. Am I right?"

Paulina nodded. June took the heavy basket. "My name is June. I am a friend. Please let me help you take this where you were going. We need to help your brothers."

Armando's Conscience was right. He pushed back the faults of his past to concentrate on what should come first tonight. "Lead us to the prison, young lady. If you would like to see your brothers freed, lead us to them."

Again, without hesitation, Paulina lowered her head and dove through the underbrush that lay between her and the prison. She led them past the gurgling mote where The River was now lapping quietly at its sides and back behind the additional foliage that partially hid the prison from view.

Groaning could be heard from inside. Obviously, the injuries of the day and hunger were taking their toll.

"Paulina, is that you?" came a voice from the dark within.

"Come quick."

"Did you bring water? Hurry."

Paulina stopped and reached up for a key dangling from an outside beam of the prison. She rushed to its doorway. A nauseating smell oozed out as she wordlessly turned the lock...but the door was held shut.

Armando took position in front of the prison door and automatically, the inmates cowered back in fear, heads down, punishment expected from the outline of a man in front of them. This was all they could see in the darkness.

"My brothers, please forgive me for taking this long to do my duty to you," Armando stated plainly in the Lowlander language.

Slowly the heads rose hopefully as Armando's voice and words penetrated the gloom. The downtrodden within stared and listened in silence. The man in front of them was not a slave master. He spoke Lowlander language and he was on the other side of the bars that enclosed them.

"You will not understand all that I say tonight but understand one thing for sure. I am here to open this door and set you free. Your obligation is to run south as fast as you can, no matter how tired and wounded you are. You must leave no one behind. More of your evil slavemasters are on their way and you must flee deep into Lowlander territory to avoid recapture. "Please do not hesitate to do as I order. Run and take your infirm with you. Run back into our homeland and disappear. My love goes with you...Now go!

Armando opened the creaking, ancient door and folded it back, out of the way. The first men out stepped forward slowly, hesitant and afraid. They looked around in fear.

Armando looked at them all but did not smile a welcome. "Go! Don't stop here! Run for your lives!" ordered Armando gruffly. "You are free! If you are caught, I cannot protect you! Get as far away from here as you can and find support from your countrymen and each other. Stay away from here!"

The Consciences of the Lowland men jumped into action, making sure that their humans ran, carried and hauled each other quickly away from the prison. "Listen and obey," the Consciences ordered. "You are men of quality, no matter how you have been treated. Save yourselves and your brothers while you can!" (These Consciences were just waking up to the strength they had not known they had. Amazing.)

There had been twenty or more men in the prison, but they were native to the territory to the south. Within only a few minutes, they had all disappeared. Paulina and her basket and clean rags disappeared with them as they all whispered their gratitude into the night.

At last, they were all gone. It happened so quickly that the small 'freedom team' was almost at a loss as to what to do next. But they could not afford to hesitate themselves. Hustling back to the drawbridge, they found their hidden horses and rode back east past the now empty prison and east into Lowlander territory.

They wanted to see the reaction of Drum's men who had come to create an 'acceptable picture of Lowlander workmen' for their benefit...but not from up close.

In the moonlight, the three 'midnight raiders' continued the way they had arrived and then climbed to the bluff of Urhonordo territory. This was where they had watched the mote construction only a few days before. So much had happened since that it seemed a long time ago.

Now, however, they had the satisfaction of knowing that they had affected the change that they had all wished they could accomplish earlier. What would they see tonight?

THE DRUMSWORTH 'INVASION'

The advance team of Drum's entourage arrived within minutes. Armando and his young companions watched as the Drumsmen stared with pride at the new drawbridge that they had never seen before.

But they did not cross it. Instead, they made the discovery. The prison that they were supposed to empty was on *the other side of the mote*. This was new.

Now they detoured around the back of the mote toward the prison with aggravation. "We should have been told about this," one hollered.

"What do we need a stupid mote and a drawbridge for, anyway?" yelled another.

They, too, were on a very strict time allotment. They needed to convert prison slaves into voluntary farm workers and do so before anyone arrived to see it happen.

"Ridiculous," they muttered to each other.

"A waste of time."

"Part of a show for the stupid royals."

There was a bit of laughing and joshing with each other over the new situation but, in only a few minutes, the laughing stopped. The prison that they had expected to empty was already empty. Its prisoners were nowhere to be found.

The bewildered Drumsmen rode around the prison searching in an obvious quandary. The observers on the Urhonordo bluff above them smiled in satisfaction and headed north.

If the three invaders had waited a minute or so, however, they would have enjoyed the behaviors of the Drumsmen as they stormed across the drawbridge toward the front door of the manor.

There were four men inside the manor enjoying the fact that they were the only ones there except for the servants. The inside servants were all Lowlander women. The men had noticed as they did every night when they were inside by themselves, how good looking these Lowlander women were.

To be blunt, they had each chosen one screaming, protesting, fist pounding woman and retired to a comfortable bedroom with a bottle and thoughts of the pure enjoyment that this activity would bring.

Their Auras were enjoying the female benefits that had been obviously left for them. An enjoyable evening had just begun when they heard a loud knocking at the front door.

Everyone was prepared to ignore it, but it kept on and grew annoyingly louder. Since the female servants had all been spirited into bedrooms, the half-naked man closest to the door decided reluctantly to tend to it, himself.

There he found four fist waving men that he recognized as Drumsmen, but there was no Drum. That, at least, was good news. "What do you want, idiots? What are you doing here in the middle of the night, pounding on the door? Go away."

"Let us in!" growled the first door pounder as he stormed through the door. "Let us in unless you want to be yanked out into the mote without your boots on. "

The Call of the Auras

The door opener did, indeed, have little on but a flimsy undergarment which left little to the imagination. The last three at the door hustled inside as well and locked the door behind them.

The Drumsmen were obviously upset, and all began talking at once, arms flailing, spit spraying from their cursing mouths in a completely incoherent jumble of possible explanations that they themselves didn't understand. "The Prisoners are gone!" the Drumsmen finally managed to say together.

"The prison is empty!"

"The Lowlanders are missing!

"You idiots left the door open!"

"Heads will roll. Who was the last one out?"

With quick recognition of the fact that their slave masters were in trouble, their Auras leapt into defense mode. How could this have happened? The Auras didn't know or care. Right now, defense for these slave masters was essential. There had to be a reason for what had happened that was not their fault.

With immediate, encircling action, the Aura's surrounded the slave masters as they stumbled out of nearby rooms to provide the excuses needed.

"We must've been invaded!" That's what happened," stated the first slave master."

"That must be what happened, for sure."

"Sneaky devils. We didn't hear a thing."

"Not our fault, though. If the slaves are gone, somebody stole 'em. By the way... why are *you* here?"

The Auras for the Drumsmen had an answer for that blame shifting question.

"We were sent by Drum to set up a sweet impression for a bunch of royals, but what do we find when we get here? Drumsworth was invaded while you numskulls were in here having a good time!" accused one Drumsman as he saw the women peeking out of bedroom doors.

The Auras for the slave masters picked up on the invented 'invasion excuse' and enhanced it with a useful embellishment.

"It was Lowlanders!" yelled one.

"Must've been the Lowlanders comin' back for their own," agreed the first Drumsman.

"They know how to be sneaky. That's the way they are."

"Sure enough! That's what happened, alright!

"We've been invaded tonight by Lowlanders!"

The four slave masters now joined the Drumsmen to put together what they now decided *must have* taken place. It was a perfect conclusion...and left everyone totally in the clear.

After the success of the day, the Lowlander slaves had been carefully stowed away in their prison with their gruel.

One man remembered getting the gruel and lugging it over the new drawbridge to the prison.

Another remembered taking the prison key off the wall.

Another had put the gruel inside the prison, and another had locked the prison door, tight.

They had all checked everything. It was clear that they had all "done their job."

The Drumsmen were content. They had simply "arrived to find an empty prison. They had done their job, too."

That left only one logical conclusion for men who had treated Lowlanders as slaves and had wasted no time in actually knowing or caring about any them. It was just as their fathers and the noblemen around them had always said; Lowlanders couldn't be trusted.

In addition, Lowlanders were obviously now 'on the march'.

The surrounding Auras of Drumsworth were ecstatic. Not only had their suggested excuse worked admirably but their humans had even made up an additional conclusion on their own.

And the Auras had another idea: If the 'invading Lowlanders' were 'on the march', then these loyal Drumsmen needed to warn someone. It was their duty.

"We need to tell Drum that the Lowlanders are on the march," yelled one of the Drumsmen.

"Tonight?" croaked another.

"I think so," sighed a third, reluctantly.

"Let us get our britches on," sighed one of the slave masters.

Together the Drumsmen and slave masters now geared up and rode north as a group, intent on alerting Lord Drum to the invasion that had occurred as well as warning him of what they were now sure was to come… a full-scale *Lowlander Invasion!* Aura leadership was proud.

MIDNIGHT NEWS- URHONORDO

It was midnight when Armando and his young companions arrived at Urhonordo Castle. Queen Luna had, of course, reached there ahead of them. The travelers filled her in on everything they had seen and done at Drumsworth. Luna was not surprised.

"There is a time when I should listen to my intuition as well as my Conscience." she had said sadly. "I felt you were right about slavery from the first time that you told me, dear," Luna said with a hug for the weary Armando. "You are always so correct in your observations. You never jump to conclusions. My Conscience was so hesitant to decide important matters without visual proof. I'm sorry that I needed that proof.

"And I must admit that you were clever, as well, Manda. Taking that shortcut through Urhonordo was brilliant. Luna sighed. "I wish I had been down there to see the looks on their faces when Drum's men found the Lowlanders missing."

"It was really fun to watch, Mother." Manda exclaimed. "You would have loved it."

"You should have seen them." added June, laughing. "When Drum finds out that his precious 'slaves' are gone, he's going to be furious. They were his farm laborers, too, working as slaves in both places. Now I don't think he has anyone to do work for him."

"He'll probably just go down into The Lowlands and round up some more of my people," Armando sighed sadly. "Tonight, we found out that many Lowlander people in south River Kingdom are being treated as slaves and they don't even know to resist it.

They don't think they have a choice. They haven't learned how to tell bosses 'no'. They're probably afraid too…maybe afraid of being locked up in that prison. I had no idea that was still going on down there. I have a lot of training to do with my people."

"Schools perhaps?" asked Luna softly.

"Only if they want them," Armando answered. "They need to know that they are their own bosses."

Manda was impatient. "Fine...schools or training might be a good idea in the future but not right now. *Right now,* everything is terrible. We have an *approved* road that is being built but will *hurt* farmers. Actually, it was being built by liars and cheaters *before* it had approval. Thorndike and Drum don't respect their King but they already got their approval signed for their road. And above all, now we have proof that they are *slave holders*. Great!" she fumed. "Just wonderful."

Armando smiled. "I was wondering if one or both of you would like to inform the King about the slavery we discovered at Drumsworth and what we did about it. You can also tell him that I will not be signing my name to the Royal River Road, thank you. I would appreciate it."

The two tired young people looked at each other. "I'm going on home right now, Sir," said June quietly. "I will definitely deliver your message."

"And I'm coming with you for back up," Manda announced. Her parents rolled their eyes and smiled.

"But that's not all," Manda continued defensively. "I need to find out what The King is going to do about it. We didn't go through all this discovery just to leave things the way they are. This road needs to be cancelled."

June grinned. He felt the same way.

After a quick nibble of bread and cheese, the pair were again on the trail north. They were tired but alone...and not in a hurry. Their horses seemed to sense their need for this quiet time.

"I wonder what it felt like to go back to Thorndike Manor for the first time as a Queen," Manda began. "I don't think Willa ever got along much with her father. She wasn't the boy he wanted. I think it would have felt strange, going back to a place where you were never happy."

"Willa is not stuck on the past now that she's a queen," June answered. "She's smarter than most women might be in that position. I think she was good to people who worked for her father, though, so they were glad to see her, and she was glad to see them. That's what I think I saw."

"I saw that, too," Manda responded. "She was good friends with that servant girl, Flo, when she was a little girl, I think. They hugged when they saw each other. Flo took us all around the manor. She was nice. She was almost like a sister to Willa, back in the day."

"I don't think old Thorndike let too many people in to visit after Willa's mother died. I think she was pretty lonely," June agreed. The two rode on for a while in silence.

"You like Willa a lot, don't you?" asked Manda quietly.

"I do," June answered honestly. "She is just our age, right? To be truthful, sometimes I forget that she's my fathers, wife."

"Me, too." Manda stared down the road. "Mother and I thought we might not like her because she was so young and we liked your mother so much, but we now we love her."

June stared straight ahead. "Me, too," he said quietly.

Manda didn't say a thing. She knew she had heard June correctly. He had admitted that he loved Willa. *Did June really say that?* she thought with surprise.

"Yes, he did," whispered Cmanda. "Are you jealous?"

Yes.....

"What are you going to do about it? It sounds like his heart belongs to another woman."

Yes, thought Manda stubbornly. *And she belongs to another man. That's not fair. He's sad and now I'm sad.*

"I guess that you are admitting now what I have thought for a long time," whispered Cmanda. "You *really like* June, don't you?"

Since we were twelve, Manda thought sadly.

They rode on in silence, but June was thinking his own thoughts.

I just admitted that I love Willa, June thought with embarrassment. *Why did I say such a stupid thing? Manda must think I'm an idiot.*

"I don't think so," Cjune whispered back. "Manda is smart and she has a gentle heart."

I know...She's the smartest girl I know.

"Why don't you tell her that you think that?

Then she would really think I was an idiot, June thought sadly.

"You *really like* Manda, too, don't you?" Cjune murmured.

Hmm...I guess I have liked her since we were twelve, June suddenly realized.

"Tell her."

No.

They rode on for a while in silence.

Manda decided on another subject. "What do you think your father will say when we tell him about the slaves? Don't you think he will cancel the road approval that he signed?"

June had been wondering the same thing. "I don't think he will," he answered slowly. Manda would probably like this idea, but he had watched his father all his life and his father was a man of protocol.

"Father signed a valid document. He will feel that his signature should not be taken back."

"I hope you're wrong," Manda sighed. "But I know your father is a man of protocol. He signed a document and now it's binding, right?"

June grinned at Manda. "Exactly," he groaned.

The ride north continued quietly. It was late and there was more thought in the two weary travelers than there was conversation. In truth, the horses found their way from Urhonordo to River Kingdom with very little guidance from their sleepy riders. The morning would be a better time for important discussions and probably more understanding of everything that they had thought tonight.

MIDNIGHT NEWS –RIVER KINGDOM

At the same time, on the River Kingdom side of The Great River, the ride to the north was far from sleepy. The Drumsmen and slave masters were spurring their horses northward at a fast clip, each rider intent on breaking the news about the missing slaves to Drum in a way that showed their own innocence.

Each would emphasize 'the Lowlander threat' they had all just manufactured with the help of their Auras. *No one* had allowed the Lowlanders to be freed, of course, but *each* now thought they 'knew' what the Lowlanders were up to. It was time to prepare for the inevitable 'invasion of these sneaky people' that had only begun tonight...but they *knew* more were coming.

The oldest of the Drumsmen, a big man by the name of Morland, nominated himself to the leadership role. He had the story straight. He would be their primary spokesperson. After nearly an hour of dust and sweat, the group spotted Drum's carriage up ahead and pulled up to wait.

The Call of the Auras

With comradery born of their current need to stick to their same story, they took a stand in the middle of the road and Morland took his place at the forefront. They all knew Drum. They had all worked for him all their lives. None of them liked Drum, but they hated Lowlanders...as they had been taught. They knew Drum would feel the same.

Their Auras were not only with them, but ready for action. This was a bonus situation that all Auras hoped for. The conditions were perfect for the forces of Auras to make a huge and unexpected gain in their surge for mayhem in River Kingdom. This "invasion of the Lowlanders" was a gold mine.

Drum's returning carriage now drew around the bend, traveling at a leisurely pace. The occupants were asleep. The driver was nearly asleep, himself. Squinting through his sleepy haze in the dark, he was at first startled. Then he saw the Drum markings on the horses in front of him and drew up.

"What say you?" the driver called out. "I see that you are Drumsmen. What say you? Why are you here? Don't you know it is the middle of the night?"

"Indeed, we do," growled Morland. "But we bring news of the greatest importance to Lord Drum. I must speak with him...now."

The driver moved the carriage closer, scowled and leaned over to Morland. "He's asleep," he whispered. "You know how he is. Are you sure you want to do this?"

Morland looked back at his comrades for assurance. Receiving the appropriate head-nodding, he spoke up clear in the night. "Awaken Lord Drum. We must have a word with him now. It is a matter of River Kingdom security."

"Well said," offered his Auras. "Nice formality...nice gravitas. Stay strong."

Drum stuck a scowling face out of the carriage door. "What do you idiots think you're doing out here in the middle of the night? Have you any idea that this interruption is inappropriate for a person of quality such as myself. You are fortunate that I do not have my men slaughter you where you stand for this invasion of my privacy."

Morland blinked but held his own. "We *are* your men, Milord. That is why we stopped you...to warn you that you may be in danger."

Now it was Drum's turn to blink. "Danger? What do you mean, 'danger'?" He glanced quickly from side to side. *It is the middle of the night and we are in the middle of nowhere,* he deduced in a flash. Quick reflexes switched his thinking into an authoritative stance, and he stepped boldly out of the carriage. *This is not a time to show fear,* he advised himself.

"Indeed", echoed Drum's Auras with the same advice. "Stay strong."

Morland climbed down from his winded steed, lowered his head and took a knee (as he knew that Drum demanded). "We came to warn you that you are in grave danger, Milord," he announced to the dirt.

"Get up, fool! Say what you have to say as a man. Speak up. I thought I heard you say that I am in grave danger. Surely, I misunderstood you." With his hands on his hips, Drum was now fully awake and in command. "What is this nonsense you bring to me in the middle of the night?"

A clatter of explanations rose up in a chorus, but Morland silenced it as he gave them an appropriate look. "We have just come from Drumsworth with important news, Milord. We knew you would want to be informed right away. Drumsworth has been invaded!"

Drum didn't react. As he had always known, his strength was not based on his real power. It stemmed from the power that others *perceived* him to have. Tonight, a shiver went down his spine as he heard the announcement that his man had just delivered. But boldly, he strode in front of his men, his hand stroking his chin as though in thought. "Invaded?" he repeated. "Invaded by whom?"

Morland stood tall displaying his own version of strength. "The Lowlanders," he answered gravely. The men behind him provided a jumbled but enthusiastic echo.

"It was a huge force that invaded the prison and stole every one of the prisoners."

"There were probably a hundred of them that snuck in and out without a sound."

"They wouldn't have dared to perform such a coordinated attack unless it was part of a bigger plan."

"The Lowlanders are on the move to take back their land."

"We're sure of it!"

Drum looked at his men and shook his head. *How stupid do they think I am?* he asked himself. *They forgot to lock the prison door and the Lowlanders escaped. And, of course, they were quiet. They're smart.*

The Auras heard Drum's thoughts and they were ready. They had eight men here who had convinced themselves of an invasion, a respectable reason for losing prisoners that were in their charge.

Now they had Drum, a man whose partner had just shown leadership that Drum resented. Things were at a tipping point. Tonight was not presenting a threat. It was presenting an opportunity.

The Call of the Auras

"This is the time for action, Drum," the Auras whispered in his ear. "Your goal is to rule the Lowlands and force Lowlanders back where they come from, right? What better reason do you need for that than a Lowlander invasion, real or imagined? Your men have delivered this situation to you. *They need to believe* what they say. *Your* belief and theirs is all you will need to convince the nobles that now is the time to make your move. The nobles *want* to believe you. This is the perfect time for you to start your plan for 'Lowlander Control'!"

Drum stopped in his tracks. Morland didn't move a muscle. The Drumsmen on their horses gradually ceased their chatter. What was Drum thinking? What would be his reaction?

Incongruously, Drum stood tall, turned back to them, smiled, and began clapping his hands. There he stood in the dark, *clapping his hands*. Did he really understand that Drumsworth had been invaded?

Drum spoke up strong and immediately they knew why they respected him. "You say there was an invasion by Lowlanders at Drumsworth? I say that the stupid Lowlanders don't know who they are dealing with. You are Drumsmen. I am Drum. We have waited long enough to put the Lowlanders in their place and return River Kingdom to its original noble state."

The Drumsmen cheered. They were not quite sure what 'put Lowlanders in their place' meant, but it sounded good. It sounded forceful. Whatever it meant, they were for it. The Auras began to glow.

Drum motioned his men from their horses. There on the bumpy old river road they held a conference. "I have one important question," Drum announced in the voice of a field general. "What is the condition of the mote and the drawbridge?"

Morland answered with pride. "They are both complete" he stated. "They are ready for use. The mote has filled, and the drawbridge is in place and functioning." The Drumsmen on either side murmured their proud confirmation.

Drum nodded. "Just as I planned," he announced loudly. "My timing is excellent, as usual". His men nodded in agreement. "Now, here is the rest of the plan for tonight. Southern nobles agree with our belief that Lowlanders have taken too much from noble and white men's rights. Each noble has a militia. Your duty is to go to them tonight and tell them about the invasion that we, the original settlers of River Kingdom, are about to endure.

"Tell them to rally their militias and get them ready. Tell them to prepare to support me and Thorndike to take back our kingdom and return it to those who deserve it!"

The Drumsmen roared their approval. Together they told each other how righteous their cause was and how much of a positive change their conquering of the Lowlanders would make in their kingdom.

With a renewed sense of purpose, most of the Drumsmen turned back to the south with their instructions firmly holding sway in their thoughts.

Drum, on the other hand, turned around and headed back to Thorndike Manor. The time was now. There was no time to delay. The big plan was ready to begin. Drum knew *he* was in charge...which was the way it was always supposed to be. Thorndike was asleep with no idea of how the leadership he thought he had shown today would fold under the direction of the real leader...himself

Drum was wide awake as his carriage rolled back down the drive that led to Thorndike Manor. The manor was quiet as it was still barely daybreak. Smoke curled above the chimney over the kitchen, indicating that Flo was already busy in the kitchen but everything else was still.

As the carriage came to a stop. Balynn shook herself awake and looked around. "Where are we? Aren't we home yet?" she mumbled. Rubbing her eyes as she peered out the window, she immediately recognized her surroundings. They were back at Thorndike Manor.

"Father! What are we doing here? Did someone go the wrong way? What's happening?"

Drum was already out of the carriage and banging at the huge front door of the manor. "Let me in, Thorndike!" he hollered. "Let me in right now! Get up! Get out here! We don't have time to sleep any more. Let me in!"

"Lord Drum? Is that you?" came Flo's voice from inside. "Is that really you? Didn't you go home last night? Does Lord Thorndike expect you?"

Drum continued to pound on the door. "Get out here, you smart-alack militia man! Get out here. It's time to do what you're really good at!"

The door stayed tightly closed. Something was wrong. Drum was here again, early in the morning after a long day yesterday. That wasn't a good sign. Flo stood back and waited.

Thorndike's chambers were directly over the main entrance. The noise below woke him with an unwelcome jolt. After endless celebratory toasts with his militia the night before, his thinking was still paddling up-stream.

His Aura, however, was ready for action. "It's your man, Drum, downstairs," his Auras hissed from the dark corners of his room. "He's back for some reason, but it sounds like he thinks it's important. It sounds like he's after your militia. Something big must be happening."

"Let *him* explain," the Auras cautioned carefully. "Don't get mouthy. You were great yesterday. Today, you should just listen. This may be what you and he have needed. Get down there."

Thorndike pulled his pants up over his nightshirt and stumbled down the stairs. He nodded to Flo who pulled the heavy door back slowly. Drum barged in, obviously in a hurry. Balynn looked on from the carriage in confusion.

Flo waved to Balynn from the doorway as Drum and Thorndike hurried to the conference room. There was no waiting for formalities. Everyone present could now sense that an emergency was underway.

As the door closed behind Drum, Thorndike's faculties returned from their deep sleep and his always-present Auras circled in readiness around his shoulders. Their presence joined with equally hearty Auras that accompanied Drum. The two men stood at opposite ends of the conference room table as Thorndike focused on the determined face of Drum, now more of a man of action than Thorndike had ever noticed before.

Drum went straight to the point. "Last night, Drumsworth was invaded by Lowlanders," he stated angrily. "They thought that I was away so they struck without hesitation when they knew no leadership would be present. This was *our* call to action, Thorndike. It's time to call up your militias and get them ready for a fight."

Thorndike was a warrior by nature. He was the son and grandson of warriors. "My militias are always ready," he answered without hesitation. "What, exactly, is the plan?"

Drum sat down and pulled a map out of his vest. This map had always been with him. Although it was creased and tattered at the edges, it was still plainly legible.

Thorndike moved closer and squinted at the map. Drum smoothed the edges of the map and announced with pride, "Behold the outline of the new home for all the troublesome Lowlanders, those who invaded last night and those who think they have rights to live in *either* of the kingdoms. *All* will be captured, and *all* will be returned to their homeland to be held in check by *your* militia...in *River Kingdom South!*"

Thorndike pulled up a chair and stared at the map with fascination. He and Drum had discussed shoving the Lowlanders back onto their own land for years, but Drum's map was a step beyond that concept. This gave a real name to a real location, something *more* than just a space, something that would need additional governance...control...*his militia!*

Thorndike's Auras swarmed invisibly around the map to digest its meaning. It didn't take them long to decide. This map was beautiful. It enlarged mere dreams of prejudice into the formation of a completely new kingdom. Thorndike's Auras were evil, of course, and supportive of all evil, but Drum's Auras were creative. This was something to be admired. Thorndike's Auras were enthralled.

Thorndike sat back and stared at Drum with awe. The two of them and the nobles from both their areas had been meeting for months...even years, but this was new.

Nobles looked down on the Lowlanders. They had resented their intrusion for over one hundred years. Lowlanders were a blot on the superiority of the fair-skinned people of River Kingdom. They were good workers for sure but never good enough to mix with the white people of The Kingdom. Their spread and, more importantly, their influence, needed to be confined. This would do it.

Urhonordo was an example of what could happen if the Lowland mix was not held in check. Intermarriage had taken its toll all the way to Queen Luna and her brown skinned husband, Prince Consort Amando. This kind of intermarriage had been given respect by the past Kings of River Kingdom since the days of King Sole one hundred years ago. In the minds of many nobles, River Kingdom had gone downhill ever since.

Drum patted his map lovingly. "What do you say about this, Thorny," he asked. "Does this look like something that you and your militia could get behind?"

Thorndike smiled a slow, wide smile and he leaned back in his chair. "When do we start?" he asked. "What is our first move?"

Drum had an outline of action...not on paper, you understand, but in his head. He had been reviewing and reforming it since he had turned the carriage around a few hours ago. It was simple:

First, a batch of Thorndike's militia would be stationed at the border between River Kingdom and The Lowlands. This would keep *new* Lowlander arrivals (the now anticipated invasion) from entering River Kingdom.

Second, the new Royal River Road would be quickly finished and made easier for militia and its equipment to travel.

Third...*a wall!* Thorndike's militia would set up a regular routine of travel to the border to erect and maintain a permanent wall between The Lowlands and River Kingdom. Local southern militias would help, but Thorndike would be in charge.

Fourth, all uncooperative or undesirable Lowlanders would be taken from River Kingdom and deposited in 'River Kingdom South' (the old Lowland Territory) where they belonged. There the Thorndike Militia would keep them confined so that only the desirable (and obedient) Lowlanders would be allowed to return. They would be put to work by nobles on whatever needed to be done and then returned to River Kingdom South when they were through.

Drum carefully outlined the plan, taking pains to stress the importance of Thorndike and his militia. Thorndike listened raptly. The plan seemed perfect. He could see nothing wrong with it. It was actually what he had always dreamed of, and he knew that the local nobles would be equally enamored with the proposal and the absolute control it represented.

Flo appeared quietly in the doorway of the dining room with a tray of biscuits and tea. Efficiently she placed her tray on the table and began to pour the tea. What had brought Drum all the way back to Thorndike territory? Had he forgotten something?

She glanced from side to side and listened as she now began to place biscuits carefully on their dishes. Something was not right. She had sensed it at the front door, but she shouldn't be trying to eaves drop.

"Yes, you should," her Conscience whispered in her ear. "You know that the two of them are never up to anything good. If they are discussing something important, it's probably not good, either. Listen hard for clues."

"Would either of you prefer cream in your tea?" Flo asked sweetly.

"For heaven's sake, no, woman! Can't you see that me and Drum are in a serious planning session right now? Important moves are being made at this table. A new country is in the works. Don't talk to us about cream!"

Spying the biscuits, he grabbed two and stuffed them into his mouth at the same time, spraying crumbs in every direction as he continued to pelt the equally excited Drum with questions.

Flo (Florence Thorndike, of course) did not have her father's disposition but she *did* have his persistence as well as the endurance of her mother. She simply nodded and backed into a corner, out of sight and out of mind ...but very able to hear every word.

Jorge had followed his mother down the hall but remained hidden in the hall. The young militia man had heard the re-arrival of Drum. That could mean the re-arrival of Balynn as well. Quickly Jorge scurried past the door of the dining room and out the front door.

And there she was. He had been thinking of her all night and she was exactly where he hoped she would be right now...waiting patiently for her father in their carriage. Slowly he approached her, being sure not to startle her.

"Good morning, Balynn...do you want to come into the manor this morning? My sister just made biscuits." he asked as he got closer.

Balynn straightened her rumpled dress and smiled. She had been thinking about him, too. "How long do you think my father will be in talking to your father?" she answered with a grin. "Those two can talk for hours when they get together."

"I have no idea," Jorge answered with a shrug, "but I don't see why you should have to sit out here alone in the morning mist. Why don't you come into the kitchen and have a biscuit?"

Balynn smiled demurely and tried to look as though she had not been wishing for this invitation. With a graceful lift of her skirts and a hand on Jackson's arm, she stepped out of the carriage. "I would love to," she answered.

The two hustled together through the cool morning and down the hall that led to the warm kitchen. The door to the dining room was still open with Drum and Thorndike bent intently over maps and papers on the table before them.

"I can already smell those biscuits," Balynn said with a grin. "I could eat a hundred of them."

"Me, too," chimed Jorge. "My brother and I try to beat everyone else to biscuits in the morning. Flo makes them for the whole militia, but the first ones are warm and gooey. They're the best."

Both young people grabbed a warm biscuit and sat themselves at the table as Jackson now ambled through the back door. "Balynn! What are you doing here? Did Jorge steal you away today like we did yesterday?" He chuckled and grabbed a biscuit. "You didn't happen to bring Mimi with you, did you?"

"Sorry, I'm afraid not. It's just me. My father suddenly decided that he had something 'really important' to share with your father...something that just couldn't wait... so we came back. The two of them are in the dining room together right now."

"That's really strange," murmured Jackson around a mouthful of biscuit. "They got so much done yesterday with the approval for the new road, I thought they would just take a day off and relax. There are probably farmers and landholders already down by The River pulling up rocks to plug up all the small levees before the heavy rain starts."

Jackson took another bite of biscuit. "Do you think we should go down to the mote right now, Jorge? They expected the drawbridge to be ready yesterday or today. Without us down there, who knows what those numskull Lowlanders will do with it if we aren't there to make sure they put it in right."

"You don't give the Lowlanders enough credit, Jackson. They're smart on their own," Jorge said thoughtfully. "I just hope the yahoos that are supervising them aren't too rough on them." Balynn munched on her biscuit and watched the two brothers in fascination.

"You just say that because Flo told you we were part Lowlander," Jackson said with a smirk. "Maybe you and Flo are part Lowlander, but I'm not. I'm all white from way back, just like Father." Jackson was trying Balynn's affection on purpose.

No one said a thing. Their "Lowlander heritage was now out in the open. What was Balynn going to say about it?

Flo now bustled into the kitchen with an empty tea pot and happily took note of what she saw. "I see you boys found someone nice to talk to while Father and Drum have their big discussion," she said with a smirk, refilling her pot with hot water. "I could only understand half of what they were saying, but it sounds serious."

She raised her eyebrows at the three of them and shrugged. "You three amuse yourselves with chitchat while I do some useful eves dropping." Smiling back over her shoulder, she sauntered back down the hall with the pot and some more muffins. "Everybody loves my muffins."

The trio at the kitchen table now sat quietly. Jackson got another biscuit, but Jorge wasn't either speaking or eating. Furtively he glanced up at Balynn. She was picking up crumbs with her fingers. Had she heard what Jackson said? If she did, was she upset? She was Drum's daughter and Drum was a Birther who hated Lowlanders. Would his daughter feel the same way?

Mimi was not here. Jackson's Aura nudged his jealous reflexes. He had decided to find out if Balynn would now look down on them for their bloodline. If he couldn't have a girlfriend with him, Jorge shouldn't have one either. But Jackson was impressed. She didn't seem upset.They were still gazing at each other.

Hmm. He would try another approach at being included. "If you two are going to give me the cold shoulder, I think I will take a muffin over to a woman I know who probably wants one," he said with a smirk.

Balynn snickered. Jackson was obviously messing with his brother. How she wished that she had had someone to joke with when she was growing up. She could join in with the joking now, however. "Poor Jackson. I don't think Jackson has a real love life to talk to this morning, Jorge, but maybe we can help him find him his woman friend to give a muffin to." She squeezed Jorge's hand.

Jorge squeezed back. "Oh yeah, Jackson? You have a secret woman friend who needs a biscuit? Where is this magic woman? We should go find her and give her a biscuit right now."

"Alright. If you think I won't, I will. Follow me," said Jackson with a grin. He picked up a biscuit and marched down the hall.

The game was on. The silliness of the situation continued to grow happily as Jackson picked his way up one deserted hall and down another, followed by Balynn and Jorge, hand in hand, taunting him with snide remarks. But now Jackson stopped. He also stopped smiling.

"What's the matter, Jackson?"

"Where is your mystery lady, Jackson?"

"Have you lost her?"

Suddenly Balynn recognized where they were. The long, dark hall stretched in front of them. Jorge raised his eyebrows. "What's the matter, Brother? Why are you stopping here? Are you afraid of a dark, mysterious hall?"

Balynn walked past the hall without stopping. "I don't think we should go down this hall," she said.

"Why not? Are you afraid?" laughed Jorge. "Come on. Is there a boogie man down there?"

"No," answered Jackson.

Balynn turned back and stared at Jackson. "You know about her, don't you?" she asked.

"I saw her for the first time yesterday," he admitted.

"Saw who?" blurted Jorge in frustration. "You two have a secret that you won't tell me? You don't...do you?"

Balynn took Jorge's hand. "Not exactly. But there is a secret something down at the end of this hall that Flo showed *me* a while back. Now it sounds like Jackson knows about it, too. How did you find out, Jackson?"

Jackson stood first on one foot, then another. "It only happened yesterday," he said, looking unhappily at the floor. "I was going to tell you about finding her today, Jorge, but I didn't get a chance."

"Her?"

"There's a lady down there behind that big door. She's old and she's scared to death of me for some reason. She started yelling at me like she thought I was going to hurt her."

"She liked me, when I saw her," Balynn added quietly. "She thought I was Willa. I really felt sorry for her."

Jorge stood back and stared at the two before him. He shook his head and frowned. "You both had a secret and neither of you told me. My new, beautiful friend and my old, rotten brother don't even tell me when they have a strange secret...that they should have shared with me." Jorge decided to pout. "See if I tell you anything interesting in the future that *I* find out."

"Alright, Brother. Since you are deciding to pout about it, I will let you peek in the door and see what we saw, but you can't go in. You look like me and you'll scare her like I did. I still have the keys with me. Will seeing the poor old thing satisfy you?"

"Maybe," said Jorge eagerly.

"Then give me your biscuit, Jackson," Balynn said with a shrug. "The least we can do while we are here is give the poor lady a biscuit. She likes me. I won't scare her like you boys will."

Quietly the three tiptoed down the hall. In front of the big door, they stopped to take a breath. Then they noticed it. The door was slightly ajar.

This was not good. None of the three knew anything about this mystery lady, but they all knew that this door standing open was not a good thing. Continuing to tiptoe in a stealthy fashion, they crept in. Immediately Balynn noticed that the fire was still crackling in the fireplace as it had been when she was there before. The hearth was not cold, but the big chair was empty. The lady was gone.

THE CONSCIENCE OF A KING

The Royal Family of The Castle gathered around June and Manda for their briefing on the issue of slavery. Their report was as they had feared it would be. Slavery *was* taking place at Drumsworth.

Those slaves had been released and were now free, but that group of slaves had already been used to create a mote and install a drawbridge. The project was nearly finished.

King Solem scowled at the report and silence reigned. What should he do now? What should be the next course of action? He had already signed the approval papers for the new road.

A fall storm had begun outside, and rain began beating against the windows of the conference room. Cracks of thunder split the silence and lightning reflected the solemn faces of those around the table.

Manda was not accustomed to unhappy silence. There was at least one happy thing to report. "But you might also be pleased to know that Drum's slaves are free now, Your Majesty," she announced proudly. "We found the keys to the prison with the help of Paulina, one of the slaves from *inside* Drumsworth Castle. She was sneaking them some extra supplies of water and food. They took the supplies and disappeared with her into the Lowland woods."

"When you arrived, were there many Lowlanders in the prison?" questioned Solem.

"Yes, Your Majesty, but they're free now...all except for the other women who work inside Drumsworth. They are still there, I'm afraid."

Solem shook his head. "I should have waited to sign until you two came back," he said sadly as though talking to himself. "I should have known that Drum was a liar."

"Trust is not a bad trait to have, however," Solem," whispered his Conscience. "Just be careful where you use it."

June was as impatient with the gloom in the room as Manda. "What was done was done, Father," he said impatiently. "The question is, what are we going to do about slavery now? Are you going to tell Thorndike to stop work on the new road?"

"Of course not," Solem answered immediately.

Willa turned to Solem in surprise. "Really, dear? You're not going to tell him to stop work right away? Why not? Drum is a slave owner. We cannot approve of a road built by slaves...can we?" Her question echoed in the huge room as the rain outside continued to pelt the windows.

"*Thorndike* is not a slave holder," said Solem quietly.

"The other day, we saw two of Thorndike's militia men acting like slave masters down at Drumsworth, Father." said June, exchanging nods with Manda. "Thorndike knows as much about Drum's slaves as Drum."

"*Jackson* Thorndike is too sweet to be a slave master," commented Mimi stoutly. "He wouldn't do such a thing."

"He *would* and he *did,* Mimi," countered Manda. "He and his brother were the two that we saw *lording* over the slaves when we watched from the bluff. I recognized them yesterday at the luncheon. I was going to tell you when I saw that you fancied Jackson."

"You must be wrong, Manda. Jackson isn't like that."

"It doesn't matter who Mimi fancies, Manda," growled June. "What matters now is that Father needs to stop work on the new road, right Father?"

Solem rose from his chair and walked back toward the dark, stormy view from the windows that rose high on the conference room walls. His hands were clasped behind his back. For more than a few minutes he stood there in thought, his back to his family.

"No," he said, still gazing out his stormy view. "The approval was signed by me. The road will be built."

Faithful Horace was standing unnoticed in a corner next to his ever-present tea pots, but the cup he was holding dropped from his hand. Horace had told his brother yesterday that he was sure the road construction would be halted when slavery was confirmed. But he was wrong. Hector would be devastated. The new road would mean the end of Hector's farm.

"You shouldn't have assured him, Horace," murmured Chorace. "You didn't have proof and now you're wrong." Horace hung his head.

June leapt from his chair, sending it crashing to the floor behind him. "I knew you would say that, Father! But you can't mean it! You can't let the road be built by slaves. You must stop it!"

"Tell him like a man," counseled Cjune. "Don't jump about and whine like a child."

June listened and calmed himself as much as he could. He began to pace again. Manda's face reflected June's reaction. She couldn't believe it. If his father had seen the conditions that these poor slaves were living and working under...

Cwilla gave Willa a poke. Willa rose and went to Solem. "I know you are a man of Conscience, Solem. June obviously doesn't agree with you about the road. I must admit that I am surprised, too. Please...tell us why you are saying what you are saying so that we can all understand."

Solem looked at Willa. She was the last person that would go against him, but here she was, obviously unhappy with his decision.

"Sometimes being a King and thinking you're right, isn't enough, Solem," whispered Csolem. "I don't understand your thinking, either. You need to say why."

Solem had been a king for many years. Having to defend his actions did not come easily, but even his Conscience was questioning him. He would try.

Slowly he returned to his seat at the head of the table. Everyone leaned in. "Yesterday I signed an official document of approval. In this Kingdom, that made the approval law. It has always been that way. I do not pass laws just to ignore them the next day."

"But we have never had to deal with slavery before, Father," countered June in his newly acquired, calm voice.

"We must deal with slavery in other ways," said Solem with sadness. "Yesterday we dealt with the approval contract for a new road. That contract didn't say anything about slavery. Slavery and the new road are two different things. They must be dealt with separately. That is appropriate procedure and protocol."

"Hog wash!" snarled June. "People are more important that protocol, Father."

"And we have not seen Thorndike endorse slavery," Solem continued. "Perhaps he didn't know that the twins were supervising slaves. We need to be sure that he knows about this before we say that *he* is a slave master, too."

"More hog wash," Manda whispered to June. Then she stood up and said her piece. "Thorndike will say whatever is convenient, just like he did yesterday... and then he'll do what he wants. The farmers in the middle of the kingdom will be left high and dry without enough water and slavery will be quietly accepted. But don't worry. The nobles will be happy. They don't like us Lowlander types anyway, so they won't worry about slaves!

"*You* know this is all wrong, Your Majesty," she continued. "*You* know it! I apologize, Your Majesty, but I no longer feel comfortable here. I am half Lowlander, as you know. I need to leave." Manda stood and left the room. She was going home.

June stood and faced his father with determination. "Since Drum is going to need labor to build his part of the new road, he is just going to go find more Lowlanders to do it, Father. He is probably already doing that.

"And *you* have given him the official approval to do so. I can't believe this. I can't support your decision." June turned and walked out into the rain. (Cjune and Cmanda quietly applauded.)

Mimi was in a quandary. What had just happened? She had never seen anything like this before. Quickly she ran to the window and peered out, hoping to catch a glimpse of Manda and June. What were they doing out in the rain? Why did they run out? She knew that it had something to do with slavery, but Mimi didn't know much about that. She wasn't even sure what slavery was.

Horace crouched on the floor, picking up pieces of his broken teacup. His mind was churning with disappointment.

Willa reached out her hand to Solem. She didn't know what to say to comfort him. What was worse, she didn't agree with him and neither did her Conscience.

But Solem was now ready for her. "We know that Drum and your father can't be trusted. We know that the nobles don't respect the Lowlanders and will make slaves of them if they can, but cancelling the approval that I made will not change any of that, Willa."

"I am trying to understand what you say, Solem, dear. What does your Conscience tell you?"

Csolem had been quiet for a few minutes. Solem knew that his Conscience was disappointed in his decision, but suddenly Csolem spoke up. "I know that you think you are right to stand behind an action you have taken, but that action does not need to stand in the way of another action to improve this horrible slavery situation.

"Think, Solem. What can you do now that will *stop* slavery but *not* cancel your approval? You make laws, Solem. Make one that will fix this."

Solem's face brightened. "That's it!" he exclaimed to Willa. "You and my Conscience are an important pair. As King, I should not cancel a good law because *something else* is not good...I should simply issue a law against that "something else."

Quickly he sat again at the table, Willa at his side. "Horace? Are you still here?"

"Indeed, Sire," Horace responded sadly. He couldn't hide the fact that he was disappointed and worried for his brother.

"Run please, Horace, and fetch me my pens, ink, and paper. I need to write a new law.! There isn't a moment to lose."

Horace ran for the supplies as ordered but Willa just stared at Solem, completely perplexed. "What are you thinking, Solem? I must admit, I have no idea what you are talking about. *I* didn't say a thing."

Solem just grinned. "You asked what my Conscience was saying," he said, giving her shoulders a squeeze. "And my Conscience gave me an idea... I don't have to cancel my approval of the new road, but I *do* have to add a law that *forbids it from being built by slaves*. What do you think, my love?"

Willa was stunned. It was a perfect answer. Solem was brilliant. She said so. "You are brilliant, Solem! Brilliant! Let me go tell the children!"

Willa ran out of the conference room and onto the rainy colonnade. Thunder and lightning crashed and rumbled overhead, lighting her way. But no one was there. The colonnade was empty. Manda and June were gone.

DISCOVERIES IN A STORM

Manda was not in the mood to discuss anything. Fetching her horse from the stables behind The Castle, she climbed on and rode off into the night. She had heard what she needed to hear, and she didn't like it. But she needed to let her father and mother know what King Solem had decided. She heard June calling behind her, but she didn't turn around.

Father would be upset, she was sure, but what could he do? The decision was King Solem's to make. One King could not change the decision of another, especially about their own land.

The weather was continuing to storm hard, soaking into the dry farmlands under her horse's feet as it was undoubtedly doing in both Urhonordo and River Kingdom. *Slavery affects both sides of The River, just like this rain,* Manda thought angrily. *Solem's decision may affect Urhonordo just as much as River Kingdom.*

"More", shouted Cmanda over the pounding of the rain. "Slavery in River Kingdom is all about control of brown skin. That skin makes Lowlanders a threat to uneducated nobles. They don't have that skin, so they are afraid of it. That's what makes it necessary to enslave and control all those who have it, you see."

The rain continued in torrents, but Manda kept riding.

Cmanda also continued. "In Urhonordo there are lots of Lowlanders and even half-breeds like you, Manda. That means that some folk from River Kingdom may even see Urhonordo as a threat that needs to be controlled. White fear and prejudice causes a desire for control so that 'the color' won't spread. "

The good people of River Kingdom don't think like that, thought Manda with a vengeance. *All of the Royal family are good. They always have been. They just don't understand the horror of slavery."*

"*Good* people don't think like that but don't be too sure about Thorndike and Drum," Cmanda yelled over the horse's hoof beats. "It might be time to think about Urhonordo as well as Lowlanders, themselves, in terms of slavery. It's possible that they are linked together in the minds of the prejudiced."

By the time she reached Urhonordo, Manda was slowly gaining a new perspective on the slavery issue. Slavery was not just about making people work without pay. That was surely part of it. Slavery was also about fear and control of people with color and she was one of them.

With a new insight circling in her brain, she tethered her horse and ran into Urhonordo's beautiful castle. She had never thought about the slavery of the

Lowlanders as an issue that might put her family and even herself in danger. She had always worried about 'them', not 'us.'

Breathlessly, she rushed into the map room, intent on revealing her new insight.

"Come in and sit by the fire, dearest," Luna said with only a small glance toward her dripping daughter. Luna's head was buried in a gathering of papers and maps and only she held the key to what was where. She was obviously very intent on something.

Prince Armando was in a corner with an oil lamp on each side, pouring over a stack of history journals that were opened to differing pages, seemingly with differing points of view. "Hello, my girl," he greeted her absently. Then he looked up at his daughter's face. "Solem decided not to rescind his approval of the new road, didn't he?" he asked without expression.

Manda nodded but looked around her. Something was happening here, but she had important insight to share. Impatiently, she flung her wet wrap to the floor and waited for recognition from her absorbed parents. They needed to stop what they were doing so that she could let them know what she had just discovered.

"Maybe your parents already know what you just figured out," whispered Cmanda calmly.

You mean that maybe they have known for a long time that Urhonordo might be in danger? she asked herself. *If this is true, why didn't I know? Am I that stupid?*

"Not stupid at all, Manda," answered Cmanda quietly. "Just young and inexperienced."

Manda picked up her wet garments and hung them on a hook." Yes, indeed, Father." she now answered. "The new road will continue."

"I thought that would be so," answered Armando. "He is strict about commitments he has made, especially to the nobles. He knows that they don't like him that much, already."

"And the nobility doesn't care about Lowlanders, so slavery is not an important issue to them. Wonderful." Manda sniped.

Luna looked up and smiled wistfully at her daughter. "The nobles *do* care about Lowlanders, dear," she said, "as long as they are getting work done...and not getting in the way...and not disagreeing with them."

Manda sat down suddenly. Her stoic face began to crumble as tears came to her eyes. "The nobles just want their road and their trip to Drumsworth on vacation."

"And their protection," Armando added from across the room. "Don't forget how important Thorndike thinks protection is. Protection from what, I have no idea, but I know he is strong on that point."

"Maybe protection from Urhonordo and Lowlanders," contributed Manda flatly.

Luna and Armando looked up at their daughter in surprise. They looked at each other and then back at Manda. *Our daughter is beginning to understand,* they both thought at once.

 Luna came around her table to give Manda a hug. "I see that you have been doing some thinking about this," Luna said. "Armando and I didn't want to frighten you with the prospects we see but what you just said shows that you are beginning to have some of the same concerns that we have."

Manda looked at her feet. "To be honest with you, Mother, I thought that freeing Drum's slaves didn't have anything to do with us personally. I just thought it was wrong. The fact that what Drum thinks about Lowlanders might affect us didn't occur to me until I was on the way home from The Castle. I just thought working Lowlanders might be in danger. I was so mad at King Solem that I wasn't thinking of anything else."

Armando chuckled. "I understand completely, Daughter. I, too, was content with freeing Drum's slaves with June and you until I got to thinking about it. Then I read these history books that your mother always has lying around. There is no doubt that Lowlanders were taken advantage of back in the history of this area. They were even locked up back then, too.

"The Lowlanders were being controlled," said Manda quietly. "It was all about fear and control, wasn't it?" She had read the history books, too, but now she understood them.

"Fear and control," Luna echoed quietly. "Those are the words that equal slavery. Slavery is just control, control based on fear. Why did I never think of it that way before?"

Armando joined them and stared at his daughter with wonder. "You have just said the words that are key to the question that I was asking myself," he said in amazement. "The slaves that you and June freed were big and strong, right?"

"Yes, Father...the ones that were healthy and not damaged from their work."

"Why then did all these big, strong men allow themselves to be made into slaves by just a few slave drivers? Why? That didn't make any sense, did it?"

The room was quiet. But Armando already had the answer. "The Lowlanders were slaves because they *allowed* themselves to be controlled. That's why. And why did they allow that? Because, of their culture. As individuals in their homeland, they never thought of controlling anyone but themselves. That is Lowlander culture. They never expected it.

"But noble fear of brown skin needed control. They were men of River Kingdom's white culture. Royalty had authority (control) over them. They decided to take control of the Lowlanders because unsuspecting Lowlanders allowed it. They didn't know they could say 'no'. Now that free Lowlanders are doing well without them, the nobles are afraid of losing their control. Lack of control over Lowlander color makes them afraid of losing *their* white culture.

"Many Lowlanders never knew that they could have *and should have* just taken control themselves as equals. They gave up their control because they didn't know they needed it. They didn't know how to just say 'no'." Many still don't."

Manda and Luna looked at each other. The truth was so simple that they had all missed it. Manda now saw slavery in a new light. "June and I gave the Lowlander slaves their freedom," she said quietly, "but we should have told them that they now had control of their own lives...control of what they did and who they did it for."

PLANNING SESSION AT THORNDIKE'S

Days had passed but Drum was back at Thorndike Manor, again. Flo dropped off another batch of steaming biscuits and stepped back into her corner of invisibility. With fascination she watched her father and Drum argue their way to and fro in their discussions, each trying to maintain a position of more importance than the other in the discussion.

Thorndike was a man of action, not one of detail and decision making. Drum was ready with specifics. The maps were back on the table and a great deal of pointing was underway. The Auras of both men were swarming with joy around their heads and the joy of evil planning was at a new peak.

Flo began to shudder as she watched. The specifics Drum was detailing were about people, not just places on a map. His ideas sounded heartless and inhumane but were stated in such a calm manner that they didn't seem to have the importance that they would surely have when they were put into action.

The Call of the Auras

"I think we should build the wall first," Drum announced after a moment of contemplation. "That's what the nobles want. That will control the influx of invading Lowlanders and that's what will make the nobles happy to start.

"We will run The Wall from Drumsworth all the way to the western boundary of River Kingdom. Then we'll round up all the unsuspecting Lowlanders and shove them behind that wall on the Lowlander side. They'll never know what hit them."

Thorndike nodded and took another bite of biscuit. "I'll love to see that when it happens. And I will post guards along that wall so that no one can climb back over into River Kingdom. My men will just slap them down when they try," he added proudly.

"Excellent". Drum leaned back and clapped his hands smugly at his own brilliance. "*My* mote is complete, and the drawbridge is in place for security. No one will get into Drumsworth unless we let them. I have prepared *a whole wing* of my Castle to house your militia in that area while they build *and* maintain the wall that will finally keep the Lowlanders in their place. Once that wall is built, no one will have to worry about Lowlander invasions."

"Better yet, *my* militia will *make the Lowlanders themselves, build the wall* that will keep them in." contributed Thorndike with a smirk. "That will speed up the process and let the militia spend more time supervising the building of our excellent road in that area."

My militia will be key to this whole plan, Thorndike gloated to himself with satisfaction. *Drum's plan is a sound one, but he has no militia. He can do nothing without manpower. He needs me more than he is admitting.*

Drum smiled to himself. "That's right, Thorny. I will be there in my castle doing what I do best...overseeing a big bunch of Lowlanders captured and ready for work." *Thorndike could never have put together a plan this ingenious,* he thought to himself. *But he will be The Commander of The Militia as we have always planned. He can handle that, while I do the thinking.*

Thorndike also has no idea how to keep the nobles in line for support. He is good with his militia, but he does not see my big picture. With the Lowlanders in place and the grateful nobility at my side because of it, The Crown of <u>all</u> of River Kingdom will be mine.

Flo needed to escape this horrible room. She needed to find her brothers. These men were planning to send Lowlanders back to their homeland and keep them there. The militia (*her father*) was to take control of them and their country and keep them there, imprisoned in a new country called River Kingdom South. What would that mean to her and her brothers?

"Would that mean that you and your brothers might be sent away, or would this just happen to *other* Lowlanders, ones that looked like Lowlanders?" asked a soft voice. Flo wasn't sure she heard anything but...

"You are now realizing where you are living and who you have been living with all this time," offered Flo's Conscience. "You have grown a lot recently and opened the truth at least part way to your brothers and your sister. But now you must remember your heritage. You *are* part Lowlander. You must remember this...with pride and concern."

Indeed, thought Flo reluctantly. Father is talking about 'them', but he is also talking about 'us'. She continued to pour tea.

"We must pull together a meeting of the nobles immediately," Drum continued, scribbling notes on a sheet of paper. "Perhaps we should get them here in three days. That will give me time to get back to Drumsworth and get things organized with the nobles down south. We need to make sure that all the nobles know where their good news is coming from."

"From prejudiced liars and cheats," said a soft voice in the rafters.

Drum jumped and looked from side to side. "Did you say something, girl?" he snapped.

"No sir, Lord Drum," Flo answered quickly.

There <u>was</u> another voice that was talking, Flo thought immediately. It was not just my Conscience, *Now it is speaking to the whole room.* She looked around.

"What's going on, Thorndike? There's that voice that I heard at the meeting yesterday. Do you have ghosts in this manor?"

"What are you talking about, Drum? I don't hear anything," answered Thorndike but he was looking around, as well.

"People who don't listen never hear anything," the voice murmured.

Cflo whispered in Flo's ear. "That wasn't me, but I agree with whoever is speaking," she said.

Thorndike stood and looked around. "Somebody is just trying to be smart," growled Thorndike. "When I find them...and I will... I will put them in chains and make them live in The Lowlands...in that prison you have down there, Drum.

Somebody just thinks they're being funny. Forget them! We will do as you suggest. We will meet in three days! See you then." Thorndike left the room.

Drum looked closely at Flo but concluded that the timid servant he saw was not the source of the mysterious voice. Quickly but continuing to look from side to side, he swept up his paperwork and hustled to the door of the manor.

Flo placed the dirty dishes on her tray and brushed the crumbs off the table. Her hands were trembling. Someone had spoken directly to her...and then to Drum and Thorndike. Flo didn't look from side to side. If it was a ghost, she didn't want to know for sure. Just the same, the voice had said what was true...all three times. Flo quickly picked up the tray and started out of the room. Then she stopped. "Thank you," she whispered on impulse.

"You are welcome," said the voice.

PLANNING SESSION AT URHONORDO

Manda woke up to the sounds of many people gathered in the courtyard of Urhonordo Castle. Quickly she pulled on a pair of leggings and a tunic. Today, she knew what she would be doing. If there were plans being made to do something about slavery, she was determined to be part of it.

She was surprised, however, by the number of people assembled on the front lawn of her castle. There might be nearly one hundred and many were preparing for travel. There was fighting equipment with those that counted themselves among the ranks of Urhonordo's tiny protection force, but everyone else had clothing, food, and huge water baskets full of water. These folks looked like they were going camping.

Manda knew almost everyone by name. What was happening with these 'campers? "Where are you going, Marla?" she asked one of the women. "What's the plan? What did my father tell you something I should know?"

"Not much," the young mother said, "but he did say that those of us who live up north here would be safer if we went south for the winter. He said that there might be danger up here in the north for the next few months. We didn't have to go but most of us decided to go ourselves. Me and the baby will go south, and my husband is going with Prince Armando...It's not ideal to separate, but probably safer for the baby." She smiled sadly and hugged Manda.

"I need to find my father, Marla. There has to be more to this than you just 'going south'."

"It's not just me," Marla confided. "I think that everyone out here right now is going to do that, everyone but the men, that is. Prince Armando said that he suggests that all of us with Lowlander blood should disappear for a while for safety reasons…if we think it's best. Strange, right?"

Now Manda was beginning to understand. If she was right, the evacuation that was happening right now was to keep the Lowlanders out of harm's way until the challenge of slavery was less. As Marla had said, it was all about safety for the families temporarily. But what was her father going to do to make life safer for the Lowlanders, long term?

Manda shook her head. How sad…safe, but sad. She rushed through the group that was moving south, hugging and smiling, but at last she found her father.

On her way, she changed her thinking. She had to accept that she wasn't a little girl anymore. She was a woman…who was half Lowlander. Whatever her father had in mind, she needed to be part of it.

Armando was loading up his horse but not with equipment used for fighting. He had digging tools…shovels and pick axes. As she glanced from side to side, she noticed that their tiny militia as well as other male Urhonordorians were loading the same things. She had assumed for some reason that her father was going to do something forceful, something that required weapons. These implements were peaceful in nature. What in the world was his plan?

The Prince read her questioning look and answered. "There is more than one way to fight a battle, daughter. Are you here to watch or to fight our battle with us?"

"I'm with you, Father. I have no idea what you are up to, but I am with you."

"I knew you would be," Armando answered as he went back to his work. "There are extra tools over there. Everyone brought theirs and put them in a pile. But don't strap any onto *your* horse, yet. We will be leaving soon. Your mother will explain."

Queen Luna now stepped to a piece of high ground and raised her hand. As was the custom, everyone stopped what they were doing and turned to listen to their Queen.

"Thank you all, Urhonordorians. You have answered the call that was put out last night. I am sure that many of you are curious about the implements I asked you to bring and why we are taking so many with us.

The Call of the Auras

"Let me tell you, dear people, that we are asking for your assistance, not demanding it. Our plan is simple. Prince Armando and I are requesting all of you who are not going south to stay here and prepare to labor at our side to accomplish two needs. The first need is to develop a system for watering plants when the current irrigation system is no longer available. The second need is to master the skill of self-control by making friends as equals with people who may not wish to be our friends. Of these two needs, the second is the most important. And, as you probably can guess, this effort will not be taking place on this side of The River."

The gathered group looked at each other in confusion. Whatever was their Queen asking of them? They had no idea. They had thought there was some kind of emergency. Was this it? It didn't sound like an emergency.

Queen Luna continued. "Across The Great River, there are many people in River Kingdom who really don't know Lowlanders very well. Many live their entire lives in one spot, without leaving their side of The River or meeting anyone new. That means that sometimes fear of the unknown developes. 'If I don't know you, should I trust you?' That's a fair question, don't you think?"

The crowd looked at each other and nodded.

"So," continued Queen Luna, "Our mission is to make friends with people that don't know us because we want them to know us as Lowlanders and trust us, even though we look different from them. In the future, this may be very important."

More head nodding.

"Now, Prince Armando figures that, if we do something positive for a possible new friend, that will help them get to know us, right?"

"That brings us back to the first need. Across The River from us, there is a section of farmland along the Central shore that will soon go dry without enough water. My proposal is that we Urhonordorians search that land to see if we can find some water for these people before they lose what they are using now. What do you think of our idea? In that way, we can make friends while helping people."

A burly gentleman in the front row of the group had a question. "Why will the people across the way lose their current irrigation system? It has worked for them for years."

"Indeed, it has, Simon. But King Solem gave permission to some of the nobles to build a new road past their area. That means that their irrigation levees will be filled in. That is what will cause their water shortage problem."

"I see the problem," acknowledged Simon with a shrug. "Why would King Solem do that, and what has that got to do with us?"

"King Solem doesn't see the problem. Maybe he's right. I can't prove that he's wrong, but I thought some of us might look into it for the farmers and see if we can help. It would be a really friendly thing to do."

Alfredo was a wiry gentleman standing next to Simon. "I don't mind helping a neighbor," Alfredo said, squinting at his Queen suspiciously, "but I do need to understand *why we* are doing this. and why are our women and children being encouraged to go south. What is *that* all about?"

Queen Luna looked seriously at the faces around her. It was time to tell all of them the whole truth. "The reason is slavery," she said bluntly. A negative rumble circulated from the front to the back of the small crowd. This was something that no one in Urhonordo had ever contemplated.

"Slavery has taken hold in River Kingdom," Luna went on, "and there is a good chance that it will spread. Maybe it already has. It is being encouraged among the nobility. Not all the white people condone slavery, but enough of the wealthy nobles with militia feel that way that Armando and I feel that all Lowlanders may be at risk, soon."

Simon looked up with a start. "That's what happened to my brother, I wager!" he shouted. "He went down to Drumsworth to see if he could find my sister and he never came back. I bet he got captured by the Drumsmen and made into a slave. I heard that they were a nasty bunch down there."

Armando came to stand by Luna. "Two days ago, my daughter and Prince June freed Lowlander slaves that were being kept in that old prison down there, but they told the prisoners in there to go south as far as they could so they would not get caught again. That is what we are asking some of you to do. If you have Lowlander blood in you, you are at risk with these people."

"We should just go over to River Kingdom, capture all of them nobles and lock *them* in that prison. Let's just see how they like it," growled Simon.

"You know that isn't the answer," whispered Csimon.

"That brings us back to what Armando and I are asking the rest of you to do. To put it simply, I would like us Urhonordorians to help the farmers across the way to find irrigation water and get to know and trust us as people instead of slaves. We would be fighting our own way...fighting white fear with brown friendship."

Manda couldn't stay quiet. "I know that the farmers over there are already worried about the water problem. Remember how angry they were at The Gathering? Hector tried to tell everyone, but the nobles didn't want to hear it...Remember?"

Several throughout the assembled had gone to The Gathering. They remembered. They also remembered how pushy Drum and Thorndike had been. Now they were beginning to see the issue that their Queen was explaining.

Armando continued. "Today I will send a messenger to King Solem. I will explain that *I* will be leading some Urhonordorian workers across The River to check out their water supply and help them find a water source that isn't The River. That is what I would like to do with the help of all of you."

"You want us to go over to River Kingdom and become their slaves just because *you* tell us to?" snorted Alfredo. "I don't think so!

"Of course not," Armando responded quickly. "I want you to help someone who needs help because they need help and *you want* to help them, not because *I* am telling you to."

"Sounds pretty close to slavery, Prince Armando. That will be a lot of hard work."

"It's not slavery," Manda cried out suddenly. "It's not slavery because *you* can say no! That is the key. No one is telling you that you *must* do this. No one should go with my father unless they feel that making friends by helping people is important.

"You have control. Father is *asking* you to help and to make friends with the farmers across the way. Then they can see how good and kind and wonderful Lowlanders are. If they see that in you, maybe their nobles will see it, in all of us Lowlanders. Maybe they will figure out that Lowlander slavery is not a good thing and Lowlanders are wonderful people. But nobody is *ordering* you to do anything. You are *not* slaves. He and Mother are just *asking for your help...and mine."*

Manda looked around. Everyone was silent. *I've said too much again,* she thought. *When will I learn to just listen?*

"Sometimes passion is a good thing, Manda. You said what needed to be said," whispered Cmanda.

Alfredo stepped forward. "I hear what you say, little woman," he announced. "Then he turned to his neighbors. "Many of us here have Lowlander blood in us. I, for one, am fill blooded Lowlander and proud of it. I think what our Royals are trying to say is that we of Lowlander descent are in danger of becoming slaves, ourselves if we don't get some more white friends to know us and not be afraid of us. We Urhonordorians don't have a big standing militia like nobles have on their side. We can't protect ourselves by force."

'Your right," contributed Simon. "Some whites don't trust us 'cause we're brown. I don't trust a bunch of them either, especially the nobles. It's definitely time for people who trust each other, to get peaceful people together while we can. I see that, Prince Armando, Queen Luna. You two are smart and the little woman here said it plain. I'm ready to try your plan." Simon turned back to the crowd. "Who's with me and Alfredo?"

One by one, the men began to step forward, leading their horses into a line. They were not smiling. They were not happy. This was not something they wanted to do but were choosing to do it because it was necessary and a good thing to do. They were ready and their Consciences approved.

Prince Armando looked down at Manda. His look said it all. Manda would be the messenger that informed King Solem of his plan. She was more than ready to go.

PLANNING SESSION AT THE CASTLE

June was conflicted this morning. The decision to build the new road had been made by his father. It was going to be allowed to stand, in spite of the slavery he and Manda had found and released at Drumsworth.

June was moping in his room. He was disappointed with his father. No new road was worth allowing slavery to continue and June was sure that was going to happen.

Drum would simply find some more slaves to replace the ones that he and Manda had freed. The new slaves would probably even be used to build the road itself. How pointless their big move at the prison seemed to him now.

June knew that Manda felt the same disappointment in his father's decision that he did. She had left for home in anger even though there had been a storm. Wondering if she had arrived home okay contributed to his feelings of uselessness.

"You are such a baby, June," Cjune said bluntly. "Did you expect to get a medal just because you did *one* thing right? Slavery is a big problem. It will not be fixed in a night, even though what you and Manda did was good and brave. It's going to take a lot of work...a lot of changing of hearts, or are you too immature to understand that?"

June shrugged. The last thing he needed now was a lecture from his Conscience. Suddenly he heard his name echoing down the hall. It was his father. He was calling from the sitting room.

He probably wants to give me a lecture about why it was important to keep his road approval in place, he grumbled to himself. *Oh well... Willa will probably be there with him. I don't think she liked the decision, either. I wish Manda was here to tell him off.* He grinned at the thought.

"Grow up, June," chided Cjune. "Try your hand at coming up with positive solutions instead of pouting." June snorted and ambled down to the sitting room.

Solem was bent over a small desk while papers littered its top and covered the floor around it. "I must make this right, "Solem muttered. "The nobles must see this as strong, but not authoritarian. This must be a positive direction, not a bossy order."

"Well said, dear," added Willa from a chair nearby. "Your last letter still sounded too bossy. I know the nobles don't like anyone to tell them what to do."

"What are we doing now? Writing love letters to the nobility about their beautiful slave-built road?" sneered June, dropping into a chair by the fire.

"You sound like Flo says my new brothers talk, June...speaking up before you know what's going on," Willa shook her head and grinned. "You left last night before your father came up with his wonderful idea. Now he's just trying to get it worded right. Why don't you just listen to what he wants to do?"

"So, tell me. What 'wonderful thing' does he want to do?"

"I am writing another new law, son, one that has never existed before now. Actually, it never occurred to me that it was needed. This new law will make it illegal to use slavery when building the new road... or anything else. What do you think?"

"Isn't that wonderful, June? Isn't that a great idea?" asked Willa smiling widely.

June sat up straight. Although it was still raining outside, the light in the sitting room seemed to brighten. "Perfect! That is the perfect answer! I am sorry I doubted you, Father. I should have known that your Conscience would give you the right answer. This is wonderful!" June jumped up and gave his father a hug.

Willa began to smile even more. The two men in her life that she loved the most were smart and doing the right thing. Their Consciences were busy congratulating each other.

But suddenly June stepped back. A serious look replaced the grin on his face. "This is a wonderful law, Father, but it isn't enough. A law that makes slavery illegal when building the new road is good, but now we need to say that slavery in River Kingdom is illegal, period. Your new law just needs to say that *slavery itself* is just, plain illegal." June stood still, as though re-hearing what he had suggested.

But Cjune would not be still. "Now you are thinking like a man, June. I'm proud to be your Conscience," he whispered.

Suddenly all became aware of a commotion in the hall outside. Agnes could be heard giving a lecture to someone who was intent on coming into the room.

"You really need to begin to act like a lady, Amanda. You mustn't just barge upstairs without an invitation as you did as a child. And, for heaven's sake, stop hiking your skirts up to your waist. It isn't lady-like at all."

June opened the sitting room door with a laugh." Throw her out, Agnes. We don't allow strange people up here announced. Throw her out." June opened the door wide for Manda while Agnes, hands on her hips, just shook her head.

"I'm sorry, Agnes. I am trying to do better. Honestly...but I just couldn't wait to tell everybody my news. It's really important...to all of us."

Willa laughed as well, but Solem, still working at his desk, barely raised his head.

Manda was surprised at the people she saw in front of her. She had expected to find Willa, but not the whole family...especially The King. "Pardon me, Your Majesty. I really apologize for my behavior but..."

"Come in, Manda," Willa ordered with a shrug. "What is it that you have to say that is so important. Don't worry about Solem. He has his nose in his *writing*. You aren't bothering him."

Manda hurried into the room and took her stance in the middle. "Wait until you hear my father's idea." she said. "My Father decided to request his fellow Lowlanders to help the nobles and farmers of the Center farmlands of River Kingdom with their water shortage issue. He said that this might build friendships with Lowlanders and help with the water problems that the Central farmers expect. It will do two things at one time. Isn't that a brilliant idea?"

Without notice, Solem rose from his chair and came forward, towering over Manda and glaring at her. "Your father wants to enslave more Lowlanders to the nobles?" he charged. "Has the man lost his mind? Why would he do such a thing?"

Willa ran quickly to Solem. "No, dear. I don't think that is what Manda said. You have been up too late, and you are tired. Sit down for a minute and listen."

Slowly, Solem backed away, but he was not satisfied. He glowered at Manda who had retreated to the door.

"Speak up, Manda,:" June said quickly, running to her side. "I don't think my father understands what your father is suggesting. I think that I do, though. Speak up but speak slowly so that we all understand."

Manda took a few hesitant steps forward. To her surprise, June stepped forward with her and took her hand for support.

Smiling slightly, she began again. "Father feels that the Central farms of River Kingdom are going to go dry because of the road. My mother believes that there may be untapped water inland from The River, such as Thorndike has found on his land. They both think that they may be able to find it if they have enough help. The Lowlanders of Urhonordo have agreed to try to help them find water. They hope this help might also help the nobles of the area get to know the Lowlanders better."

"As slaves?" questioned Solem.

"No, Father," contributed June. "As people...helpful, nice people. I get it. I understand...I think. But who will pay the Lowlanders for their help?"

"No one," responded Manda. "They are all volunteering."

"But they will be working without pay. Isn't that the same thing as being slaves?" growled Solem.

"Not if they are working because they *want to* and are *not being forced* like they were at the mote," added Willa quietly. "They will be in control of what they do... They can say no at any time. At least that is what I think."

"Exactly," said Manda with a thankful nod. "These Lowlanders will be under their own control. They can quit or keep working. There will be no bosses. They just want to let the River Kingdom farmers get to know them by doing something nice for them."

"And your father has a bunch of them that say they are willing to do this?" Solem asked incredulously.

"Indeed," answered Manda proudly. "Father just needed to request your permission to do this on River Kingdom land. He will not proceed without your approval."

Solem sat back down at his desk. There was a long period of silence. Quietly, June reached over unnoticed and grabbed Manda's hand again for support...at least that was what she thought it was for...

Solem rose from his workspace again, but this time he slowly smiled. The sitting room in The Castle suddenly became a warmer, more inviting place as he walked from behind his desk. Willa watched carefully, but she could tell at once that Solem's attitude was now positive.

This time he approached Manda gently and looked into her eyes with what only could be viewed as gratitude. "Your father and mother are geniuses," he said plainly. "Their thoughts on this difficult issue are just what I needed to complete my new law...which I was trying to write when you came in. Now I know how to complete it.

"My Law will prohibit slavery of *any kind* in River Kingdom but will encourage appreciation and a fair wage for all who volunteer their services." (The local Consciences cheered quietly on every shoulder in the room.)

PLANS AND QUESTIONS TOGETHER

Drum was in a good mood this morning except for the fact that his favorite servant, Paulina, didn't come when she was called. A couple of the other servant girls came running right away, however, ahe was not about to be sidetracked. He had work to do now that his road had been approved. He wasn't worried about the Lowlander slaves that had been "stolen" the other night. There were plenty more Lowlanders where those came from. His Drumsmen were already out rounding up a new, unsuspecting batch of men.

The new slaves would not be working on his mote as the old ones had, however. The mote was complete, and the drawbridge was working flawlessly. These new men would be filling in the holes of the levees for the new road as it led up to Drumsworth. That was necessary, of course, to provide a beautiful smooth surface leading to Drum's end of The Royal River Road and River Kingdom South.

Drum had never worried about irrigation before, and he was not about to start now. The surface of this road was what was important to him now. It was the key to his future rule and The Crown that still waited for him inside the wall of The Crown Room...speaking of 'walls'...

The Call of the Auras

The first thing on the minds of his neighboring southern nobles was The Wall as they hurried to answer Drum's call to a meeting. He had sent word last night that the building of the new road was officially approved and was a sure thing this time... but he had also let them know about The Wall.

The Southern nobles were a wealthy but fearful class in the political body of River Kingdom. Although their lives were comfortable and frequently excessive, they were insecure and jealous as individuals and always in competition with each other. In addition, they were furthest away from The Castle and the protection of The Royal Castle Guard.

So, because of all their precious indications of wealth, the nobles of each grand manor felt that they needed and deserved the most protection. And there was another issue. Their estates were closest to the ungoverned, mysterious Lowlands and the most dangerous individuals of all, the Lowlanders. For this reason, this group of nobles were focussed upon something even more important than the new road... The Wall.

The Wall would keep the nasty Lowlanders *out of River Kingdom*, period. To them this was the most important piece of construction next to the road itself. In the minds of the nobility of River Kingdom, this protection was even more important than improved transportation. The dark-skinned Lowlanders had always been considered to be a major threat. Among Drum's southern neighbors, they were the main topic of conversation in all their social and economic gatherings.

"Lowlanders are nothing but low-life savages."

"You can tell by the way they look and the clothes they wear."

"They don't know how to behave like civilized individuals."

"Anyone can see that."

"And they all want to marry someone of the superior white race."

"There ought to be a law against that."

"You can already see what has happened in Urhonordo."

"It's disgusting."

"And there are probably a whole lot of them right over there on the other side of those trees."

This morning, emotions were high with the nobles. This was their time to put into motion what they had discussed with Drum over the years. The Wall was necessary to keep the Lowlanders in their place, once and for all.

Soon most of the nobles were seated. Balynn had gathered up the servants to prepare for them since she couldn't seem to find Paulina who usually did this.

The servant ladies (all of whom were Lowlanders) had brought in the tea and scones required for an early morning meeting. The discussion was lively. When Drum entered the room at last, there was a round of applause and the servants disappeared...at least from view.

Drum joined in with the clapping as he entered, as usual, pointing out this one and that one of the nobles with special acknowledgement as was his habit. Those acknowledged in this way always seemed to covet the attention.

Drum was not just another noble, anymore. He was a celebrity who had spent time with Royalty up north and was able to make that Royalty give them the road they wanted. He was special.

Drum's first speech was a blow-by-blow description of how the final road approval had been gained. Strangely, Balynn noticed, there was no mention of the day trip to Thorndike Manor or even of Thorndike himself. Drum's focus was on himself and the completion of the mote and the drawbridge.

This focus was required for this audience, Drum decided. These creations were important because of The Wall that these nobles were particularly interested in. This wall would run from the Drumsworth Drawbridge through to what would soon be known as 'The River Kingdom South Security Gate' of The Wall. The Wall would then stretch from that gate on to the most western boundaries of River Kingdom, blocking all access to the Royal River Road but, more importantly to these nobles', were their 'endangered manors' to the north of it. The Wall would, at last, keep their piece of River Kingdom safe and keep all disgusting Lowlanders in their place on the other side.

The nobles were ecstatic. For the first time in a long time, they felt that their need for white supremacy and safety had been heard by The King and there was a chance that River Kingdom could return to its old superior way of life.

They could make their own decisions, decisions that didn't need any consultation with the likes of Urhonordo or The Lowlands. *Superior, white control of* River Kingdom would be back, at last.

After the initial cheering and self-congratulations, however, reality began to sift into the discussion. Montrose Manor's noble, Lord Montrose, shared a border with The Lowlands on the southwestern end of River Kingdom, just beyond the more central Drum parcels. His was one of the smaller holdings so he had little in the way of militia and his landholder's farm hands were almost all Lowlanders.

These were 'good Lowlanders', however, ones that his landholder had handpicked and kept under lock and key in a small shed on his property.

Lord Montrose scratched his head. "Just exactly how is this going to work?" he asked carefully. "If the Lowlanders are all locked up on the other side of The Wall, who is going to do the work that needs to be done? To begin with, who is going to build The Wall?"

The nobles all nodded to each other. Lord Montrose (Monty) was asking what they had quietly been wanting to ask, themselves.

Drum was ready for them. "That is the beauty of my plan," he chortled. "The Lowlanders are going to *build their own wall*. What do you think of that?"

The nobles erupted into cheers and laughter. What a great idea!

But Monty was not convinced. Shaking his head, he continued to question. "So, you say that the Lowlanders are just going to voluntarily build their own wall to keep *themselves* out of River Kingdom? Is that what you're saying?"

There were more laughs, but it was a very good question.

"Of course not," explained Drum confidently. "You all know Lord Thorndike from up north, right? Well, he has a huge militia that has been training for this. A few days ago, *I* talked Thorndike into assigning some of his militia down here to guard The Wall.

"They will be here soon to round up a whole 'work team of Lowlanders', if you know what I mean. Those big militia boys will "supervise" the Lowlanders and keep them 'in check' building their own wall. How's that for a great answer?

Drum sneered as he gazed around the room. The nobles concentrated on the answer, processed it, and then began to sneer as well. The Lowlanders would be building *their own cage*, like the wild animals they were. It was the perfect answer.

"And you can keep your own 'pet Lowlanders' that you've had for a while if you want them. You can even get more for special needs when you want them. All you will have to do is go to The Wall and pick what you want."

More cheers went up around the room. Drum had all the answers. The man was a genius. The meeting ended shortly with full attention focused on the wonders that Drum had announced. By noon, the nobles had dispersed to their own, private manors with no discussion at all about the irrigation of their farms.

The nobles had decided roughly that they would each contribute some of their own field hands and 'private Lowlanders' to the required remodel of The Royal River Road that was near their own property.

There was no concern with required closures of levees. The finished road would easily handle all the increased travel of militia for building The Wall now and Drumsworth vacationers next summer. The concept of next year's irrigation problems was far from their minds. They did not work in the fields. Their land holders did and they had not been invited to the meeting. Irrigation was not the nobles' issue.

Balynn was impressed. Her father had conducted this meeting with a flair, enjoying the credit he had garnered for the successful layout of his plan. He had answers for everything. He was in a leadership position with these nobles. And she could tell that this position made him happy.

For a moment at the end of the meeting, her mind flashed back to a distant recollection from her youth. It was her memory of The Crown, carefully hidden as it was in its secret vault behind a picture of her great grandfather. *I wonder if he still cherishes that crown as a wonderful heritage...or is it more,* she thought to herself.

But now, everyone was gone, and she was alone, again. She wandered through the corridors of the manor by herself. Where was Paulina? She had not seen her at all lately. Balynn had been allowed to attend the big meeting this morning, but now she needed someone to discuss it with.

"You can discuss it with me," offered Cbalynn. "Your father said a lot of terrible things this morning. Why don't you discuss them with me? Maybe we can figure out what to do about them."

Balynn jumped. *What do you mean 'terrible'?* she thought. *And aren't you only supposed to help **me** do things right, instead of my father? I haven't done anything wrong lately so what is there to discuss?*

Cbalynn. "You have a point," she admitted, "but I do see an upcoming need for you to make difficult decisions in the future because of what your father is putting in place. Maybe you should think about these possible future decisions now before you have to make them."

Like what decisions? thought Balynn to herself. *All the decisions are going to be made by Daddy and the nobles...not me.*

"Like how you feel about slavery?" Cbalynn began.

"I hate it, of course", Balynn snorted.

'Good answer," said Cbalynn promptly. "So, who's going to build your father's wall?"

"Lowlanders."

"Lowlander slaves?" Cbalynn added.

Oh dear, Balynn thought suddenly,

"And who will be their slave masters? Jorge or Jackson or both?"

Balynn pulled on a shawl and walked out to the front of the manor. Would the boys she now knew actually do something like that? She decided to ignore the question.

"Haven't they already been slave masters when they were working on the mote?" asked Cbalynn.

I guess you're right, Balynn thought sadly. *Buy that's over now. I really don't want to think about that.*

"Indeed...And then, there's your father. What are you going to do about your father?"

My father is just fine, Balynn thought with a smile, thinking with pride at the respect the other nobles had shown him this morning. *All the nobles look to him for direction. They have faith now that he can get The King to do what they want him to do. And when that new road is built, citizens from all over River Kingdom will flock down here easily. They can even swim in the mote.* The thought was delightful.

"Will the slaves that built the mote be able to swim in it?" asked Cbalynn.

Why do you keep talking about slavery? Balynn thought angrily. *I have nothing to do with slavery...and neither does my father...anymore.*

"Then who has your father sent the Drumsmen to 'round up'? Who is going to be building The Wall? Who is going to be "supervised" by Thorndike Militia while *they* do all the work?"

Balynn walked to the edge of Drumsworth Lake. Her world had been so simple only a few months ago.

"Simple but ignorant," Cbalynn acknowledged, reading her thoughts. "And your life was boring, too, don't you think? Which would you rather have? An interesting life or a boring one?"

I don't like all the mixed emotions you are asking me about now, Balynn pouted silently. *I am just a girl. I shouldn't have to be bothered by all these questions that don't have any good answers.*

"I understand, said Cbalynn sympathetically, "but these hard questions come with growing up. What would you rather have...no hard questions or... Jorge?

<u>"NO SLAVERY"</u>

Up north at The Castle, more heavy-duty thinking was going on. Willa and Solem were busy with the scribes of The Castle putting finishing touches on the "No Slavery" message that Solem had finally completed. Many exact copies would need to be delivered quickly to every noble in the kingdom. The good news about the Royal River Road approval would be the first part of the message, but the law forbidding slavery would be delivered with it so that no time would pass in between. All slavery was forbidden in River Kingdom for the first time.

June and Manda wandered out to the garden to wait for the finished message to be copied. Manda would take her copies to her parents and the rest of Urhonordo. She was sure her parents and Urhonordo would want to be included. June would hand out the announcements to all nobles and commoners throughout River Kingdom.

June and Manda both felt the importance of what they were about to do. They would be announcing major changes to their world where major change seldom took place. They would be bringing good news to many but unexpected and even unwanted news to others. How would their information be received?

Both of their Consciences were on alert. "Are you worried about what some River Kingdom citizens might do when you tell them there must be no slaves working on the road or anyplace else?" mused Cjune to June. "Many people in River Kingdom may think they have servants, but they actually have slaves. They don't pay their help and don't let them leave. They may get mad and object to this new rule."

Cmanda was eavesdropping and whispered to Cjune. "But maybe you should tell June that he and Manda should deliver announcements together so that Manda doesn't get hurt.

Cjune agreed. "Maybe you and Manda should go together to deliver your Royal Announcements," June's Conscience advised. "Manda may need protection if people get angry."

June laughed. "If I try to protect Manda, someone will have to protect *me*." he joked.

"Who are you talking to?" Manda asked.

"My Conscience. He thinks you may need protection if people are asked to treat their servants like servants and not slaves. What do you think?" June responded.

"*I* need protection? Your Conscience has a short memory, hasn't he?" snickered Manda.

"I told him not to worry about you," June laughed, "but our Consciences do make an important point. Not everyone is going to greet our news with happiness."

"I am not worried about Urhonordo people," Manda sighed. "No one there ever thought of slavery as an option. You, however, are not going to be greeted with joy when it comes to the slavery issue and the nobles. Maybe you should take some of The Guard with you...unless you want *me* to come with you for protection," she added with a snort.

Both young royals were quiet for a moment. "Tell me," June asked seriously. "What exactly is the difference between a slave and a servant? Why was Drum wrong when he said that he was just using workers and not slaves to build his mote?"

"Because his workers weren't willing, June. They were kidnapped, kept in a prison, forced to work, and not paid. That's why."

"*We* have servants here at The Castle. Some days they don't want to work, I'm sure." June continued. "Why are they *not* slaves...or are they really slaves, too?"

Manda looked hard at June and suddenly she knew something she had not known before. She loved him. He was thoughtful. He was concerned. He was thinking about others rather than himself. He was...wonderful. Without hesitation, she moved closer to him and took his hand.

"That is the best question you could have ever asked me, June," she whispered. "You asked because you are thinking about people you care about and you have heart for them," she said simply. "That makes you very special."

June just blinked. What had he said that produced this wonderful reaction?

Manda explained. "The difference, June, my friend, is control vs. self-control. I just figured this out myself. The servants in your castle, Horace and Agnes and the rest, work hard all day but they are there because *they* have chosen to be there, not because they are forced to be there. They are free to leave The Castle if they want to go, right?"

"Right."

"Slaves can't do that. Slaves are captured and held in place by fear or force. They are required to do what they are ordered to do and they can't say no. The issue that makes a slave a slave is control. Slaves have no control of where they work or what they must do. Slaves have no control of their own lives at all."

June sat very still. He tried hard to understand what Manda was explaining to him, but she was holding his hand. He was having trouble thinking at all.

Finally, he managed a question. "Why do smart Lowlanders let slavery happen to them?" he asked. "And why is slavery never a problem in Urhonordo?"

Now Manda had to question herself. "I'm not sure, but think the answer again is control. Urhonordorian Lowlanders are there because they came there and decided to stay. They can stay or go if they want to. There is no control one way or the other."

"There are many Lowlanders in River Kingdom that come and go, "said June quietly. "They have control."

"Right," agreed Manda. "Free Lowlanders in River Kingdom are not slaves. But many Lowlanders in River Kingdom were captured from the Lowlands a long time ago or even recently. They were never controlled by anyone in their homeland so they didn't know they should stand up against it. Now, when they are captured, they don't know how to tell their captors 'no' so they just accept their situation as though it's required.

"I think the nobles always feel they should have as much power as The King, but they don't. Instead, they decide to control their slaves which makes them feel powerful. Power loves its power. That's one of the reasons the nobles have slaves. The other reason is that they want a lot of things done and don't want to pay anyone to do them."

Manda sat down flat on the cold garden bench and stared at the muddy grass before her, still wet from the recent rain. June sat as well, only a few inches away. "She's smart," Cjune whispered in his ear. "She thinks about things more than you. She's making sense."

June was sitting very close to Manda. He dug his toe in the mud in front of him. "I finally think I understand," he said slowly. "This idea of slavery has one hundred-year old roots among the nobles. I never really thought about what control of one group means to another."

"And then there is the color issue," Manda muttered.

"What color issue?"

The Call of the Auras

Manda stared at June. Then she picked up his white hand in her brown one. "This color issue, June. Do you see the difference? I am brown. You are white. Haven't you noticed?" Her eyes were wide with disbelief.

June squeezed her hand and laughed. "Yeah, I see the difference, so? What difference does that make?"

"Haven't you also noticed that all the slaves that we set free were brown?"

June shook his head. "I guess I never paid attention to what color they were. They were just people."

Manda just smiled incredulously. He was being honest. Color was not important to him.

She took a deep breath. "You are right, June. All people are just people. But color is important to the nobles. It makes it so much easier for them to see who should be controlled. And, since Lowlanders don't fight back, I think nobles have decided that 'the brown people must not be as smart as the white ones.

Now it was June's turn to stare. "Really? Do you think nobles really think that?"

"I do and the color difference makes it easy to see who should be controlled."

"Color? My god, I never thought about color, but that's true, isn't it isn't it? The people who are being made into slaves are mostly dark skinned aren't they? I hadn't thought of that."

June reached out and retrieved Manda's hand again, pulling her close. Slowly their arm folded around each other. It felt natural, as though they had been planning to do this since they were twelve. "I love you, Manda. I think I always have."

"I know," Manda answered. "It just took us a while to figure it out. I love you, too."

After a few warm moments, they were holding hands again and pushing the mud in front of them back and forth with the toes of their shoes. When the copies of the announcement were ready, they would be going in separate directions to deliver them to very different kingdoms. The way that these announcements would be received would be very different as well, but, for now, the two young royals were learning about life from each other.

*****

The next week was one of significance in River Kingdom. June led the delegation of couriers throughout The Kingdom with the announcement of the approved construction of The Royal River Road. The nobles from north to south cheered mightily at the prospect of an elegant new highway that would link the northern part of their kingdom with the south. This major road was the first end-to-end road among all known kingdoms, and it made them proud.

They could see themselves traveling throughout the kingdom, now with The King's Royal Guard available to protect the roads, all of which would connect them with each other more than had ever occurred before. And of course, Drumsworth would be even more easily available for extended vacations. Next year's social season promised to be spectacular.

Townspeople were thrilled, as well. Transportation to all parts of The Kingdom would mean that there would be more connection with other areas for more trading. Obviously, the new road would be able to provide the protection of the Royal Guard to all portions of River Kingdom more easily, should the need arise, but the best news was that there would be more places for individuals to sell and trade their wares. New towns and villages along the new road would thrive.

In the Central portion of River Kingdom, the new road was greeted with forecasts of doom, however. One-by-one, the word reached the inland farmers and scattered land holders and was greeted universally with projections of crop failure and disaster. Together the farmers journeyed north to have a word with Thorndike and request that no 'road improvements' be made in the Central area until they could find some way to mitigate the disaster that would be brought about by the plugging of their only water supply.

As could have been predicted, Thorndike was more interested in the completion of the new road than he was in any lowly land holder's problems. What made matters worse was the fact that none of the nobles that owned the Central farmlands were concerned at all. They did not work the land and had merely concluded that the complaints and fears of their landholders were completely unnecessary.

The announcement of the new law regarding "no slavery", on the other hand, had a completely opposite reaction among all parties. Landholders used Lowland labor in their holdings as workers in all phases of the operation. They always had. To them, the law was not necessary. They had a good relationship with Lowlanders for the most part and didn't need a law to make it happen.

The nobles, however, were, horrified. To them, this law was actually *thievery!* In the minds of many nobles, slaves were *their property*. No King could just pass a law that took something away that they *owned*. How dare he even think such a thing, let alone put it into a law.

The Call of the Auras

The nobles *owned* their slaves. In some cases, they had purchased them from other nobles, but, in most cases, their militias had gone down to the Lowlands and kidnapped what their nobles needed. Many of the nobles had owned their slaves for a long time. The nobles owned their children and grandchildren, as well. Who did King Solem think he was to take away a good noble's property?

June had been flabbergasted by the nobles' responses.

In every case, Prince June made his announcements bravely, handed the formal paperwork directly to the noble of each manor and concluded with a respectful bow. In some cases, the noble received the papers solemnly. In others, the noble threw them to the ground. In many cases, the nobles had one message to deliver before they turned abruptly and marched into their homes. "We shall talk to The King about this. This will not stand."

*****

Manda's dispersal of the Royal announcements in Urhonordo was much easier and less dramatic than June's. Since most of the citizenry could read and none had slaves. Manda simply posted the announcements on gate posts and buildings in the villages. She handed them to villagers and farmers that she ran across. She didn't take time to explain. No one seemed concerned or surprised about the Royal River Road which was not even in their Kingdom. Their way of life was easier. They didn't feel the need for a fancy road or worry about the one in River Kingdom...if it didn't cause flooding on their side of The River.

Manda took no more than a day to make her deliveries but arrived at Urhonordo Castle to find it a hive of activity. A cohort of river-flow experts dispatched by Queen Luna had quietly roamed the central farming area of River Kingdom in the last few days and they had discovered some sources of water inland that might supplement river water for River Kingdom's central crops. They would not know for sure until they did some digging, so now, the Urhonordo citizens were preparing to travel across The River. They were willing to see if they could help now that the announcement told them that the new road was a sure thing.

Manda looked around her and was proud of what she saw. Dozens of men and women were loading horses and satchels, preparing for the journey. They would travel across the shallows down river, the area that Manda had shown her father last week.

Crossing there would place them directly into the Central farming area of River Kingdom. King Solem had approved of this adventure and was grateful for the assistance since Central nobles were apparently oblivious to the need. Hector would be there to meet them and introduce them to the other landholders and farmers.

Manda found her mother standing on a porch that overlooked the busy scene on the front lawn. "I delivered all my anti-slavery announcements, Mother," she said happily as she approached. "Urhonordo was not surprised or upset. I was pretty sure that was how our people would feel about slavery and the road. And now I see everybody ready to help look for water. That is so wonderful.

"King Solem was thrilled with the Water-search idea. I think he heard you and I when we brought up the water problems of the poor Central farmers, but I also think he felt pressured by the nobles. If we can find another source of water inland for those farmers, that will solve a lot of his problems."

"Some but not all," mused Luna. "And what do you mean "we"? You're not expecting to go with "the water search group" are you?"

"Of course, I am intending to go. Hector is my friend. I want to look *with* him for the water he needs."

"You just got home. You are tired. Your horse is tired. There are a lot of strong people already ready to go now. Your father is going to lead them. I don't think they need you to do this."

"But Mother..."

"You know I won't forbid you to go, but I do suggest it. Your father will give your love to Hector for you."

Manda knew better than to argue with her mother's 'suggestion.' Reluctantly, she watched as her father raised his hand as a signal to all those around him. He was leaving. The Water Search Mission was under way.

Again, despite her disappointment, Manda watched as a long file of Urhonordo citizens filed past her and off to the 'shortcut' that would land them where Hector would be waiting. She was again aware of who was in this group. Many of them were of Lowlander-Urhonordo mixed blood or pure Lowlander heritage. Together they were marching toward the cutoff where they could cross, ready to help people they didn't even know.

Cmanda knew what she was watching and said so. "All of these people in front of you have wonderful Consciences," she said softly. "They are going to help people who need help, people they have never met, just because your father and mother suggested it. You are very lucky to have such thoughtful parents and community members."

Manda stopped watching with a start. Her Conscience was saying that she was lucky. Why? Because she had good parents with good Consciences. She had never thought of things that way. But now she understood. She *was* lucky. She continued to watch as the group of 'people with good Consciences', being led by her father and *his* good Conscience, disappeared out of sight.

THE CONNECTIONS

Hector was a landholder and had been up delivering messages with his family all night. During the afternoon, two Urhonordorian messengers had arrived at his home just south of Thorndike lands. They had requested that he spread the word south and inland to the rest of the Central landholders of River Kingdom. The Urhonordorians were coming to help find water for their farms.

Hector was a proud landholder as his father and grandfather had been before him. He knew just where Thorndike property ended and his began. Hector's property was groomed and weeded. His wheat had been harvested in a timely way and sent to market to be sold for a good sum. His noble had paid his taxes with the proceeds and furnished Hector and his family sufficiently in terms of their own food stuffs, housing and equipment.

Hector had always felt satisfied...or almost satisfied with this system that had existed for centuries, so why did he need help with his farm?

He was just worried about his levees and so were his fellow landholders.

Thorndike's property was less well kept, less tidy and less productive than Hector's...but it was larger. It produced larger quantities of everything. Thorndike was a major producer. Plus, he had a huge militia housed on his property. No wonder he had been able to push through the building of the new road that would ruin Hector's levees.

Hector's concern about the levees was near panic stage. The levees were a lot of work and they were the major function of Hector and his family and those in the adjoining plots of land of the Central farms. But these levees were what kept their crops alive.

Each plot of land was owned by a noble, but the nobles didn't work the land. Their landholders such as Hector did. That is where the nobles and land holders had differed at The Gathering, and this is where they still differed. The nobles had no idea how the levees worked or how plugging the levees with the new road would affect their land.

The levee system was a constantly changing system of hand-dug canals that coaxed river water inland by digging tributaries in from The River. The manpower needed was constant. On good years, farm hands were able to create deeper ponds that would last longer into the year. On drier years they would have to dig those ponds longer and deeper. It was a hard life, but it worked.

Hector and his family had received two notices, first from his brother Horace (who was always the first to know) and second from a courier sent over The River by Urhonordo. Both notices had said essentially the same thing. King Solem was aware that the new road might cause trouble with the levee systems, and he was happy to have approved Urhonordo's offer to help with this possible problem.

Hector was not at all sure that assistance from Urhonordo could do anything more than he could do, but he was grateful for their concern. Hector had never been to Urhonordo. He had never even met a citizen from that kingdom. He had seen them, of course, at The Gatherings but he had never actually talked to one.

They had a beautiful Queen. He had definitely noticed that. Maybe she would be the one that would visit. *No,* Hecter concluded realistically to himself. *Queens never do such things. But why is anyone from Urhonordo coming here?*

Then Hector stopped short. *I forgot,* he thought suddenly. *I do know at least one Urhonordorian. I know Manda. I have known her since she was a little girl. And she was on our side at The Gathering. She already knew about the problems we here in the Central farms would have with the new road. She even said something then. She also brought Queen Willa here to see the problem.*

With increased the enthusiasm he got from that memory, Hector plowed from homestead to homestead to deliver his message, and now he was less concerned. The Urhonordorians were coming, and he actually knew one. She was smart and thoughtful and understood their problem. An Urhonordorian visit might be a good thing.

By early morning the next day, a small contingent of land holders had gathered at the spot where the Urhonordo messenger had indicated they would arrive. *At least someone is going to try this with me,* he thought gratefully.

The would-be helpers would arrive by way of Manda's shortcut and Hector knew where that was. It was the quickest route and the most unknown. It didn't go near Thorndike's estate at all, which was a good thing. The little welcoming committee of farmhands and land holders was abuzz with hope and curiosity.

At last, they saw the expected Urhonordorians. They were just entering The River on the far side and were picking their way across the boulders that formed the shallow spot that Manda had discovered.

Even though the group was expected, their appearance was not. What they noticed immediately was frightening to the farmers. There were dozens arriving at once, an invasion force of large, strong, dark-skinned men. In addition, they were equipped with a fearful array of what appeared to be weapons...long, heavy objects that were surely javelins and clubs and swords!

Hector squinted fearfully, searching for the familiar form of Manda but he could see nothing but serious, determined looking men. And they all had brown skin with even darker hair sticking out wildly from under the hoods of their travel garb. They weren't Urhonordorians at all. They were *invading Lowlanders!*

The day was damp and foggy along The River and completely without the comforting rays of sunshine that might have made the visitors' appearance less ominous. Now trembling with fear, the little farmer group began to back away from The River's edge. They had not brought anything to defend themselves with. They were getting ready to bolt back into the land.

"Yo! Hello! Hector! Is that you?" yelled one of the large men that was leading the 'invading horde'. Hector blinked and strained to get a better look. With shock, he recognized the caller. It was the Royal they called 'Prince Armando'.

This man is from Urhonordo! Hector realized with a jolt. *He's the most important man in Urhonordo, the husband to the Queen... and he is the one that is here to visit us?* Hector realized all this now with amazement. *The Queen has sent her King to help us. Wahoo!*

With the joyful understanding of this realization, Hector rushed among his cohorts and explained the significance of what they were all seeing. King Solem had sent 'the king of Urhonordo' to help them. That indicated that King Solem *really did* understand that the levees were in danger and *was* at last on their side.

Gradually, the dripping contingent from Urhonordo straggled up onto the shore of River Kingdom and climbed off their horses. Hector stepped forward. In front of him, Prince Armando stood tall, as if at attention, his hand on a sword that hung on his side.

He nodded politely and held out his hand. Hector held out his hand as well, even though it was shaking. The rest of the surrounding men, brown and white, wet, and dry, watched cautiously but breathed a sigh of relief.

Armando decided to speak first. He could see that the little farmer group was all white-skinned and quivering with fright. He knew that his men were all very dark and were not who the farmers had expected. His voice was soft.

"Thank you for coming down to The River's edge to greet us," he began. "I know that you expected Urhonordorians to be arriving this morning. I am Prince Consort Armando of Urhonordo, and all of my men here are also Urhonordorians, though probably many of us were born in The Lowlands. You were surprised by the way we look, right?"

Silence............

Armando took a deep breath and continued. "How many of you have ever talked to a Lowlander before?"

Silence.........

Armando now smiled at Hector. "You must be Hector," he said. "And you know my daughter, Manda, don't you?" Hector nodded. "Do any of the rest of you know Manda?"

Surprisingly, several hands went up in the group including Hector's. Armando looked around. "Since some of you have met Manda, you should know three things. The first is that she sends her love and wishes her mother would have let her come here with me as she wished to do.

"The second is that she and my friends here with me have read some of my mother's maps and they think we can help you find water.

"Third is that you who raised your hand and said that you know my daughter can now say that you already know at least one Lowlander because Manda is half Lowlander, herself. Do you know that?" Again, Armando looked around slowly.

Hector stepped forward bravely. "I know Manda. I didn't know that she was part Lowlander. I never thought about it. But I always trusted her, so I trust her father. It will take me some time to get used to trusting all the rest of you, though," he added honestly.

A man from the back of the *Urhonordo* group snickered, "Me. too. You people are so white that you look like you live under a rock. I will have to get used to you, too," he quipped.

The farmers nudged each other.

"This man thinks he's funny."

"He *is* funny, George. What he said sounds like *you*."

"Maybe we *can* work with these guys."

"If you weren't looking at skin, you would think they're acting just exactly like us."

Hector and Armando both listened. They would not move quickly but they both now wanted to make this work. And they both had a plan. Hector's plan was to hear Armando's plan.

Hector had no idea how he and his fellow landholders would survive when spring came and the water from The River was blocked. Armando's plan was simple, but it would take some explaining. Carefully the Prince selected a rock from the shoreline and maneuvered it to a sandy clearing. He sat down and waited.

Without further instruction, the men from the farmer group began rolling rocks toward Amando's rock and setting them in a circle near Armando. The farmer that had been standing next to the Urhonordorian that yelled his quip stepped out a way into The River, grabbed a rock and plunked it down in the circle. Withing a few minutes, most of both groups were seated and nervously facing Armando.

The Consciences of both white and brown men hopped out onto shoulders and heads and looked around, trying to make out what they were seeing and how they felt about it.

"What Is that head man going to say?" whispered a white Conscience.

"He seems like he's an okay guy," whispered another.

"I've known him all my life," added a brown man's Conscience who was listening. "You can trust him."

"Well, I'll be darned," said the first white Conscience. "You brown people have Consciences, just like us white guys do?"

"Of course," snorted the brown Conscience. "We're human. All humans have Consciences."

"Yeah," retorted another brown Conscience. "But a lot of humans don't listen to theirs, I don't care what color they are."

"You got that right," agreed a nearby white Conscience.

Armando looked around at the group in front of him. They were sitting next to each other. They weren't fighting. *If only I can explain this right,* he thought nervously.

"You'll do fine," Carmando assured him. "Consciences of both colors are getting along, already.

Armando began. "Being born in The Lowlands, I think I know what you farmers are worrying about today. It's all about water, I understand." A lot of white heads nodded.

"Water is what Lowlanders are always worried about. They love their grassy plains and cacti, but they are always worried about the water that trickles into The Lowlands at the southern end of The Great River.." (All the Lowlander men nodded now.) Because water is always scarce in The Lowlands, Lowlanders have spent a lot of time looking for water in places where most people wouldn't think to look.

"Some Lowlanders have dousing sticks that can sense water where you can't see a thing. Other Lowlanders look for sign in plants that turn green in a dry area. Others look for round rocks like those we are sitting on. They are round sometimes because water has washed over them in the past."

The farmers looked sideways at the brown faces near them. Lowlander faces were smiling. Their heads were nodding. The Lowlanders were taking stock of the white men next to them. So far, so good.

"Water searching is what Lowlanders have always done. That is what we Lowlanders *who are now also Urhonordorians* would like to try here. *We* think... we *hope* that we can use the skills we used in our homeland to find the water that you folks need here. That is what we would like to do. What do you think of this idea?"

Farmers and landholders were not experienced in much of what went on beyond their small portions of land and were, by nature, suspicious of that which they had never experienced. This 'invasion' of these men who looked so different from anyone they knew was a definite challenge. They were not sold yet.

Hector, however, was a bit more experienced. He had spoken directly to The King himself, at the most recent Gathering. Now he stepped forward to take advantage of that experience.

"Hey, you guys. You have all seen Manda around here on her horse for years. She spoke up for us at the Gathering and so did her mother, the Queen. They are both at least part Lowlander...if you look close you can tell.

"Because of this, I say that these Lowlanders and 'part-Lowlanders' are on *our* side and have come here today to help us. That makes them 'good people' to me. I say that we should accept their offer and be grateful for it." He hesitated a minute and looked around. "At least that's what I think."

 He sat down and waited. Armando and his cohort waited. At last, the big farmer that had spoken up before made an announcement. "I just heard from my Conscience that the Lowlander-types have Consciences just like we do. To me, that means that they are about as good as we are. I say that we make use of their offer. That's what I say."

There was a short round of quiet rumbling, but within a few minutes, the heads of the farmers were nodding. Hector was pleased. He looked Armando in the eye and said what he thought he should say...in his current position of 'leadership'.

"Thank you for your very kind offer, Your Majesty," he said with true respect. "We here in the center of River Kingdom have never received such a kind offer, not even from our own Royalty. We gratefully accept your kindness. We didn't expect it and we are not well prepared for so many visitors. We will have trouble housing you all while you help us, I'm afraid."

Armando grinned in relief. "Thank you so much for your acceptance of our humble offer," he replied, "but don't worry about housing us. We are all nomads by nature. If you will just show us to a place of dry, unused land, we will provide our own accommodations."

"Come with me, men," the man who had spoken before said, standing up and looking around. This man turned out to be Thomas, a long-time landholder and one with the respect of his neighbors. Thomas was a hard worker. Everyone knew that.

"I say that there is a piece of clean grass just a short way from here on my land. It's close to The River for drinking and bathing. These new 'friends of ours' (he said this carefully) can all hang out there. It will be a good place to stay for a while and to meet. What do you say, men? Is this a good place to start our..." He stopped for a minute. What was it that they were going to do together?

"We can call ourselves 'a work crew' like all the crews we are all a part of when we need to change or reposition our levees," offered Hector. "That's what we can all be...hopefully," he added, looking around at the multicolored faces in front of him.

"Right. Good group name...Fits the situation," said Thomas without fanfare. "Come on, all of you 'work crew people.' Let's find you a good spot." Without further discussion, Thomas turned and marched off into the inland.

Little by little, the motley 'crew' of white and brown followed him, some with true commitment to this new idea, some still hesitant but willing to try.

Armando looked at Hector with satisfaction. "It looks like you have a group of good men here who are willing to meet new people," he murmured, "even if they don't look the same as the people they have always known."

"I'm just learning to do that, myself," murmured Hector. "Manda is the only 'half-Lowlander' that I've ever known, and now, you, of course," he added with a half-smile.

THE GLORY OF THE WALL

The importance of this meeting could not be overstated. Tonight, Drum and Thorndike were going to announce the construction of the most prized dream of the nobles...The Wall...a barrier that would keep the obnoxious Lowlanders in their place once and for all. Drum knew that this barrier was the key to his success with all northern nobility.

Thorndike was here tonight for only one thing, however. He was here to collect what he expected the announcement of that wall to inspire in the nobles... an enormous militia, a combination of all the northern militias of The Kingdom under his supreme control. He closed his eyes...

" You will need them for Urhonordo", a voice whispered.

Thorndike looked around. He had been hearing that voice all day. It was disconcerting.

Drum was here for two reasons. The first, of course, was Thorndike's control for an enormous, combined militia. He had already sent his Drumsmen out to inform the northern nobles of the 'Lowlander Invasion' and encourage the need for their militias. If Thorndike controlled all the militias in the north as well as his own (and Drum controlled Thorndike) there would be no one in position to tell Drum what he could or couldn't do. He would have more power than anyone...particularly more than King Solem.

The second reason for Drum's attendance tonight was to enjoy the support of the white nobility itself. Drum had complete control of the nobles in his southern region, but lately he had really enjoyed his support from the north. This was Thorndike's territory, but Drum thought he had captured the affections of this end of the kingdom, too. Since The Gathering, Drum had reveled in the support of these nobles and their response to him. He was now positive that this appreciation of him was what was key to his goal...*his* royalty in Urhonordo.

Thorndike positioned himself at the head of the table with Drum at the foot. Thorndike's Auras had suggested this. He should take the leadership position. This was *his* house.

Now he stood and addressed the nobility that surrounded the formal table and crowded into additional chairs behind them. Everyone was here. It was crowded. He was confident in his mission, but Thorndike began sweating.

"Since the reign of King Drum," Thorndike began. "Lowlanders have infiltrated areas of influence in River Kingdom" he said. "Urhonordo has become completely over-run, and Lowlanders have even married into the Royalty there. This has been disgraceful, of course. It is time that someone do something strong so that this atrocity doesn't happen here!"

"So, this is all about Urhonordo, isn't it?" a voice asked.

Thorndike looked around. It was that voice again. His mind tipped sideways. *Who said that?* he thought. *Who is here that's trying to confuse me?*

The Auras heard the voice and swirled around his ears. "Don't listen to any voice you think you hear," they hissed. "Speak up! You need to talk about The Wall!"

Thorndike blinked and continued. "We don't want what happened to Urhonordo to happen to River Kingdom, do we?" he continued.

"But you want Urhonordo for yourself, don't you?" snickered the voice.

Thorndike frowned and searched the faces around him. Who was here that was speaking out of turn? It was disconcerting.

Grum hadn't heard the voice down at his end, but he knew something was wrong. "Of course not," Drum answered Thorndike forcefully. "What we want is control of the Lowlanders. We want them only when we want them, and we don't want them when we don't want them. Right Thorny?"

The nobles cheered. "We don't want another Urhonordo! No sir!"

The cheering seemed to put Thorndike back on track. "Our elegant new road will help us avoid that terrible outcome just by being a beautiful way for good, white people to get to and from where they want to go, right? But tonight, I want to share with you the *new* construction that Drum and I will be adding to that road that will keep all of River Kingdom safe from the horrors of Urhonordo."

"But Urhonordo is waiting for you," whispered the voice.

Thorndike shook his head and tried to eliminate the voice. "Urhonordo will never happen to us!" he cried as he shook his fist in the air.

"What's all this about Urhonordo?" the nobles whispered to each other.

The Call of the Auras

The Auras around Thorndike rushed to the other end of the table. "Thorndike has lost it. Take over!" they hissed to Drum. Drum could see that Thorndike was becoming distracted.

"Right you are, Thorny!" Drum shouted enthusiastically, rising quickly with a confident smile. "What an introduction! Thank you so much, Thorny. Let me get right into the reason for our grand meeting tonight. This meeting is a celebration! As you know, my grandfather would never have put up with the horrible dilution of the white race that has taken place in Urhonordo. Am I right?" Drum began walking around the room and clapping.

"Right!" shouted the nobles. Slowly, they began clapping as well.

"This is why Thorny and I put our heads together and figured out a solution to that very problem, a solution that we are able to announce tonight only because of our recent victory in obtaining the approval of our beautiful Royal River Road!"

 Nobility cheers went up again. The crowd of nobles was back and cheering. The Auras watched in amazement. How does Drum do this?

"As you also know, River Kingdom was invaded recently by the very people we are talking about tonight...brown-faced Lowlanders! (The crowd booed on cue.)

"I know...I know...That invasion was a sure sign to me and Thorndike that these people are dangerous. That is the reason that Thorny and I are constructing something right now that will put these invaders back under the white control that my grandfather had back in the day. We are building *The Wall!*"

Flo and her brothers were just down the hall, bringing pitchers to pour ail. 'The wall?' They stopped to listen.

The room cheered again and then burst into curious chatter. What wall? Where would it be? How would that work?

Thorndike nodded his head and looked like he knew the answers to these questions, but he didn't speak. He was trembling. Where is the voice? He wondered, looking around.

Drum couldn't tell where Thorndike's mind was, so he continued. The nobles seemed to be buying what he was saying...and they needed to. Tonight was important.

Drum held up his hand regally. The nobles quieted. "Our new wall will be built on the border between River Kingdom and The Lowlands and will be built there for only one reason....to keep Lowlanders behind it and out of the way... of the noble River Kingdom population!"

The room erupted in cheers. The nobles were standing and slapping each other on the back. This is what they had always wanted.

It took a while for the room to settle down. Flo and her brothers circulated among them with pitchers of refills, as the nobles toasted each other in a state of pure supremist satisfaction. Lowlanders were going to be controlled as they had been in the past, they thought. They were thrilled!

Drum, having now created this room full of joy, strode among the nobles happily clapping and accepting their congratulations. But after a few moments, he spotted Thorndike. Somewhere in a world of his own, he was not as happy as the other nobles. He was being ignored.

"Fix this," ordered Drum's Auras. "You need his militias."

Drum nodded. He was enjoying himself, but the Auras were right.

"Thank you for your enthusiasm," he began. "When this wall is ready, all Lowlanders will be removed from the nuisance they cause here in River Kingdom and taken back to live behind The Wall in their own country. They will not be allowed to stay there unless they are needed by worthy Nobility," Drum now recited with dignity. "Those nobles who have their own special pets may keep them, but, if they become old or disruptive or uncooperative, they may be returned to *their place*... behind The Wall!

"Lowlanders will not be free to go against their masters or say no. They will do as they are told. When anyone wants a Lowlander of any particular type or skill, nobility will be able to come down to The Wall and pick out what they want. They can bring him or her back when they are through with them. What say you, my friends? Is this Wall not what we have been talking about and wanting for years?"

The nobles were overcome. This was better than they had ever dreamed. They rose to their feet and began to chant...Drum. Drum, Drum...

Thorndike's Auras got up in front of Drum. "You're taking all the glory," they chided firmly. "If you want your *combined militias*, you had better give some of this glory to our man Thorndike before you lose him. He seems to be drifting mentally."

Reluctantly, Drum could see that this was true. He needed the extra militias. His manner changed immediately. "Thorndike is the man that has been putting this Wall construction together for you at his own expense. This has been for the use of *all* of River Kingdom. Now he has something very important to share with you... *and* ask of you. Tell them, Thorny!

The Call of the Auras

Thorndike rose from his chair, his focus returning. "Thank you, Drum for your stirring announcement. The next phase of our plan must begin immediately. Here at Thorndike Manor, the clearing and filling for the new road to The Wall has already been begun by my men as I am sure you noted as you arrived. (A smattering of weak applause indicated that the nobles only wanted to talk about the wonderful Wall some more. The road was already old news.)

Thorndike was not deterred. His desire for a grand, *combined* militia now rose up before him. "The rest of the road will need to be worked on immediately, of course. A good road that is wide enough for military vehicles *and the constant exchange of Lowlander workers* when needed must be completed and be made ready for use. This road will be for *all of you* and our militias to come and go as we see fit!" (A bit more applause returned.)

"The construction of that road will require a great amount of manpower, of course. Are your militias ready to work on it?" (No applause.)

"Come now, people. You didn't think the road to The Wall was going to build itself, did you? (No response)

Flo and the twins were listening from the hallway. Immediately, the siblings noted a change in response from the nobles. They had stopped applauding. The boys who had spent hours working on the road already, rolled their eyes.

"What do you mean, Thorndike?" the nobles groused together. "*You* are the biggest landowner in the northern part of The Kingdom. Your militia is already working on the road. What do you need our militias for?"

Thorndike had been waiting *for* this moment. More militia meant more power, first for road building, then for military conquest!

The Auras re-twined around his sweating brow and hoped he had recovered from whatever had ailed him.

Thorndike stood and glared at the nobles. "You expect to use the new road, don't you? You expect to get your Lowlanders to and from where they are locked behind their Wall when you want to use them, don't you? How do you expect to travel quickly to do that?"

The Auras sighed in relief. He was right on target.

The nobles shrunk back. They hadn't thought of that. They had to admit that he now struck a chord when he mentioned their need to go and come using this new and wondrous Wall.

Flo and her brothers also shrunk back but for a completely different reason. They heard the part about getting Lowlanders from "where they were locked behind their Wall." This small piece of what their father said was larger than anything else. Lowlanders were going to be 'behind their gate.' Translated, that meant that Lowlanders were going to be *locked up behind* this new wall. The three shivered. Flo and the boys were part Lowlander.

 Without any knowledge of his frightened offspring in the hall, Thorndike's successful set of questions and his glare had now penetrated the selfish minds of the previously cheering noblemen. The fact that *they* might have to do *anything* for the wonderful outcome Drum had outlined had escaped them entirely, but Thorndike's questioning was becoming successful.

Drum decided to be helpful again. "*My* militia and servants will be building the southern section of the road that leads to The Wall," he said, although he had no intention of letting that happen. *That's what slaves are for,* he thought smugly... *the strong backs that my Drumsmen are now rounding more up for the job, even as we speak.*

"We in the south are taking care of *our* obligations," he continued contritely. "Don't you nobles up here think that this road and the use of The Wall is partly your responsibility, too?"

The Auras curled and swarmed around Drum's legs with pleasure. He had rescued Thorndike before, and he was doing it again. They were impressed.

But Drum wasn't just being helpful to Thorndike, his Auras knew. The militias that Thorndike wanted were part of Drum's ultimate goal, as well. He was thinking ahead. One huge, combined militia under the leadership of one supportive Commander was what he might need soon on his way to becoming a king.

"What would you have us do, Thorndike?" asked one noble with a more humble tone than he usually employed.

This was Thorndike's opening. "Keep it simple," whispered the Auras.

"If you would allow your militias to join with mine, we will be able to finish the road and use the benefits of The Wall almost immediately" he stated.

"Indeed," Drum responded pointedly. "I know all of you will wish to join us ... *so that no one will be excluded from the use of The Wall and the Lowlanders behind it...*so that we, with our combined militias, will control the Lowlanders once and for all. "

The northern nobles nodded. They got the message.

Drum produced the paperwork quickly. Thorndike yelled down the hall for another keg of ale. The nobles began signing. Flo and the boys went for more ale, and then it was done. All the manpower in River Kingdom was now under Thorndike's control. What appeared to be Thorndike's momentary loss of focus was forgotten.

"Total control of River Kingdom Militias is mine," he mused with satisfaction.

"But Urhonordo is still waiting," whispered the voice, "and Drum doesn't care."

<u>HOME AGAIN</u>

Willa was alone in the sitting room. Solem was thoughtfully reviewing the reactions June had received when he delivered his pair of announcements, especially the one about slavery. Willa was feeling negative for some reason. She had tried reading, but now was admitting to herself that she still had mixed emotions about The Royal River Road and her father's interest in it.

She understood the reason that Solem had stood by his agreement to sign the road's Royal Approval despite Drum's undeniable use of slavery for his mote, and she was thrilled that the anti-slavery law had accompanied his approval. *That should make me happy,* she thought stubbornly. *Now the road cannot be built by slaves, and neither can anything else. It is a wonderful law, and I am proud that Solem signed and announced it before all (or most) of the road construction was started.*

So why was she not content? The fact that her father had already started construction of the new road *before* any approval had been granted still bothered her. This was an indication of dishonesty, and this was *her* father. There was no doubt that he had not operated in an honest fashion. Hopefully this dishonesty would not continue, but, for now, the doubts persisted. She was part of *his* family and his dishonesty made her feel negative in a way she couldn't seem to shake.

"You can't be responsible for anyone else's actions, Willa," commented Cwilla. "You are only responsible for your own."

"I know. I know. But *my* father has cheated regarding things that are crucial to the happiness of the whole Kingdom. And he is still involved with that horrible Drum, I am sure," she moaned quietly.

"Who are you talking to?" asked a cheerful voice from the doorway.

Looking up from her book, Willa discovered her eavesdropper. It was Mimi, dressed for an outing and bubbling with pent up energy. Willa laughed. "Where are you going today, Mimi? You look like you are ready for something big. Have you been invited to some festivities that you haven't told me about? By the way, you look splendid."

"Thank you, Willa. Do I really? Do you think it's too much? I really tried to look my best."

"You look fine, Mimi, but where are you going? You look *way too good* to waste that preparation on lollygagging around The Castle all day."

"Exactly correct," sang Mimi as she waltzed across the room. "And you, poor thing, are sitting here looking miserable and talking to yourself. I have come here today to rescue you from *all* unhappiness. We are going on a day trip."

"Oh really?" Willa asked with a smile. "Just where are we going on this day trip?"

"We are going back to Thorndike Manor, of course."

"And just why are we doing such a thing?"

"To view progress on the new Royal River Road," chortled Mimi with excitement.

"And to get a glimpse of a certain young militia man who might be working on the road?" asked Willa knowingly.

"Exactly." responded Mimi. "Come on, Willa. You know you're just sitting here feeling negative. I can tell. You know that there is nothing planned here today. The sun is shining. It may be getting chilly, but heavy weather will be upon us soon. We should take advantage of this perfect fall day and be happy."

"Just the two of us? We don't have a carriage ready or anything."

"I already arranged it. There's a carriage waiting for us at the colonnade," announced Mimi smugly, "And *you* can drive. I know that you know all about horses. It will be just the two of us. Doesn't that sound adventurous?"

Willa laughed at the young royal. She was so full of energy that it was contagious. She also felt sorry for a young girl alone in a big castle. She had lived that lonely life, herself. "Alright, Princess Adventure, I'm with you. Just give me a moment to prepare and we will be off."

Within only a few moments, the pair were on their way. Mimi had been right. The day was clear and bright but very chilly. Winter was on its way but hadn't arrived yet. It was a perfect day for travel. Mimi was alive with commentary and speculation, and she chattered happily as they rode along.

The Call of the Auras

Willa enjoyed her time at the reins of the tiny buggy she was driving, but a negative thought crept in. She was aware of the disaster that had befallen the queen whose place she was taking. She had not even informed Salem personally that they were going. She had left word with Agnes to tell him later, afraid that he would have said no.

"Don't focus on bad things that have happened in the past," Cwilla advised. "That has been your problem all day today and that is just plain negative thinking. Focus on the future. Think positive."

As the buggy rounded the curve that led to the Thorndike entrance, Willa slowed to a stop at Mimi's insistence. Up ahead there was a hive of activity. Dozens of men were hard at work on the road in front of them and one of them might be Jackson. They both peered through the crowd eagerly.

Immediately Willa was reminded that her father had begun work early on the new road. *He will do what he wants to do when he wants to do it,* she thought sadly. *That's what has caused problems lately. He is just not honest, and I fear that will continue.*

"There you go again, thinking negative thoughts," sighed Cwilla. "You have developed a very bad habit of focusing on the past mistakes of your father. I'm sure he has done many good things. Think about those."

Willa sat up straight. *I will think positively. I am determined,* she thought staunchly.

"Your father must have started this part of the new road before June made his announcement," Mimi observed innocently. "This part of the road looks almost finished. I don't see Jackson, though. Do you see him?"

Willa shook off her negative thoughts. "No, Mimi, I don't see him, but I'm still looking."

Mimi was in a pout. "I was sure I could see him working here today. Now I'm mad. It looks like we came all this way and got all dressed up for nothing."

Willa was determined to be positive. "Maybe not, Mimi," she offered with a grin. "Maybe he never went to work today at all. Did you ever think of that? Maybe he is still at the manor...or at least in the barracks behind it. Do you want to go see?"

Mimi nearly bounced from her seat. "Oh, Willa! You are so good and smart! Of course, that is just where he might be. Will you take me? You'll only need to wait for a minute while I snoop around and look for him. I promise."

Now Mimi stopped short. "I have an even better idea. You can go inside and talk to Flo again, Willa. She's really your sister, right? That was so strange...her being your sister all that time and you never knowing it. But you like her, right? Do you want to talk to her again…for a little while?"

Willa looked straight ahead but smiled and kept driving. "Of course. And it's not that strange. I always loved her then and I love her even more today. I just wish we had known that we were sisters before now."

Willa's smile was covering her thoughts, however. *What kind of man is it that doesn't tell his lonely daughter that she has a sister? This is just an early example of my father's dishonesty,* she groused to herself, her negative thoughts threatening to overtake her again. *Not telling us that we were sisters had us living a lie. It was wrong. Didn't he have a Conscience at all even back then?"*

"Everyone has a Conscience," offered Cwilla quietly. "That was then and now is now. Focus"

As they neared the manor, 'magic happened' according to Mimi. Jackson and Jorge were spotted out in front of the manor with rakes and shovels, repairing the driveway.

Mimi could barely contain herself. "Hello gentlemen," she said waving demurely. "What are you two doing out here? I thought you would be working on the new road,"

"We were," answered Jorge, sprinting toward the buggy. "But we had a passle of nobles here last night for a meeting and their horses tore up the driveway. Thorndike told us to get it back in shape before we went back to work on the road, so here we are." He reached up to Mimi gallantly. "May I assist you down, Mimi?"

"Indeed, Jorge," Mimi responded, daintily holding out her hand.

Jackson walked up and edged in front of Jorge with a grin. "Now that this unmannered oaf is out of the way, please allow 'a man' assist *you,* Your Majesty." he said with a bow and taking Mimi's hand.

"That one is the Queen," he whispered loudly to his brother. "She should be assisted first. Please excuse him. Your Majesty"

"I am not a queen here, Jackson," Willa whispered, laughing at the two trying their hand at manners. "I'm only your sister. Don't ever forget that now that we know it."

For a moment, the young siblings stopped and shuffled their feet awkwardly, reviewing the new situation. What Willa said was true, but it was so new. Neither of them had known this until recently. Neither had even suspected that they had another sister and now that sister was a queen. They were not quite sure how to behave.

Willa decided to take the lead. "We have a lot of catching up to do for the years we didn't know about each other, don't we?" she asked gently. "Who knew we would find out something this crazy after so long? I'm really glad now that I can get to know you. You are my brothers." She smiled at them warmly.

The boys began to bow again and then suddenly rushed forward and smothered her with hugs. *There was more than one reason for this day trip, after all,* Willa thought in amazement, her negative thinking vanishing.

Now she had an idea. "You know, boys, I have driven all the way here. It's not that I minded it, but now I think I would like to give up the reigns. I noticed that the new road is quite long and wide already. Would someone like to show the progress on the road to Mimi while I go in and have a chat with Flo?"

"Jackson, here, would love to do that, wouldn't you Jackson," Jorge answered with his eyebrows raised with meaning. Mimi blushed appropriately.

"Are you older than us or younger," Jackson asked as he noticed both Willa's youth and her beauty. "You turned out to be pretty good looking." Then *he* blushed. She *was* The Queen, after all. He probably shouldn't have said that.

"Thank you so much, brother," Willa laughed. "I think that I am slightly younger than Flo and older than you two, from what I understand. I'm not sure, though. I'll ask her." With that Willa glided to the doorway of the manor, aware that she was being watched and evaluated by two new brothers.

Jackson climbed up into the buggy. "Shall we go see what the rest of my militia buddies are doing. Mimi? They are going to be so jealous of me, driving around with a beautiful princess." Mimi grinned with joy. This was even better than she had expected. Jackson snapped the reigns, and they were off down the driveway.

Willa entered the manor alone as Jorge had decided to finish his chores outside. The manor's cold ambiance descended on her again, but she chose to ignore it. The front hallway had changed, however.

When they had arrived for the luncheon a few days ago, the hall had been clean and shining, glistening with the polish that she remembered from her youth. Now, however, it was lined with suits of armor, standing guard in readiness for their next battle. *Why is that?* she wondered. *No one is going to war. Why is my father so charmed by the ugly equipment of war?*

The chill of the hallway was as she remembered it, however. Slowly she passed the conference room and dining room, all neatly arrayed as Flo had undoubtedly insisted upon. With a shiver, she recalled the many meetings that she had walked past in this corridor, meetings of shouting and plotting and Drum occasionally leering out the door in her direction or Flo's. *I knew evil was being plotted here at the time,* she remembered.

"Indeed, you did," echoed Cwilla staunchly. "That is why I requested a special Conscience Alert to warn other Consciences of the evil that was being plotted here. And I wasn't wrong, was I? I still feel it. The Auras that were so much a part of the thinking that was going on then, are still here. I can't see any right now, but I can feel them."

"What is an Aura?" Willa asked absently as she continued toward the kitchen. The delicious smell of Flo's cooking began making her stomach growl.

Flo was at the hearth, as usual stirring a large pot that would probably be the evening meal. She looked up in surprise and then in joy. "You're back! You've come to visit! How wonderful" She dropped her heavy ladle and crossed the kitchen, arms open wide

"Willa, my love, you're here. This is such a surprise. How did you know that I was wishing you would visit today? Are you alone? What prompted you to come? Are you alright?" The questions fairly flew from her lips as she hugged Willa and then held her forward to examine her closely.

Willa laughed and returned the hugs with a new fondness that she didn't know existed. She was hugging a sister, a real sister, someone she had always loved as a childhood friend but now loved even more. It was the warmth of kinship, the warmth she had longed for as a child.

Gratefully the two young women sat down at the kitchen table, still holding hands. Willa looked around with fondness at the one place she had felt at least some of that warmth as a very little girl. "I remember this kitchen so well", she said softly. "Your mother was always so good to me. She almost treated me like a daughter."

"Until your father called her out for it," Flo added sadly.

"How so?" asked Willa, anxious for tales of their combined past. "What was Father angry at her for? Didn't he know that I needed a mother's touch and love? Why do you suppose he was he being hateful?"

"That was just his way at that time" answered Flo matter-of-factly. "My mother was fixing your hair the way your mother used to fix hers. You were about ten and you already looked like her. I guess that really surprised him and made him mad."

"How strange," Willa murmured. "I would have thought that he would be pleased that I looked like her. He must have missed her a lot, though. Thank you for telling me that. At least I know that my father had a heart back then," she added, still determined to think positively. "Right now, I don't see much of it."

"Me either," Flo said bluntly. She shook her head and went back to stirring her pot. "He's very busy with The Plan, like always" she added.

"But it sounds like he loved my mother a lot before she died," Willa continued, determined to squeeze out some positive information. "There is so much that I don't remember. What was my mother like? She was nice, right? You're a bit older than me. Do you remember her?"

"Indeed, I do," Flo answered at last with enthusiasm. She came back to the table with a smile. Willa was her friend as well as her sister. She deserved to know how wonderful her mother had been. "Your mother was beautiful and kind. She had long, golden hair thar curled when it hung down like yours does. She looked a lot like you."

"Did my father love her? Did he adore her beauty? Did she love him?"

"I'm sure your mother loved your father a lot," Flo answered carefully, avoiding the first two questions.

"I wish I could have known my father back then," Willa said brightly. "I wager that he was much jollier than he is now. Did he play with me and love me?"

How should Flo answer. She had promised her Conscience that she wouldn't lie any more. What should she say to Willa that wouldn't hurt her feelings?

"When in doubt, tell the truth," Cflo advised quietly. "That's always better than a lie. She is a big girl now. She can handle it."

"I don't think your father was ever that pleased with either one of us," Flo answered frankly. "He only wanted boy children; you know." Willa's hopeful smile faded. "We didn't take it personally though," she added quickly, "I told you not to feel bad so you wouldn't be sad. I knew that boys were what he favored. So did my mother."

"Oh", said Willa with a sudden adult understanding. "Father had two wives at the same time then, didn't he? I had no idea." She shrugged and shook her head, trying not to think about it further. She was a big girl now and that was a long time ago.

"Not really," Flo answered. "Our father had one wife and *one slave*. That's what he had. That's what he always had." Flo covered her face with her hands. How could she say all she needed to say to a little sister who was now a Queen? What would Willa think of *her* if she knew the real story about her mother?

Willa was not focused on family ties now, however. She stared at Flo. "What do you mean, 'slave'?" The thoughts of mistrusting her father that she had been grappling with earlier returned with a vengeance. "There are no slaves in northern River Kingdom. I have always believed that. And there never have been. Why do you say such a thing about your own mother, Flo?"

Flo kept her head down but continued. "Because it's true," she answered in a voice muffled by her hands. "I am not proud of it, Willa, but it's true. I thought you should know at least that much. Our father *used* my mother until he got what he wanted...his two boys. She was only a slave, and he did what he pleased with her."

"But look what he got for his efforts," came a voice from the hall doorway. "He got two beautiful sons for the price of one. That man always seems to get what he wants."

The Call of the Auras

The two sisters were caught unaware. Jorge stood in the doorway without shame or concern. He almost wore a look of pride. "I didn't know all this slave stuff until recently, myself," he admitted as he strolled toward them nonchalantly. "And I'm not worried like Flo is. Father won't send any of *us* to The Wall. He is too dependent on all three of us and we're all blood. We're safe."

With these words of confidence, he plunked a basket of turnips down on the table in front of them. "Father says he wants these in his stew tonight. Do you want help us slaves with the peeling, Sis?"

The girls stared at Jorge. They were in the middle of a discussion about family slavery! Willa was descending back into her dark crisis and Jorge wasn't the least bit concerned about anything except turnips.

At last, Willa recovered but with fury. "You consider yourselves *slaves now?*" she questioned harshly. "You *both say that you* are slaves? We have all lived in the same home for years and all this time you think you have all been slaves?"

"Indeed," sighed Flo. "We are all slaves here, or at least we are the sons and daughter of a slave. I don't know if that counts with the new plans and definitions under the new law, but it's always been true."

Willa stared at them both. "I had no idea that there were *any* slaves around here *at all.* All this time I was afraid of what Drum was doing down south and slavery was going on right here in my own house? ...and with my own father as 'slave master'? How could that be?"

"I think it started when the first Lord Drum moved down to Drumsworth," Jorge related as he calmly peeled his turnip. "I've been asking some of the old Militia guys about it. They say that Drum was close to the border of The Lowlands back then and he saw free labor all around him. Our grandfather was one of *that* Drum's main men in those days, so he just grabbed up a bunch of slaves like Drum did and brought them north. We've been here ever since.

"All the northern nobles have slaves. Some of them will probably be sent down to The Wall now, though, if they're not real useful. The nobles have been wanting to get rid of all the 'extra Lowlanders."

"Sent down to what wall?" squeaked Willa. "Sent down? Shipped off? What are you talking about, Jorge? You're talking about people, not packages."

Flo tried to signal to Jorge that he should stop talking. Willa was obviously upset. They had said too much.

But Jorge wasn't paying attention to Flo. His attention was focused on a pretty Queen who was his newly announced sister, and *she* was listening to him intently. He could see that she was staring at him. This was all new to her. He smiled and went on.

"That's why I was surprised when I heard the new anti-slavery law," Jorge continued with his new feeling of importance. "Most people around here are not going to pay attention to it. They have had slaves for a long time and dismissed the whole anti-slavery idea as a stupid idea of The King. Nobody is going to give up their personal slaves, they say. They'll just get the Lowlanders they don't want hauled off to The Wall where they're not in the way.

"Almost everybody is really excited about The Wall that Drum and Father are constructing", Jorge continued with enthusiasm. "They see that wall as the dream of a lifetime, a way to limit the number of Lowlanders that they think are cluttering up *white* River Kingdom. "Me and Jackson will probably be part of that."

Willa just stared at her two siblings. Jorge was talking slavery...real slavery...and total prejudice!

Flo was nervous. Willa stood and walked around trying to clear her head. The northern nobles had slaves that Solem didn't know about, and free Lowlanders were going to be walled off, 'out of the way.'

Willa sat down again at the table with her siblings and began turning a turnip over and over in her hands. "Tell me about The Wall, Jorge. What is The Wall?   Where is it? What does it have to do with slavers, yours or anybody else's?

"We were both listening last night to the nobles that were here." Flo answered sadly. "They were very excited about the new road, but they were even more excited about this thing that they all want called 'The Wall'. It's apparently going to be built down by Drumsworth on the border between The Lowlands and River Kingdom."

"Why are they going to have something like that?" Willa asked slowly. "Lowlanders go back and forth between their country and ours to work for nobles all the time. A wall wouldn't help anything."

"That's the whole point of The Wall," Jorge added. "If there is a wall there, the Lowlanders won't be able to go back and forth. That's what the nobles want the wall for...to keep the Lowlanders on the other side."

"That doesn't make any sense. The nobles want the Lowlanders to work for them, but they want to keep them on the other side of a wall? Isn't that crazy?"

"Not from their point of view," Flo sighed. "You see. If all of the Lowlanders are walled back into their homeland, the nobles can pick the ones they want to use and leave the other ones behind the wall where they won't 'get in the way' here in River Kingdom.

"What that means is, they won't have a say-so about anything River Kingdom wants to do and, most of all, they won't marry up with whites. That's one of the things they're most worried about."

"But that's pure prejudice and slavery!" cried Willa in horror.

"Now you understand all of what Jorge is saying," Flo grumbled in shame. "And 'The Plan' is not only to keep slavery but to create a very organized way to return whiteness to River kingdom for good. It is their way to get and keep good brown faces for what they want them for but without *ever* seeing faces like Jorge's and mine. We're not really that brown, but we're not really that white, either. They want to control Lowlanders because they need them, but they don't want to mix dark people with their whiteness. It's that simple."

Willa couldn't stand. The negative thoughts she had had this morning seemed trivial. Those thoughts had now grown into a crisis that she could never have imagined and wouldn't have believed unless her own brother and sister told her. But now they *had* told her. What should she do with this terrible information?

The Consciences of the three now sprang into action. None of them had pulled all these facts together in the way that had just been done. They felt like they had been asleep. All the major horrors that they had thought of as terrible had just been summarized over a basket of turnips. With shame, all three Consciences decided it was time to go to work.

"I'm sorry that I didn't understand your worry this morning," said Cwilla. "That wasn't just worry. That was *intuition.* Intuition is powerful. I should have seen that. These two and the one that is with Mimi are good people. They are being truthful. Your father and his cohorts are engaged in evil. Do not disregard the knowledge of your siblings. Act on it."

Willa had never had such a serious message from her Conscience.

"Thank you so much for confiding in me," Willa said to Jorge and Flo sincerely. "It is so wonderful that I have you and Jackson as my family. You two have told me things that I needed to hear. I am very lucky."

"You still think you're lucky, now that you know you are related to Lowlander slaves?" asked Jorge, a tone of disbelief in his voice.

"She respects you for telling the truth now, Jorge," growled Cjorge. "But she won't respect you if you don't stand up for what you know is right. Do you think you don't have to worry, just because Thorndike is your father? What about other Lowlanders?"

Jorge stopped peeling turnips and looked at the floor. He hadn't really thought about that.

"You are as smart as your sister," hissed Cflo. "You have told her about your history and *your* mother. Are you finally going to confess the rest of your secrets... the secret about *her* mother?

Flo hung her head. *I can't,* she thought in misery.

"Yes, I still think that I am lucky," Willa answered. "At last, I know what is really happening here in River Kingdom. Thank you both."

When Willa and Mimi arrived back at The Castle later that afternoon, they found a very disgruntled King Solem waiting for them. As Willa had suspected, he had been worried about them. After all, they were driving on the same territory that Queen Linda had been on when her accident had killed her. Willa and Mimi could both understand.

Willa had not said anything to Mimi about the slave issue on the way home. Actually, Mimi was so full of the joy of her time with Jackson that she didn't have room for such worldly discussions. She was in love.

She was also observant, however. She noticed that Willa was unusually quiet, even though she tried to appreciate the joyful narrative of her young stepdaughter. After the first gushes of romantic incidents, Mimi relaxed into her own private thoughts.

It was only after several miles of silent travel that Mimi offered up a different kind of discission. "I was really surprised by the way the work was being done on the road," she mused. "I never realized that we had so many Lowlanders up here in the north. They sure are hard workers...I mean *really hard* workers."

"Jackson explained it. He said that all the inland nobles had sent their servants to the construction site to help Thorndike's militia men and a lot of them were Lowlanders."

"That's interesting," commented Willa casually." Were they all working hard together?"

"Not exactly," Mimi reported upon reflection. "The Lowlanders looked like they were doing most of the hard work while the militia men were just standing over them and supervising. To be honest, they almost looked like slave masters. Jackson didn't see anything wrong with it, so I didn't say anything."

"I guess Jackson thought it was okay, right?" Willa asked calmly.

"I think so. And he was pretty impressed with how far they had gone in improving, flattening, widening...everything, even up to the entrance to The Castle. They were making a mess of the levee systems, though. That was clear to see. Everything that the farmers and landholders have been using for the last one hundred years is flattened. I know Manda is worried about that."

"Indeed," Willa responded absently. The Royal River Road was disclosing many of the problems that had been forecast ...irrigation, prejudice, slavery. Her sweet husband had given his approval for the road, and she had, in essence, agreed with his decision. Now she felt guilty.

"Hogwash." exclaimed Cwilla. "You are thoughtful and kind and honest. You have been doing what you thought was right. You had no way of knowing what you know now."

Hogwash, yourself, Willa thought. *Being a good person isn't good enough, sometimes. Lots of good things are going wrong. I have been blind. We have slavery and prejudice all over the place here in River Kingdom and, in the spring, all the crops of River Kingdom will be drying up.*

"If you think that's true, Willa Thorndike, then you had better do something about it." announced Cwilla.

Very helpful, snarked Willa to herself. *The question is what should I do?*

THE GOOD INVASION

Armando and his fellow Lowlanders were up before dawn. Immediately they noticed that many more Lowlanders had joined their ranks. These Lowlanders were what the local nobles called "domestics". Armando had never heard of this term before, but he understood the concept.

These Lowlanders were people who had migrated up from The Lowlands, looking for work. And they had found it here in Central River Kingdom.

But these Lowlanders were free men who were working for a living as much as any of the white men around them. A few of them were actually landholders.

They were as concerned as their neighboring white men about the impending drought that would be caused by the new road.

Armando now looked about at the 'work crew' that was at least twice the size he had come with. It was an army and they were all looking to him for leadership. Did he have it? Was his water-search theory correct?

"Of course, you have the leadership," assured Carmando, "but remember. You are only working on a theory. This is not a sure thing. The first order of business is for Lowlanders and whites to work together and see each other as equals. If we find water, that is a bonus."

But we must find water, Armando thought to himself. *We can't go away dry.*

Prince Armando held up his hand for attention and magically it worked for him the way it worked for Queen Luna and King Solem. Quietly the large group of Lowlanders became quiet and shuffled themselves into position facing him. They were here on Thomas's land as Thomas had suggested. They were next to the old road, a few levees and The River, all the important elements.

Hector, Thomas and a small host of white land holders and farmers stood in rows together toward the back of the group. Amazingly, there were a few white noblemen among them...*noblemen!*

"Fellow Work Crew Members, I welcome you," Armando began. "Today we come together on a mission. Our mission is to find the water that will be lost when this area's precious levees are filled in and lost forever. Heads nodded. This part of the mission was clear.

Armando spread out a huge map that Manda had made for him, but it was not like the map that Hector had presented at The Gathering. This map went from The River all the way to the western border of River Kingdom but only included the borders with the northern farmlands on the north and Drum's estate on the south. Armando's 'audience edged forward to look.

"This map includes all the territory of the Central Farmlands of River Kingdom. From what Queen Luna has read in her research, this area has been the most dependent on levees in the past and so will be the most injured by their closure. Am I right?"

There was universal nodding on this question, not to mention some audible agreement. The crowd was paying attention. Now Armando spread Hector's old map on top of the new one. "If you look closely at this map, you will see that there are two small rivers drawn on it. These rivers are on Thorndike land and Drum land. That means that they will probably have the water they need. These rivers are special for another reason...they are flowing *down* to The River not up from it. That means that their areas will need to catch spring water to use before it gets to The River...and you will have *no levees*."

Hector couldn't stand it. He had to chip into the conversation. "That is what I have been saying all along," he bellowed from the back of the group. No one listened to me. The King just went ahead and blindly approved the stupid road without giving it a second thought. And now all his noble buddies are cheering for their beautiful new road down to Drumsworth!"

There was more head nodding all around. Armando waited patiently for his audience to settle. "You were mostly right on what you were saying, Hector and I understand why you are angry. Today, however, according to my wife's research, we all may be able to solve the water problem that we have here the same way that the north and south have by creating our own inland river. At least that is our hope."

Carefully Armando rolled up Hector's map and gave it back to him with respect. He returned to Manda's drawing of the area. Very clearly, Manda had followed her mother's directions. Most of the map was yellow in color, but along the center of the area there were indications of small mountains or hills that were colored green or blue. Armando stood back and pointed. "These green and blue areas are areas that Queen Luna's research indicates to be possible areas where this central area can find water.

"She indicates that there may be an under-ground river there that runs from Thorndike's land down to Drum's. We don't know how deep it is or even if it exists at all, but you Lowlanders know what you use when you need water for yourselves. That, my friends, is the reason that I have asked you to come here. The road, much as we might not want it, is going to be built. It will be designed to be wide and dry with no levee holes.

"Hopefully, it will have two bridges, however, bridges that we will all build to cover the mouths of the rivers that *we find* flowing from *your* hills."

Silence.

Armando looked around. Had he missed something in his explanation? Had he said something offensive? Then Thomas stepped forward and came to stand right in the middle of Armando's map. The map was clear with its yellow lands and green and blue dotted spaces where water might be. Thomas scratched his chin. Little by little, the white landholders and nobles picked their way through the seated Lowlanders and joined Thomas.

Thomas looked Armando in the eye. "I see what your wife thinks we might have here. The fact is, we never looked for it before so we don't know what's true and what isn't. All in all, though, it sounds like it makes some sense. I think we should give it a try. What do you boys say?" he asked, looking around him.

Gradual head nodding began. The Lowlanders now stood up with small but noticeable smiles. They could tell that the white men here were beginning to buy Armando's idea. They were ready to try, at least.

Armando wasted no time for reconsideration. He stepped forward to Thomas and made a request. "Thank you for considering my proposal, sir," he said. "Would you be so kind as to ask your fellow landowners to step to the side and hold up their hands so that the Lowlanders know who they are? Thomas nodded without a word.

"Fellow Lowlanders, look at the map here and locate the areas where *you* think you might find water. Then us worker types will look for the land holder in that area, grab up our picks and shovels and get ready to search for water in those areas. We will meet here tonight and every night until we find water."

The Lowlanders nodded and began to mill about, looking at the map and then wandering among the land holders with questions and offers.

Negotiations went on. Arms were waving, fingers pointing, chins being scratched. No one knew when or if water would be found but the mood was slightly optimistic.

After only a few minutes, all the various teams had dispersed to their areas. Even Armando had decided on an area that he thought showed promise. The landholder of that area, one that was on higher ground and quite a way back from The River, was Clarence, a stuffy and usually quite isolated soul who liked his privacy. He was one of those least thrilled with this invasion, but he *was* impressed with this impromptu offer of assistance.

"These Lowlanders have good Consciences", Cclarence informed him. "They are just willing to help someone who needs help."

Yes, indeed...thoughtful for no reason...very admirable, Clarence thought to himself. He motioned to Armando to follow him. He had noticed a patch of green grass on a hill on his property. Sometimes he had a holding pond there. This was one of the things that the Lowlanders seemed to fancy.

Clarence would lead him to it. *You are a good Conscience some of the time,* he mused. He also motioned to the Lowlanders that had worked on his lands for years and they came along willingly. They knew Clarence was a good man, even though gruff and solitary.

THE COLLECTION CREW

The work crews had moved out quickly, but the meeting area had not gone unnoticed by a completely different group. There were Drumsmen in the weeds and tall grasses on a hill just slightly to the south.

The Drumsmen had made note of Lowlanders working in the central hills before (probably Clarence's farmhands). They were now on a hunt for replacements for the slaves that had escaped (or been stolen) from Drum's prison. And here were a bunch of them...but they were walking away.

The leader of these Drumsman was Roman, a strong individual and second in command only to Drum himself. He had a military mind. One of his heroes was Lord Thorndike. He had even contemplated joining Thorndike's militia at one time, but he liked his current position of leadership and Drum favored him.

Roman might now be in a position of potential good fortune. He and his men were looking down on a gathering of so many Lowlanders in one place that all Drum's needs for road work on his property and The Wall would be answered at once. If they had seen this earlier, it would have been a dream come true, but...

"Oh my god," growled one of the Drumsmen. "Do you see what I am seeing, Roman? There were dozens of big, strong Lowlanders down there in one spot just a minute ago but now they're all disappearing inland!

"They have captive landholders! This was a Lowlander invasion and we were just a minute late. Should we charge down there now before they all get away?"

"Yeah! Grab your swords, boys!" yelled his companions.

"Let's go! Let's rescue the central landholders!"

"Let's show 'em what southern Drumsmen can do!"

"We will be heroes!"

"Wahoo!"

Roman shook his head. His men were sturdy and ready for action but not too smart. "No, men! Halt! Calm down," he said slowly but in a respectful tone. *Enthusiasm shouldn't be wasted even if it is stupid and misplaced now,* he thought. *It will be needed soon enough.* Roman would treat this as a learning moment.

"Look closely, men. Look at this group that you think of as an 'invasion force'. Do you see any weapons on the Lowlanders? Do you see any resistance in their 'captives'? Look closely."

The enthusiastic stared down at their 'invasion force'. Gradually, one by one, they became aware of what Roman was pointing out. Everything was very peaceful. There were no weapons. They were disappointed but had to agree with him that there really was no invasion taking place. Slowly they stepped back and shuffled their feet.

A cadre of Auras had loaded themselves onto the arms and other preparations that the Drumsmen had brought with them on the Lowlander-hunt, however. Gathering up innocent, unsuspecting Lowlanders to make into slaves was a worthy endeavor for any Auras and they now decided to do what they had come to do...inspire.

Roman was chosen for their strategic support. "Think of this spot below as a meeting spot, the Auras began. The Lowlanders down there are equipped with digging tools. They are going to be looking for water. That means they will probably report back here with what they find. That means that all these Lowlanders will probably be back with their findings. A good general might use this time right now to prepare to gather them up when they are tired. What do you think?"

Roman's eyes widened with the idea. *Why haven't I thought of this before? I am a genius,* he thought with sudden satisfaction. Quickly he gathered his cohort around him and explained.

"This amazing group of Lowlanders is going to return to this spot to report to each other on what they did today. What *we* will do is divide into two fighting forces and wait for them. At my signal, we will swoop down and attack, sweeping all the Lowlanders together at one time. It will be amazing. They will never know what hit 'em."

They did have one question, however, being the thinking white men that they were. What would they do with the white landholders and nobles that might be mixed in among them?

"Let the landholders go home, of course," answered Roman with a shrug. "They are white. We have no quarrel with them. They will probably be glad to get rid of so much riffraff."

The day wore on.

Then it happened, the return of the Lowlanders that the Drumsmen had been waiting for. The group looked dejected. Apparently they hadn't found water, the Drumsmen concluded. They chuckled to themselves. This is the result they had expected. They pitied the white landholders that had thought these Lowlanders would actually help them find water.

The Call of the Auras

Two groups now returned to the sandy circle. How many groups were there? They had only seen a few this morning as they left. There were ten Lowlanders here. That must be all and the Drumsmen were impatient. With a whoop, they charged from both sides into the circle, surrounding the surprised Lowlanders and their landholders on their horses with shouts and fury.

"Separate the whites, guys!" shouted Roman. "Drum don't want *them*. We've got what we want with these Lowlanders. Surround them all and let's go!"

The Drumsmen herded the unprepared Lowlanders into a circle, grabbing the reigns of their horses and pulling them together. In the space of only a minute, they were off down the road to the south. Drum would be thrilled.

A half hour later, the rest of the Lowlander groups began to straggle in. As each group stepped into the circle, those that had come before looked up with anticipation. Some said they had found areas of promise, but darkness was falling, and they had decided to ask some others to go with them to these spots and dig tomorrow. Within an hour, all were ready to call it a day and return tomorrow.

But that was before Prince Armand arrived carrying Lord Clarence on his shoulders in triumph. They had found water!

Great whooping and joyful dancing about ensued among two groups of men who seldom danced, let alone with ones of different races. The men in this circle had now formed a bond, however, a bond fostered by need on both sides. One side needed water for crops. That was essential and understood. The other side needed acceptance and understanding. Today both seemed to be given without reservation.

All the weary, shivering heads nodded but one. A small Lowlander stepped up to Armando quietly. My brother and his group are missing, he said. "I talked to their landholders. They were too scared to say anything to you."

"Well, what did they say to *you*, Enrique?"

"They said that a big bunch of white guys invaded this circle before the rest of us got here. They stole about ten Lowlanders and left the two white landholders."

Armando fell to his knees in sadness. The day had been so joyful and successful until now. He couldn't believe this had happened...and why? And Who? Who had done such a terrible thing?

"Could you ask the two landholders to come to me? I am not angry with them. They didn't do anything wrong."

The young Lowlander brought the two forward. They were frightened because of the invasion that had stolen the Lowlanders and they were also afraid of Armando. He was big and brown, and they didn't know him that well. They cringed and peeked at him from under the hoods that they had pulled on for warmth.

"We're sorry, Your Highness," they offered in unison. "But ten of your people that was with us got solen. We couldn't do nothin. There was a bunch of white guys here altogether."

"This was not your fault," Armando said, trying to keep his voice from showing his anger. "But I was wondering if you have any idea who the men were that stole my countrymen?"

"I'm not sure," one managed to squeeze out. "They didn't say nothing to us. They just said that Drum wouldn't want us... Just the Lowlanders."

Armando drew up straight. "Drum?" he echoed. "They said that Drum only wanted Lowlanders?"

"Something like that," croaked the other in support.

Armando bowed slightly to the two quivering landholders. "Thank you, so much," he said sincerely, but his eyes were angry, and his mouth was clamped with fury. Drum was kidnapping Lowlanders. It was his standard practice. *He is a true slaveholder,* Armando thought sadly.

THE REST OF THE STORY

Manda was lonely and frustrated. She had allowed her father to talk her into staying home in Urhonordo and not following the Lowlander work crew over to River Kingdom.

That was a mistake, she thought to herself. *I can't be helpful with anything here in peaceful Urhonordo. I am just wasting time.*

Mother would say that I should just 'sit for a spell and read'.............. Ah! There's an idea. I'll just go talk to mother. She's alone today, too.

Manda jogged down the hall to her mother's reading room. There she was, as usual, reading. Manda gave her a quick kiss. "What are you reading about, today?" Manda asked, not really wanting to know.

Manda settled herself comfortably in a nearby chair with her feet tucked under her.

But now Manda noticed something that she didn't usually see on her mother's confident face. It was a look of worry. She was sitting with a large history book in her lap. Bookmark place holders protruded from several places in the book.

Luna looked up from her reading at last but with only half a smile. "I am so glad that you didn't go with your father yesterday," she said. "I know that you now wish you were there with him, don't you?"

"Indeed," answered Manda. "June and I were glad to deliver The King's announcements, but I am dying to know how June was received in River Kingdom."

Luna smiled at the mention of June, but her look of worry returned. "I doubt that he received the same positive reaction to both announcements that you got here...or that King Solem expected." She looked down at her book. "History tells us that people do not change their habits easily. Upon review of history, I see that Lords Drum and Thorndike of old were very unkind to Lowlanders. They captured many of them. Thorndike even locked them in prison to keep them out of the way. The history books don't call that slavery because it was a war action, but I think that it was slavery."

"There was slavery back in the old days, Mother? I didn't know that. Why didn't I know that?" Manda asked with surprise.

"Because Urhonordo has never had slavery here and nobody in River Kingdom called the nobles' actions slavery, at the time," Luna answered, her nose still in her book. She flipped to one of her bookmarks. "But it *was* slavery. It says here that back in the day, a noble named Dipswitch built what is now Drumsworth Castle. He used enslaved Lowlanders to dig the original Drumsworth Lake and he kept them in cages to please Thorndike's ancestor who wanted to attack Urhonordo and takeover at the time.

"Thorndike's men were under the control of a man they called 'King Drum' but they were actually early versions of the Thorndike Militia which still exists today. It seems that Thorndike wanted to attack Urhonordo and keep it for himself, but he was afraid that the Lowlanders would warn the Urhonordorians. It took a long time to get a fighting force down the rutted river road between the north and Drumsworth...as it still does.

"The Lowlanders would certainly have warned Urhonordo back then. They still would today...unless the road was wide...and smooth...and could be quickly traveled by a militia..."

"so all the Lowlanders were locked up in the prison back then," Manda added.

Manda's eyes opened wide. She had read about some of this when she was in school, but this story had not seemed important until now... until she knew some of the people that her history books talked about.

Manda remembered her father saying something about 'history repeating itself' when they had watched the Lowlanders being treated like slaves and imprisoned behind the Drumsworth mote. This was what he had been talking about.

Manda sat up straight. Her mind began reviewing what Thorndike and Drum had said at their meetings. They had been *very* interested in the new road, from the first. The road seemed to be what they thought was key to a successful future for River Kingdom.

Manda and others had thought of this as the selfishness of nobles who wanted to get to Drumsworth Lake more easily. With a jolt, Manda now realized that the *extremely selfish* Thorndike would never have wanted to help Drum and Drumsworth just to be nice. Thorndike's eyes were on Urhonordo!

And Drum must have other plans, too, Manda's revelation continued. *He is very popular with the nobles, all of whom have disagreements with King Solem. Does Drum have designs on Drum's grandfather's position of old? Are these the reasons for the worry on my mother's face.*

Manda sat back in her chair. "What are you thinking right now, Mother?" she asked plainly.

"The more I read this history and the more I review the way Drum and Thorndike have acted recently, the more I am concerned that these two have had something much greater than a new road in mind from the first," Luna answered sadly. "The way I see it, both our Kingdoms are in danger and Lowlander slavery is designed to be law in both."

"Do you think Urhonordo's in danger right now or do you think this scheme will take a while to get started...that is, if there *is* a scheme like you believe?"

"What do *you* believe now, Manda? You are observant. You are a young woman with a brain. You have been involved in this push for the new road from the beginning. Do you remember the nobles cheering for Drum and being ready to come to blows at The Gathering? Is this only for a new road being built? What do *you* think?"

"I think that I am embarrassed that I didn't figure out much earlier what you and father have been seeing. How could I have been so blind? How could I have *not* known that there was more to Thorndike and Drum's insistence than a stupid road?"

"You are young. You are more trusting. You haven't experienced as much as we have. That's all," sighed the Queen sadly.

"She's right," agreed Cmanda. "You and I *both* missed all the cues. I even had a warning from people who went to the Conscience Alert, but I didn't connect everything I should have. The question now is, what are we going to do about it?"

Luna began thinking out loud. "I wonder if King Solem thinks much about his family's history." she mused. "Have you ever heard him speak of that era when the old Lord Drum became acting king for a while?

"Never. King Solem is a man of present-day protocol. I don't think he ever thinks about the past. Bur why would someone have been an 'acting king' anyway?"

"The history book says that King Sol, King Solem's grandfather, was off to the east on a Dragon-quest. He was gone for nearly a year. The old Lord Drum was just a stand-in, but he made himself very popular with his fellow-nobles and actually acted like a real king for a while."

"Did old King Sol kick old Drum out of the castle when he came back?"

"I don't know. That part of the history isn't very clear. I guess that Drum just retired peacefully, and King Sol took over again. Maybe the old lord Drum didn't fight to stay king. I don't know. Maybe that's why King Solem doesn't even think about that history at all.

"But now it strikes me as odd that these same two lords are involved together again in something that I don't quite understand," Luna continued. "Suddenly, they *must* have a new road and they need it right away. Why? And Thorndike is impatient and pushy. Why? Just like before, these Lords Drum and Thorndike are very popular with the nobles, particularly with those who think whites are superior to Lowlanders. It all just seems very...connected."

"Hmm," Manda murmured. "So now *you* worry that history might be inspiring what is happening today, right?

Luna looked at her daughter and grinned. "It's just an idea from an old history book that I have been reading..."

"And we don't think King Solem has spent too much time reading history books," Manda continued. "He has been too busy with today's pesky nobles that never think he is doing things right...

"...and who don't think Urhonordo should have anything to say about anything in River Kingdom." Luna finished. "No. I don't think that King Solem sees any connection at all between his problems now and the treachery of the past."

"Are you going to say anything to him?" asked Manda quietly.

"No," answered Luna. "It is not my place. He is King of his Kingdom, and I am Queen of mine. My suspicions are only my own. The Queen of one country should not interfere with the political issues of The King of another, especially when there is no proof."

"But you do that all the time at The Gatherings," blurted Manda.

"Yes, and the nobles hate it. Armando and I only speak up about water issues because we all share The River. Speaking about possible insurrection or anything like it in River Kingdom would not be tolerated. I must keep my suspicions and worries to myself."

"Hog wash! You won't tell a friend that someone is planning to hurt him?"

"No. I won't."

"Then I will!" Manda shouted aloud.

"I was hoping you would say that," smiled Luna.

THE UNSEEN

Willa had not yet said anything about her visit to Thorndike Manor. She had left that to Mimi who was more than happy to speak of everything she had seen there in detail. She told of the twins working on the front driveway of the manor when she arrived. She told of her romantic buggy ride to the Royal River Road's work site. She told of the road's wideness and smoothness and how much hard work was being done by the Lowlanders, not to mention how many Lowlanders were actually working, compared to farmers and landholders.

King Solem listened and smiled at the progress. Nothing she said caused him any concern. He was busy working on other kingly paperwork when Willa tapped on the sitting room door. Without raising his head, he called for her to enter. He was in a good mood and working hard. Her presence was always a welcome interruption.

"Where have you been all morning, Willa?' he murmured absently. "From what I hear from Mimi, you two had a good visit at Thorndike's. Did you get another chance to talk to the young lady who says she is your sister?"

"She *is* my sister, my love. She is not only a sister, but a friend."

"Wonderful…I know that means a lot to you. We should begin asking her and your brothers to come to parties in the future. We haven't had a party in a long time. I was thinking of scheduling something in the winter, perhaps a winter ball. What would you have to say to that, my love? Would it please you?" Solem looked at her and smiled.

Willa had placed herself next to the fire. She was gazing into it and thinking. She was not smiling. Solem noticed instantly. Something was wrong. Carefully he put down his pen and turned to her directly. "What is it?" he asked simply.

"I don't know how to tell you, Solem. Yesterday was a day I will never forget, and I don't know what to say about it. I heard some things that will surprise you, I think…and maybe even shock you. I was stunned by information that I was told by both my brother and my sister." Willa turned and faced Solem, searching his expression for a way to begin.

"What did they tell you, my dear." Solem encouraged. "I can see that something is bothering you. You know you can tell me anything."

Willa took a deep breath. "The most important thing that they told me is that the nobles of our River Kingdom have many slaves today that you don't know about and none of them are going to give up the ownership of those slaves even though it is against your law." Her statement gushed out in one sentence. She had stated it and now she waited.

Solem sat back in his chair and looked at her closely. "Really…What makes you think this is true?" he asked calmly.

"My sister and brother told me this," she answered defensively. "They told me that they are the children of my father and their mother and that she was a Lowland slave. She was only one among many who are *still* slaves right there at Thorndike Manor. My siblings have just come to realize that they, too, are slaves… Lowlander slaves.

Solem didn't move. Willa glanced in his direction. Like a statue, he appeared to have been cast in stone in one spot, sitting in his work chair right in front of her. She reached out and touched his knee.

At last, he lowered his gaze to Willa. His look chilled her to the bone. His look didn't see her. It was cold and impersonal. "How do your brother and sister know that the nobles will not give up their slaves?" he asked with a frightening lack of emotion.

Willa blinked and tried to retrieve her voice that suddenly became stuck in her throat. "The nobles and Drum had a meeting with Thorndike after the announcement of the law where they declared this to be true," Willa managed to answer. "They were thrilled with the approval of the new road, but they swore that *their* slaves were *theirs.* They will only give up slaves they don't want and they want to lock their rejects away behind something they call 'The Wall.'

Solem came suddenly to life. "Behind The Wall? They want to put slaves in *my* kingdom behind a wall?" he bellowed.

Solem had never lost his temper with Willa but now he stood and paced first one way and then the other. He looked at her accusingly. Then he looked away. He shook his head and threw his crown to the floor. Then he stared at her angrily, again.

She shrunk back. *Is he angry with me? Does he think I am lying? Doesn't he trust me anymore? Should I run?* she thought.

"He believes you and trusts you, Willa," Cwilla counseled quietly. "That's the problem. He believes you. Hold your ground. He's not angry with you. He's just angry, period. Give him time to process what you have just told him, but he doesn't want to believe it."

Solem stormed out of the sitting room and down the main stairway. He charged out onto the colonnade. Willa followed him. *What should I do?* she thought fearfully. *Where is he going?*

Cwilla was with her. *I don't think he's going anywhere. He's just looking for answers. I think he is having to deal with decisions he has made as a King that he now thinks were wrong. He is mad at himself.*

Nearby Csolem confirmed what he heard Cwilla saying. "You're right. Solem is mad at himself, but also fearful. Now he is worrying about his whole kingdom. If he has slavery going on in his own kingdom, what else is going on that he doesn't know about?"

Csolem focused on his human. "Take a deep breath, Solem. Start thinking instead of reacting. You are a King. Act like one."

At last, there on the colonnade, Solem's reaction to Willa's news began to return to a more normal position. He stood in one spot, his hands clenched at his side, his head bowed. "How could I have been so stupid?" he asked the columns around him. "How could I have been so blind?"

Csolem continued. "You did nothing wrong here. You trusted people that were not trustworthy. That's all. I went along with you. We thought it was the best thing to do at the time."

"Willa knew something was wrong. Armando knew. Even Manda and June knew that there was a problem that I wasn't seeing. Why didn't I see it?"

"Because you're human," Csolem continued quietly.

"I am so sorry to have to tell you about the slavery in our Kingdom," Willa said as she finally felt free to take his hand.

Solem gave her hand a sorrowful squeeze. "How could I have missed something as terrible as slavery when it was happening so near to me? Was I blind?"

Willa remembered something that Jorge had said. He hadn't known he was a slave until recently. Why not?

"My brother, Jorge, didn't even know he was a slave, himself. He didn't know that he couldn't leave or say no to an order. He never tried. He wasn't told he was a slave, and he was a favored son. He was born at Thorndike Manor," answered Cwilla without hesitation. "His mother always knew that she couldn't leave and was forced to do what Thorndike told her to do. Back then, she *knew* she was a slave. That is the difference."

Willa now thought some more about this. *If my brothers and Flo can leave when they want or say no to Thorndike's orders, are they still slaves?* she asked herself suddenly. *This is an interesting question. Are the 'slaves' that Jackson and Flo know about really slaves or are they still taking orders because of habit, because they don't know that they are free?*

This question didn't have time to find an answer. Manda was riding onto the colonnade at a full gallop.

Solem and Willa, still hand in hand, turned to greet their guest. They both cared for Manda and valued both her friendship and her perspective on things, but this was not a good time for a surprise visit. They would be polite. Perhaps she could spend some time with Mimi. Mimi would be more than happy to tell her tales of her romantic encounter with Jackson. "Hello, Manda," said Willa politely.

Manda noticed the 'politeness' of Willa's greeting. She was intuitive and by now she knew Willa well. There was a problem here. She could sense it. On a normal day she would have respected the tone she was hearing, but this was not a normal day.

"Hello, Your Highness...Willa," Manda said carefully. "I see that you both are out here on the colonnade alone. It appears that something different is happening and I don't want to interfere."

"Mimi is in The Castle with some time on her hands," said Solem kindly. "I know she would love to see you today."

"He is just being kind," whispered Cmanda honestly. "He doesn't want company right now. You must decide whether or not you think what you have to say is important enough to break through his attempt to get rid of you. Do you believe that you are on a mission or delivering only and interesting piece of history?"

Manda hesitated for only a moment. "I am sorry to have arrived here unannounced, Your Majesties, but it is both of you that I came to meet today. I feel that what I have to say is important." Now she added a piece of information that she knew would be heard. "My mother knows that I am here."

That information was, of course, important. King Solem bowed his head slightly and offered his free hand to Manda. "Come inside with us, my dear Amanda. If your mother feels that what you have to say will be important to both of us, then we shall hear it."

Willa motioned to Manda now with a slightly warmer smile. Manda could see that she was unfortunately interfering with 'a situation' that was currently taking place and not resolved. She had hoped for an unconflicted reception for her story, but...

"You made your decision to come and speak," Cmanda commented tersely. "Now tell your tale well. Don't waste this moment. Your mother is depending on you."

King Solem guided the two women into his conference room. It was too big for just the three of them, but it was closest and most convenient. Horace saw them enter and nodded. He knew his duty. He would bring tea and scones. (Protocol)

Solem took his place at the head of the table with Willa sitting uncomfortably at his right side. *This is such an odd situation,* she thought. *Here we are in the middle of a crisis that involves slavery and Manda is almost insisting on an audience. And we are in this huge conference room, for heaven's sake.* Willa placed her hand on her husband's knee for support and waited for whatever was coming next.

Manda was, for the first time, unable to speak. She had known what she wanted to say when she climbed on her horse a few hours ago. Her mother had waved an encouraging goodbye. But now she was here. Had she done the right thing to come here like this?

Horace came to her rescue. "Would you care for some tea, Princess?" he asked with a comforting smile.

That's right. I am Princess of Urhonordo Kingdom and I am here on an important mission, she told herself reassuringly. "Yes, please, Horace" she responded.

"At least your voice was working," snarked Cmanda.

"I am really sorry to barge in on you like this," she began. "My mother feels that she found important information that Your Majesties should be aware of, but she feels that a formal presentation from another head of state may be inappropriate in terms of protocol."

One corner of Solem's mouth turned up.

Manda continued. "As you know, my mother is a strong supporter of education. As such, she reads all the time...everything... old and new...important and not."

Now Willa was smiling. "Speak up, Manda. What did your mother read that she thought was important to tell us?"

"She read that the Thorndike and Drum of the past abused Lowlanders and tried to overtake Urhonordo with an armed militia...one that still exists with the current Thorndike," Manda blurted in a rush. "She said that since the old behaviors and current ones seem so much the same, she feels that the current Thorndike and Drum may have the same objectives as the old ones used to have. She is worried."

King Solem rocked back in his chair. Willa covered her mouth with her hand. No one said a thing, but minds raced. Manda's announcement was so short, clear, and concise that it took another moment for the implications of her declaration to be understood. When they were, Solem and Willa leaned forward and stared.

"What are you saying?" asked Willa. "Are you saying that your mother thinks that there is a current plot to take over Urhonordo?"

"Are you saying that Drum and Thorndike are plotting against you?" asked Solem.

Manda backed up in her chair. "Be careful what you say here," warned Cmanda.

"No," Manda squeaked. "That's not what my mother is saying at all. Please... That's not what *I* am saying. Maybe I shouldn't have said anything at all. Maybe I was wrong to come here." Manda backed up another notch in her chair.

Cwilla saw Manda's discomfort. "I think that she just wants to say something that her mother thought was important. You are jumpy right now. Relax. Let her speak."

Horace placed a full cup of tea carefully before Manda and stood back. "If I may be so bold..." he began.

King Solem looked up with a start. "Horace?"

"I *do* beg your pardon, Highness, but I believe that *I* actually read a book a while back that made me think that today's actions *are* very reminiscent of times in the past. It was a history book about the time of 'King Drum' many years ago."

"There was never a 'King Drum'," growled Solem. "That is just an old fairy story that the nobles like to tell when they're angry with me. They say that this noble rose from their ranks and did all the wonderful things that they liked. But it's not really true. There never was a King Drum."

Horace looked around. *Should I say anything more? I'm already speaking out of turn, but this is important.*

"Speak up, Horace," whispered Chorace. "This is what your brother was talking about at the Conscience Alert."

Horace spoke again, this time with more authority. "Maybe *you* think that there was no King Drum, Your Majesty, but my brother is the landholder for a noble who believes that 'King Drum' not only existed but was the best king that ever lived. That's why they like the current Lord Drum so much. The nobles think that he is just like the old guy. My brother's noble and all the other nobles love him. My brother and I have been worried about that for quite a while now."

King Solem was shocked. Servants never entered discussions in the first place. This was definitely against protocol.

On the other hand, Horace was a good man...honest and trustworthy. What he said had merit. It couldn't be ignored.

Solem decided that it was time to limit protocol. This subject was too important, and it revealed something important. It revealed that there were great portions of *his* kingdom that he didn't know about.

There was pain in this discovery. Willa's gentle hand was again on Solem's knee for support. Was she just now learning about these failings he was seeing in himself, or had she always known them? Whichever was true, her support was important at this moment because now he felt completely inadequate as a King.

"You are a good King," Csolem chimed into his thoughts. "You try to do what is right at all times. What is right now, however, is to listen to others to find out what you don't know... *and* you must be willing to learn from it."

Easier said than done, Solem thought to himself.

Horace was backing out the door, afraid that he had said too much. Manda was crouched fearfully in her big chair in front of him.

"Thank you, Manda...and you, too, Horace. You are both brave to speak out as you did. I know that you don't feel that I have heard you completely yet so please sit here with me and Willa. We need to hear what you have heard that we have not... and also what you think."

"I don't know what you are discussing now," said a voice from the doorway. "But I have something terrible to add to your discussion."

It was Prince Armando, disheveled and exhausted from what had been an all-night ride. "Please excuse my interruption and my appearance, but I must admit that I come here in a state of emergency. Ten of my men were just kidnapped by Lord Drum!"

King Sol rose from his chair immediately and went to his friend and fellow monarch, bringing him a chair.

Manda looked up at her father in surprise and ran to him.

Horace scurried to get his provisions of scones and tea but Willa remained frozen in place. How many crises could be delivered at one time?   "Everything you are seeing this morning may be tied together," whispered Cwilla.

Armando leaned back in his chair and closed his eyes. Everyone around the table waited...and waited...and waited. Horace placed a cup of his tea in front of him. "Would you like cream, Your Highness?" he asked tentatively.

Armando opened his eyes and sat up. "Thank you, Horace. I just needed to gather my thoughts. I have been riding alone with my misery all night. Ten of my men were captured yesterday.

"The men I was with and the Central Landholders had just discovered one of the water sources that we were looking for, but when we returned to our group, we got the news. My men were gone without a trace." He bowed his head in pain at the recollection.

Manda wrapped her arms around her father, tears for his sadness rolling down her cheeks. "I am so sorry, Father. How did it happen? Do you know who stole them? Do you have any idea why this happened there in the center of the kingdom?

"All I know is that the farmers that were with my men at the time heard the kidnappers say that Drum wanted Lowlanders only. The farmers said that the robbers looked like they were *hunting* for Lowlanders. Can you imagine that? My fellow countrymen are now being hunted like wild game."

"Our world is becoming a very strange place," said Willa sadly. "Who would ever have thought that Lowlanders would be stolen away like animals. Why would anyone do such a thing?"

"Because the thieves needed to make up for the Lowlanders that we just turned loose from their prison," Manda suddenly blurted. "That' it, Father! That's what the Drumsmen were doing. They were trying to replace the slaves that June and I turned loose!"

"What did I do now?" asked June with a grin as he ambled through the door. "I am hearing some commotion down here and the next thing I see is a full-blown meeting that I wasn't invited to and Manda blaming something on me."

For a moment, the seriousness of the situation held everyone silent, but June just kept smiling at them.

Horace reheated his tea supply one more time and finally small smiles crept onto the faces around the table. But slavery, kidnapping and even potential war were all on the table.

Horace also went to tell Agness that the conference room members were going to need lunch.

THE AURAS RISE

Drum ambled across his front lawn and gazed with pride at the construction occurring just to the west side of his new drawbridge. Over a dozen Lowlanders were hard at work on the beginning of what would soon be The Wall. This barrier would run for miles from his drawbridge to the very western border of River Kingdom, sealing off The Lowlands from River Kingdom for good. It would be magnificent.

"Finally," Drum's Auras told him as they swirled around his feet. "We have been working on this plan since you were a boy. It's way past time to put action in front of words. Let's get *all* those Lowlanders back onto their homeland where they belong. That's what northern nobles want...purity for whites...no brown skins allowed except for work."

Only a day ago, the Drumsmen had been complaining that they could find no new Lowlanders to replace the ones that had been freed from Drum's prison. They had come up with only four and one of them was a girl.

Then a miracle had happened. On an emergency hunt to the central part of the Kingdom, the Drumsmen had discovered ten Lowlanders all in one spot. What a bonanza! They were not armed or on guard and had been taken easily without a fight. They had been driven south immediately on their own horses.

"Excellent!" Drum had declared as he had spotted the exhausted Lowlanders proceeding toward him. "What were they doing when you found these men together in a bunch like that?" he asked.

"I think they'd been working on widening the new road for one of the Central nobles," offered one of the Drumsmen, looking a bit guilty.

"That's *that* noble's problem now," Drum laughed. "Put 'em to work on The Wall.

The Wall was an essential piece in Drum's personal plan, the one that he only discussed completely with himself. His father had expanded Drum family holdings and increased the size of Drumsworth Manor to that of a fortified castle, but he had failed to develop the strategy needed to replace an active king and assume the throne.

Drum, however, was now in league with the Auras. He knew that his strategy was complete. The nobles loved The Wall and it had been his idea. This alone would maintain his popularity as Royalty…after the Thorndike Militia took over The Castle.

 Drum watched as the new batch of brawny Lowlanders toiled under the watchful eye of his Drumsmen. Huge posts were being driven into the mud on the west side of the drawbridge to support the stands of rock and mud that would form The Wall. It was a mighty undertaking that would not be completed overnight. The good news, however, was that these Lowlanders were now building *their own* enclosure, *themselves*. The irony was delicious.

Balynn now joined her father on the lawn. The construction was impressive. The Lowlanders were sweating and breathing hard even though the weather today was cold. The Drumsmen stood over them with fists holding what looked to be whips. They shouted and snarled while the Lowlanders below them cringed and appeared to try to work harder.

Is this what Jackson and Jorge were doing when they were working on the mote, Balynn asked herself. She was now seeing it up close for the first time. These men look like slave masters. This is horrifying.

Balynn tucked her hand into her father's arm. She must say something. Surely this was not what it seemed to be. "Look, Father," she began. "I see a whole lot of work being done on the other side of the drawbridge. It looks like it's going to be something big."

"Indeed," Drum responded happily. "They are building The Wall. It's going to be huge."

"How exciting! What is this wall for, Father?"

Drum thought for a minute. He hadn't explained this to her before. She didn't need details, either.

"Right" counseled the Auras. "Good thinking. Details are a nuisance for young people."

"The Wall is for protection," Drum responded warmly. "I always want to be sure that you are protected."

'How lovely," Balynn said as she squeezed his arm. "Your Drumsmen look like they are being really mean, though. They look too harsh. The Drumsmen almost look like slave masters standing over those poor, hardworking Lowlanders."

Drum stepped away and turned to her with a stern look. "Who told you that? Where did you get an idea about slave masters? Have you been spending too much time with The King's daughter?"

Balynn stepped back as well. "What do you mean Father? I didn't say that your men *were* slave masters. I only said that they *looked like them.*"

Drum's face turned very pink. He was obviously upset. "We don't have slaves here, girl," he hissed. "We have Lowlander workmen."

"Oh," said Balynn quickly. "I guess I was just commenting on the way they *looked.* My mistake..." she skittered into the manor.

"Your daughter may become a problem in the future," an Auras whispered to Drum.

She is beginning to be a nuisance, Drum thought to himself. *She's pretty and I thought for a while that she might be a nice princess when I become King, but I don't know. She asks too many questions.*

 Drum scowled as he took note of his threatening Drumsmen who were now actually using their whips. *These Lowlanders are too slow. I need to get some more white men down here to browbeat these Lowlanders and hurry this wall along. I think I'll just go back up to Thorndike Manor and get some more manpower.*

The Call of the Auras

Balynn scurried into Drumsworth Manor. "Those Lowlanders are slaves," she said angrily. "How stupid does Father think I am? I know slaves when I see them. Surely Jorge was never as bad as these men are."

"You should also know *your father* when you see him," commented nearby Auras. "If your father wants to call those Lowlanders 'workmen' instead of slaves, what difference does it make to you? *You* are not the one who has slaves. Don't worry about it."

"If you don't stand up against slavery, then you are as wrong as your father," snarked Cbalynn as she crawled out onto Balynn's shoulder. "I know that going against your father won't be easy," she counseled sympathetically.

Cbalynn looked around. Sure enough, Cbalynn could see Auras swirling happily through Balynn's hair.

The Auras could not see Cbalynn. That was her only advantage as a Conscience. But the Auras seemed to be thick everywhere. She was outnumbered.

"Slavery is bad, and you know it," Cbalynn now spoke up plainly in Balynn's ear. "Jorge is a good man, and you care for him. He is part Lowlander. How would you like it if *he* was made into a slave?"

The Auras crawled up onto Balynn's lap. "Your boyfriend was supervising Lowlanders when you first met him, remember? You don't think *he* was a slave master, do you? He was only supervising Lowlanders, right? Your father and your boyfriend know what's right. There is no slavery going on out there in the construction area...just strong supervision."

"Hogwash!" screeched Cbalynn. "Auras are just trying to trick you! You can't approve of slavery. Jorge may not have known any better before but now he has you. He wouldn't do what he did before now that he knows how bad it was. I won't let you fall in line with this. You are a good, kind person."

"Don't worry about the stupid slavery idea, Balynn", crooned the Auras. "Your father is going to be a king someday. Kings are never wrong because kings make the rules. What you have just seen won't be against *his* rules. Don't worry any more at all. Just go look at your wardrobe and decide what new, wonderful dress you want to wear the next time you see Jorge."

Balynn smiled. She knew her father had always wanted to be king and usually got what he wanted. She turned now and marched confidently upstairs. *It wasn't as bad outside as it looked*, she told herself. *And I will want to look good for Jorge and Father when Father becomes king.*

Cbalynn sighed. She would have to develop a stronger counter offensive. Balynn's Auras were smart... and effective.

AN EARLY START

Thorndike was ecstatic. His Auras were ecstatic. Everything was going pretty much as they all had planned. The Auras were finally seeing the fruits of their labors, the completion of The Royal River Road that they had envisioned months ago at their meeting.

The northern nobles had muscled their landholders and militia into the Thorndike Militia to widen and smooth the northern third of the territory. That road would be large enough to carry all Thorndike's militias (which now included all River Kingdom militias other than The Royal Guard) and their battle equipment... eventually to Thorndike's ultimate goal, Commander of River Kingdom...and more!

He could see it now. Drum would soon be King of River Kingdom South. With Thorndike's help, he would then move up to take over The Castle and his rule of all of River Kingdom. That was Drum's goal and Thorndike knew it. He was excited.

"Since Drum will be content with the rule of River Kingdom, he should give you the support you desire to take over Urhonordo, at least," crooned the Auras at Thorndike's feet. "Then you will be able to complete the mission of your grandfather that was foiled so many years ago. He deserved it then. *You* deserve it even more now."

Thorndike paced the halls of his manor and his mania began to resurface. The mention of Urhonordo always did that to him. The victory he envisioned as Commander was getting closer but he was frustrated. It was not happening fast enough. He wanted it now. He needed action. What could he do *right now* to speed things along?

His Auras were ready. "Here's an idea for *now, Thorndike,* if you are bored. Start a small task here...*now.* You can even involve your militia. Drum got a lot of positive reaction from the nobles when he told them about The Wall and what you were going to do with the Lowlanders when it was completed. You can practice by capturing Lowlanders right here, right now if you want to. It will thrill your local nobles. They have been wanting to get rid of the invasive Lowlander population here in the towns of River Kingdom for years. You could do it now, *for* them."

Thorndike stopped pacing. That was it! This was something that he could do now that would make him popular with the nobles...maybe even more popular than Drum. This was a great idea. Without hesitation, he summoned his militia.

Within an hour, the militia had their orders and began gathering. They were to go into a town of northern River Kingdom and locate Lowlanders wherever they were.

Then they were to capture those Lowlanders and bring them to a vacant pasture on Thorndike property. From there they would be driven to River Kingdom South and banished to a life behind The Wall. The plan was so perfect that Thorndike decided to supervise it himself. It would give him something to do.

Like an excited child, Thorndike scurried to his room to 'dress up' for the occasion. He would wear his grandfather's leadership garb, complete with insignia. He would carry a sword and a shield. He would look authoritative.

"It's fun to play dress-up, isn't it?" came the pesky voice again.

Thorndike whirled around, sending his sword and shield clattering to the floor.

"Who's there?" he growled. "Where are you? This isn't funny anymore. Who are you?"

Silence filled the room. Thorndike searched the corners, the rafters, under his bed. Slowly he gained his composure. *How ridiculous I am,* he thought to himself. *I am just excited, so I think I hear things. Excitement will do that sometimes.* He picked up his equipment and headed out the door.

"Banishing people that you don't want is a mistake," said the voice softly. "Haven't you learned that yet?"

Thorndike kept walking. "*I hear nothing",* he said to himself.

"You hear *me,*" the voice said clearly. "And you will hear me more." Thorndike kept walking. His mania was under control.

The militia was gathered on Thorndike's front lawn and was abuzz with excitement. They finally had a real mission. They were not going to be rousting landholders and their pathetic militias to work on the new road or even supervising them as they had done the past few days. They were going to capture real Lowlanders.

"We're going to do what?" Jorge asked Jackson.

"We're going to do what we are ordered to do," Jackson answered stiffly. "We are Thorndike Militia."

This militia considered themselves an elite force. First, they were all white. Second, most of them were direct descendants of one hundred years of militia history. Their grandfathers had been on the march that should have taken over Urhonordo. These descendants were very aware that they *would* have succeeded if they had not been thwarted by the early return of King Sol back in the day. Like Thorndike, they still harbored a grudge...most of them.

Also, like their leader, The Thorndike Militia was ready for action, and they were excited by it. They had been brought up to believe in the superiority of their position...*and* their whiteness. This assignment would be something they would enjoy.

Auras had not been used in such a large group situation before. This was new. They had no idea how it would work but they were excited, too.

The twins, however, had very mixed emotions...and no attending Auras. They scurried into Flo's kitchen immediately to tell her the news.

Flo, as the boys had anticipated, was devastated. "I can't believe this is really happening. And I can't believe that the two of *you* are going to be part of it," she groaned. "The people you will be capturing have done nothing wrong. They are just Lowlanders. *You are both Lowlanders*. How can you do this?"

"We don't have any choice. We are Thorndike Militia. We are our father's sons. We have no other choice, Flo."

"You always have a choice," commented their Consciences in unison.

 Auras now found them. "Come on boys. Are you little girls in the kitchen or are you men? Your father is waiting."

The twins chose to ignore their Consciences.

The militia moved out quickly with Thorndike proudly taking the lead. The Auras rode along in a new adventure for them. They had never rounded up Lowlanders... or anything else... before. There was no anger or fury to support. They were perplexed.

There was much good-natured banter among the troops as they headed toward the village. "This will be like herding sheep," they joshed to each other. "Let's split up when we get into the town and go on a Lowlander Hunt. Let's see who can capture the most." Now the Auras caught on.

When the group reached the village, their ranks began to resemble a swarm of bees. With their lighthearted challenge in mind, they swooped from Lowlander to Lowlander, hauling them, one by one into the town square. They had brought ropes to tie up their conquests but soon realized that they didn't need them. There was no resistance.

In shocked compliance, the Lowlanders simply allowed themselves to be assembled in the middle of the village. Many still held vegetables that they had just purchased from a vendor. Several had children at their knee.

"What should we do with the children?", one militia man shouted to Thorndike.

"Just bring them along," he shouted back. "They'll just grow up to be Lowlanders like their parents."

The process of 'cleaning the town of Lowlanders' as the militia was calling it, didn't take very long. Within only a few hours, a large collection of people of all ages filled the town square.

Bystanders who were not Lowlanders watched in amazement as customers and vendors alike were rounded up and encircled by the militia. There was no doubt in their minds about what was happening. The local Lowlanders, their neighbors and friends, were being removed from their midst. They were astounded. No one had anticipated anything like this. These were good people. Why was this being done?

The town folks' Consciences began to question their silence.

"Why is this happening?

"Why don't we say something?

"I don't think this is right."

"These people weren't doing anything wrong."

"Where are they going?"

But the militia looked very intimidating, and Lord Thorndike was at the lead, issuing commands and answering questions from his troops. It all looked very official. The townspeople determined that asking questions was not a good idea. They did watch and wonder, however.

The roundup continued all afternoon. As evening began to descend, Thorndike was satisfied that he had nearly all of the Lowlanders in this town. Tomorrow would be another day. He gave the order, and an odd procession of confused humans was gradually set in motion toward Thorndike Manor. It was a very strange sight, almost resembling a parade.

One young female looked up to her left at a pair of very handsome militia men on horses beside her. "Hello, kind sirs," she began. "Can you tell me what this is about? What are we doing right now and why have we been gathered up?"

"Don't speak to them," cautioned Jackson.

"I don't see why not," Jorge responded. "We have been ordered to gather up all Lowlanders so that they can be taken to The Wall, to live in The Lowlands," he answered truthfully.

"Why," asked the young woman. "Why would you do such a thing?"

"I guess it's because that is where you Lowlanders are going to live from now on, that's all," Jorge answered lamely.

"I was born here in River Kingdom," the girl said indignantly. "I have never been to The Lowlands. Why would I want to live there?"

Jorge couldn't think of a good answer.

The young woman had another question. "Are you going to go back to The Lowlands to live, too? You two look like you are part Lowlander," she asked innocently.

"I told you that you shouldn't talk to them," whispered Jackson.

Jorge stared straight ahead. There was nothing more that he could say. The Auras drifted off. This was boring.

It was evening and the weather had turned blustery as the procession passed by the warm lights of the manor and headed toward the pastureland beyond. Thorndike was still leading. There was no hesitation. He knew where he was going and where he was going to stash the crowd behind him.

Jorge was not happy. Jackson was now experiencing more mixed emotions. He was the closest of the two to their father, but he was now unhappy, too. Maybe his father had not thought something through. Maybe there was something that they didn't understand. Would he really be content to leave all of these dozens of human beings out in the cold tonight?

With determination but proper decorum in front of his fellow militiamen, Jackson rode to the front of the line and drew up next to his father. "Excuse me sir, but may I ask what your plan is for tonight? It is getting cold and there are children in this group."

"What is your problem, Jackson? These people are Lowlanders. They are used to sleeping outside in their homeland," Thorndike answered with a grin.

"But it's warmer down south and they weren't prepared to be out in the cold here at night, Sir."

"There's a barn out behind the pasture," Thorndike snickered. "Some of them can go in there if they kick the sheep out. They will be heading south tomorrow. Don't worry about these people, tonight...unless you want to join them." He looked at Jackson with meaning. Thorndike's Auras chuckled.

Jackson felt the chill that Thorndike had intended to send with that answer. His father was holding him in check and letting him know that he knew who Jackson was and what was in his family tree.

Jackson returned to his place in line and remained silent. Jorge watched the exchange and listened. He, too, got the message. The twins rode on in misery.

From inside the manor, Flo watched the Lowlander procession slog toward the back pasture. She saw the bedraggled mass of humanity struggle between flanks of militia men and was sure that her brothers were with them. Shame and sorrow welled up inside her but accelerated into anger when Thorndike himself strode through the back door of the manor. "Don't fix any dinner for the boys tonight," he commanded with an authoritative swagger. "They were softening up with our prisoners tonight. They have to learn to tough it if they want to stay in *my* militia."

He looked closely at Flo for a minute. "That goes for you, too," he said. Then he marched toward the dining room where he knew his ale was waiting beside the fire. The Auras marched with him.

WHAT NOW?

King Solem was in his sitting room again contemplating his strategy. In a matter of only a few months, his calm world had been turned on its head. He had discovered that *actual slavery* was taking place in his kingdom without his knowledge! Two of the strongest nobles in his kingdom had lied and tricked their way into approval for a wide, new road that would leave his kingdom with unresolved irrigation problems for crops.

On top of it all, there were rumors that civil unrest was strong enough among the nobles that they might want to replace him.

Solem had always known that there was a strong wish among many nobles for what they thought had been 'the good times' when white men alone had dictated the way things should be. Back then there had been no need to consider 'lesser individuals' such as Lowlanders or even other kingdoms such as Urhonordo. Solem realized he had not paid enough attention to this or learned from the history of it.

His sitting room now seemed to be closing in on him. He walked to the window and gazed out on the peaceful colonnade below. Although it was chilly outside, the sun was bright and cheerful. He could easily ignore the issues before him, but his Conscience was speaking in his ear. "I don't think this is a time to relax, Solem. On the contrary, it may be a time for you to prepare for war."

Solem had lived for nearly fifty years. He did not panic easily. "What makes you think this current collection of problems could result in war? Aren't you being an alarmist?"

Csolem paced Solem's shoulder. *How can I explain this to a human?* he thought. *Humans don't know much about Auras and Auras are key here. I guess I must just boil it down to basics...*

"There are evil forces in the world," Csolem began. "These forces look for negative ideas in weak people and exploit them. We Consciences call these evil forces 'Auras'.

"I have never heard of Auras."

"That's because Auras are invisible. Only Consciences can see them, but they *are* everywhere. Their goal is to create more evil in humans, almost the opposite of Consciences who are trying for more goodness. Understand?

"Not really," sighed Solem

 "Well, try harder, Solem. What you need to know is that River Kingdom is full of Auras right now. Consciences on both sides of The Great River have been trying to fight off their influence but now Auras are working hard with two humans, I hear. Can you guess who?"

Solem smiled patiently. I guess so," he responded.

"Don't bother to humor me, Solem. I am telling you something important. These two nobles have even developed a following of nobles with their own Auras. In a recent survey of River Kingdom, I have discovered that all good Consciences now realize that they, themselves, are 'at war' with these Auras daily."

Solem shook his head and smiled a sad smile. "You expect me to believe that Consciences I can't even see are at war with something else that no one can see?"

"Yes...I know you can't see me, either but you know me."

Solem had trouble taking this so-called 'war' seriously, even though he usually trusted everything that Csolem said. He smiled indulgently and was about to walk away from the window when something caught his eye.

The Call of the Auras

Approaching The Castle rapidly was a line of horse-drawn wagons, some fine but most common and utilitarian, the type that one would see carrying produce to market. The wagons were followed by even more lone riders carrying men with very serious looks on their faces.

What now, Solem thought to himself. *Here I am with serious issues to confront and now the local townspeople are here to express their grievances. Just what I need.*

Solem straightened his robe and left the sitting room. He was not in the mood to quibble with townspeople over petty issues, but he was their King, and they deserved his attention.

Horace was already at the door to The Castle as the first townsman climbed off his horse and strode up the colonnade with purpose. "Allow them to enter and escort them to the conference room," Solem called from the top of the stairs.

By the time Solem entered the room himself, it was almost completely full, he noticed with some pride. *All my subjects feel welcome in my castle. I am proud of that,* he said to himself.

"Please seat yourselves if you can," Solem began as he strode to the head of the table. "I am sorry there are not enough chairs for you all."

"Chairs are not our problem today, Your Majesty," declared a large man who had apparently been chosen spokesperson for the group. "We are here to discuss the Lowlanders." The crowd around the room nodded.

Oh dear, Solem thought dismally. *More complaints about Lowlanders.*

He sighed. "Indeed? What seems to be the problems with the Lowlanders this time?" Solem asked as politely as he could.

"We have no problems with the Lowlanders, your Majesty," the big man answered politely. "We just want to know where they went and when they'll be back." The crowd around him nodded again, this time adding their own commentary and questioning...

"Yeah, where did they go?"

"Why didn't you tell us you were going to take them?"

"How do you expect us to get along with half our town missing?"

Solem held up his hand. As usual, the townspeople hushed themselves, but they were all staring at him expectantly.

Solem decided to sit for this. Obviously, they were talking about something he knew nothing about. The crowd backed off a bit and made room for their spokesman to sit near The King.

Solem began with a question. "Did you just say that your fellow townspeople are missing? Is that what you are all saying?

"You're darn right, Your Majesty. *All* of our Lowlander folk are missing, to be exact. Even some that work for the nobles. We all just got together yesterday to make a plan to come here today and ask for 'em back. Our town can't do a thing without 'em. The nobles want their servants back. We just want to know why you took our townspeople in the first place and then we want to know when you'll be bringing 'em back." More head nodding.

Solem swallowed hard. Csolem took to Solem's ear quickly. "They all think that you took the Lowlanders out of their town for some reason. Tell them that you didn't do that. Right now, they are just asking questions. Don't wait until they get mad."

"Let me understand, good people", Solem managed to say. "It sounds like your Lowlander population has gone missing for some reason and you want to know why. Am I right?"

"Indeed," snorted the big man. "We saw your militia take 'em and march 'em off somewhere. Them people make up nearly half our town these days. They're our friends. We depend on 'em. We all thought you liked 'em too."

Solem looked around the room. Their faces were all saying the same thing. They wanted answers.

"I don't have an answer for this except for my usual," Csolem sighed. "Tell the truth and be honest. That's all you can do."

King Solem did not feel very kingly, but he stood and followed Csolem's advice. "I know you will find this hard to believe because you town is so close, but I must tell you all that this is the first I have heard of this. *I* did not take any Lowlanders from your town. *I* would never do that. And I have no idea who did."

"But we saw your militia ride into town and take 'em!"

"We saw it with our own eyes!"

"They even took my grandchildren," cried an elderly man...and they were only half Lowlander! Why?"

"I don't know why," Solem said sadly. "And the militia you saw was not my militia."

"Whose would it have been, then?" the big man growled. "It was a big militia like yours."

"Thorndike Militia," whispered Csolem.

Solem looked around at the angry, worried faces. Their town had been completely disrupted. He had been feeling the same way about his whole kingdom just a few minutes before.

"I am very sad that you have all had this terrible thing happen to you," he continued. "I didn't do it, but I also must say that I didn't protect you from it, either. I am your King. You should be able to plan on my protection from something like this."

The group grew very still. Their king was speaking to them humbly...almost as an equal. This was a first.

The spokesman squinted at King Solem. He thought he could see that Solem was being truthful. He stood next to Solem and looked into his eyes. "What should we do, then, if it wasn't you that done this?"

"I am not sure what *you* should do here, but I am sure that *I* must do *something* immediately," Solem said firmly. "What has happened here as you describe it was an unlawful act against all of you, the Lowlanders and the whole kingdom. Unlawful acts will not stand. I will go to work on this immediately to find your townspeople and bring them back where they belong."

Silence reigned again. Then the group's anger seemed to melt. They nodded sadly and eased their way out the door of the conference room. They had done what they had set out to do. Now the King had promised to do what needed to be done. Within a few minutes, Solem was left standing alone in the Conference room, thinking about Lowlanders.

SLAVE RECRUITS

Drum was not thinking about Lowlanders. He had just left The Crown Room.

He had taken the crown out again to dust it and admire its gleam. How it must have pained his grandfather to put it away in hiding all those years ago. He, too, must have felt the call of greatness that was now so strong in Drum today. How proud he would be if he knew that Royalty would soon belong once again to the Drum family line.

Drum walked out to the drawbridge and again surveyed the progress being made on The Wall. He did this every morning now. The Wall was growing rapidly but not rapidly enough. The Wall would need to be miles long to reach the west side of The Kingdom. He hadn't really realized just how much labor that would take or how much time. He was impatient. Planning had taken so long. Now he was into the implementation phase of his plan...but not the part that he was most interested in.

"Tell those lazy Lowlanders to work harder," he yelled to his Drumsmen. In the distance he could see some of his neighbors' smaller militias and landholders pulling rocks into position to fill the levees on their land. They were not happy about it. They knew that the water that flowed by them would now simply continue to 'flow by' without watering their crops.

They had seen *pictures* of a little river that supposedly flowed on Drum's land and could be directed to some of them, but they had not seen an actual little river yet. They grumbled and groused as they worked. The nobles that owned their land were still unconcerned.

Suddenly Drum saw something else in the distance coming toward him. It looked like a parade. How strange. What would a parade be doing on the road now?

No! No! On the contrary. This must be an invasion force! This was truly a frightening thought! As a plotter of future invasions himself, paranoia was active in Drum right now.

"Attention you there on the work crew!" he shouted to his Drumsmen. "Grab up your shovels and get ready for war! We are being invaded!

Confusion reigned for a moment in Drumsworth but finally subsided when a forward scout sent by Drum came back laughing. "It's not an invasion that anyone should worry about, Sir," he announced. "Maybe we should even put out a welcome mat. This 'invasion' is a whole town full of Lowlanders from up north. Some of them are young but most of them will make great additions to our workforce for The Wall."

Drum now began to chuckle, himself. This was just the news that would give him what he had dreamed of...hundreds of Lowlanders working on The Wall that would fence *them all* out of The Kingdom. What a bonanza! Thorndike was really showing his worth.

Slowly the weary Lowlander procession made their way to the beginnings of The Wall that would soon be keeping them inside it. The drawbridge was down and Drum stood upon it triumphantly. They filed through the already existing gate attached between the drawbridge and The Wall and out into their homeland, a land that many of the young Lowlanders had never seen before. The Thorndike Militia followed behind them riding on horseback, herding the stragglers harshly.

Drum smiled his approval, closed the gate and waved the militia over the drawbridge and onto his grounds. "Give your horses to my servants," he said jovially, and I will show you where you will be housed. Everything is ready for you."

The militiamen obeyed willingly as they had been prepared for this by their Commander. It would feel like a vacation, they had been told. They would be staying in Drumsworth Castle, itself, a large step up from their usual quarters. All they had to do was work in shifts and keep the Lowlanders penned up behind The Wall. There was plenty of space to lock them up in the prisons out back.

Happily, the militia men looked around them. This is where they would stay while they supervised the Lowlanders that would eventually be sorted out and sent where they would be needed in River Kingdom. It was all in the plan.

Once the militiamen had been shown their quarters and assigned their duties, Drum hurried into The Crown Room again. This is where he had secretly kept and improved his *real* plans for Royalty. These plans were now ready for implementation. All his preparations were paying off.

Balynn had wandered down earlier, aware that there was noise and commotion coming from the road that led to her home. She tiptoed past The Crown Room, careful not to disturb her father who had locked himself in there a lot recently.

Bundled up against the winds of winter that were blowing from the north, she slipped outside. Maybe the boys were coming down to help with The Wall. She was longing to see Jorge.

Balynn was shocked at what she saw. The entire southern side of The Wall was teeming with Lowlanders of all ages. The youngest and strongest were already at work on The Wall while women and old folks mingled behind them. Children straggled through the crowd looking bewildered and crying.

The soft heart that rarely had a reason to surface sprung to life within Balynn. She could understand fights between adults and her father, but it was wrong to be cruel to children and old people.

Cbalynn climbed into Balynn's ear and wasted no time in underlining her sympathetic feelings. "This is what I have been warning you about," she hissed. "Now, do you understand the evil that your father is proposing with his plan?

"This is what Flo warned you about months ago. Your father wants to not only control Lowlanders but *eliminate* them all from River Kingdom. He thinks this makes him popular with the nobles. You have seen The Crown in the Crown Room. Your father wants to wear it as King. What do you have to say about *that?*"

"You're wrong," whispered Balynn to her Conscience. "You must be wrong. The King has approved the new road. Father and I stayed *inside* The Castle. Father would never do something so disloyal!"

"I'm pleased to see that you feel that insurrection is disloyal, Balynn. I am also pleased to see that you don't like the mistreatment of the Lowlanders. Now I am wondering what you intend to do about it?"

I am going to prove you wrong, Balynn thought with determination. *Maybe the twins are with the militia that brought the Lowlanders here. They will tell me the truth. They will tell me that my Conscience doesn't know what she is talking about.*

Balynn stormed out of the manor and over to the drawbridge. She climbed up on its railings to peer over the heads of the crowd searching for a familiar face. She didn't spot anyone she knew but *they* spotted her.

"Mistress Drum. Is that you? "Don't come any closer," hollered a young militiaman. "This area back here is full of Lowlanders."

"I see that", answered Balynn. "Why is that so?"

"Because they are Lowlanders, of course," the young man answered.

"Okay. I understand that, but I was actually looking for Jorge. Have you seen him?"

"Nah," sneered the militiaman. "The twins are Thorndike's favorites. Thorndike's keeping them home for the main action. He's not stashing 'em down here on guard duty."

Balynn climbed down from the drawbridge slowly. *My Conscience was right,* she thought with dread. *I should have known it. I should have seen what was happening. Why didn't I see it?"*

"You love your father. Sometimes love is blind," whispered Cbalynn. "But it's important, now that you understand, to do what is right...to think the thoughts that you know you have inside you. Think about what is kind and honest and good. Do what you can to use those good thoughts for good actions."

"Don't worry about the stupid Lowlanders," a nearby Aura said in its most convincing tone. The Auras sensed that her resolve was weakening in terms of following her father's lead. They were not about to let that happen.

"You are an intelligent woman. You already know that you like Prince June. This is the man that your father has in mind for you when he takes over the leadership that has always been his destiny. When your father dies, you will be Queen of River Kingdom and Prince June will be your partner. It will be perfect. He has been thinking of you, all along."

"Hog wash!" whispered Cbalynn. "Don't listen to the bad thoughts that others may put in your mind. Your father's thoughts have never been for *you*. They have always been for himself. *You* have always known that. Listen to your heart, Balynn. Your heart is where I live, and your heart and Conscience are both telling you the truth."

Balynn trudged through the door of the manor and was greeted by the servants that had been Paulina's friends in the past and actually hers too. She had been missing for a while and was very shy for some reason. Nudging each other, one of servants finally stepped forward. Balynn noticed something she had forgotten long ago. They were all Lowlanders.

Quickly their spokesperson said her piece. "Can we have your permission to go into food storage and get food for the Lowlanders? They are all nearly starving and much has been stored in advance to feed the militia. While we get food for the militia, we can also get it for our people...

 Paulina said that you were nice."

Balynn hung her head. *Am I nice?* she wondered. *Do I think that I should follow my father's plans or are his plans wrong...bad?*

"You know the answer, Balynn," Cbalynn said confidently.

"Come with me," Balynn said to the Lowland ladies around her. Delicately she removed the key to the storage that hung by The Crown Room and handed it to them. After accepting hugs from the ladies, she passed them the key fearfully.

"What's that I hear?" Drum bellowed from inside The Crown Room."

Balynn jumped. "I'm giving the servants the key to food storage... so that they can prepare food for the militia," she answered truthfully.

The Lowland girls grinned gratefully and disappeared.

GOOD DEEDS MODIFIED

Armando had returned to finish the job on the Central farmlands. Now he stood back and looked at the 'improved' Royal River Road that ran along the Central farms. His Lowlander volunteers had even helped to complete this section of the new road and it looked smooth and wide. It was beautiful. All the nobles whose farms were on the west side of the road had finally come to realize the benefit that had been brought to them by these helpful strangers. Even the nobles were now standing here to celebrate and say goodbye to new Lowlander friends who had helped them.

Two small springs had been found upland by the Lowlanders and their superior water-finding talents. Small rivers had been dug from them down through their farmland and then under the two bridges that would span these smaller rivers as they emptied into The Great River beyond.

Throughout the winter, farmhands here would re-dig the irrigation streams that were necessary to water their crops. Not only had these nobles finally realized the work that was needed to do what Drum and Thorndike had insisted upon, but now they had new friendships with and respect for the men who had helped them. These Lowlanders were not only hard workers. They were geniuses...and now friends.

Several of the nobles begged them to stay and live there along the Central coast of The River, but most had family and obligations in Urhonordo and they were homesick. Waving goodbye and promising to come back and visit, Armando and his band of Lowlanders picked their way back across the low part of The River, now beginning to freeze at the edges. Winter was setting in.

Armando and his Conscience were pleased to see that his theory about 'developing friendships by being helpful' worked. He could hardly wait to tell Luna. He wanted to tell her as well, that her theory about the location of water in the highlands of Central River Kingdom had also proved to be correct. She was so smart about so many things.

For these reasons, Armando arrived at the door of his castle with a smile on his face. His wife, however, was not in a good mood. Lowlanders had arrived from scouting missions that *she* had requested and they had not brought back good news. Armando sighed,

"Relax a minute and listen to your husband," Cluna advised Luna quietly. *His* good news will give *you* perspective on what you have to say to him."

Grateful that he was home and safe, Luna listened to her Conscience and hugged Armando close, begging him to tell her of his mission and what had happened. Cluna was right. She needed to hear something positive.

Armando began with the good fortune his men had had using her ideas with their natural water-finding skills. When he had told her everything, including the offers his men had received to move to River Kingdom, Luna laughed for the first time in days. She hated to change the subject to the information she knew she now had to share. It was devastating.

With a sigh, Luna began what she needed to tell him. "While you were gone, I became worried about the similarities between now and the last time that Drum and Thorndike joined forces. I remembered that the history books tell us that Lowlanders were captured to keep them from warning Urhonordo about invasion back then so I sent a few men on a scouting mission to see what was happening at Drum's property now.

"My scouts have just returned with even worse news than I feared. Not only have they captured local Lowlanders here in the south but also they have captured *all* the free Lowlanders from the town above Thorndike Manor in the north."

Armando hung his head. His homecoming happiness no longer seemed important. He also remembered the ten who had gone missing from his work crew along The River. *I should have gone after them*, he thought with shame.

"You were on a different and important mission on River Kingdom land," Carmando whispered sadly. "You had to inform The King. That was essential to getting them back."

But I let my friends be taken, Armando thought with misery.

Luna noticed his sadness, even though she didn't know the personal loss that made that sadness even more acute. She hesitated about telling him the rest of what her scouts had discovered. She would wait until he had rested a bit.

"This isn't a good practice," whispered Cluna. "Your husband needs to know everything...now."

"One more thing, dear," Luna now added with pain. "Your countrymen are now being held behind a wall down at Drumsworth. That wall is designed to keep *all Lowlanders* segregated back in The Lowlands forever. They were banished from River Kingdom and driven on foot to Drumsworth. They are now prisoners there."

Armando was stunned. He could say nothing. He couldn't believe how quickly his good news had flipped from happiness to misery. The Center of River Kingdom now loved his people while the northern and southern portions now wanted to cage them away forever.

And there was no doubt about it. Evil was on the rise in River Kingdom. Was King Solem aware of this?

Manda had heard her father's arrival. " Father, you're home! I am so glad! I was so worried when Mother told me about The Wall." Manda hugged her father hard. "Mother predicted something like this. She read up on the history of this area and she said just what you said the other day. This is like history repeating itself. Remember when you said that?"

Armando looked at his wife. She was so strong...and so smart...and now she was so right. But his eyes were sad. His daughter was also right. His countrymen were already imprisoned and soon Urhonordo, itself would probably be in the danger.

He and Luna had lived a life of peace in Urhonordo and preparation for war had never been a focus. Now he regretted that. His honor guard was small, and he actually had no militia at all. Armando looked at the two women in his life and knew what he must do.

"I must leave again, immediately," he said, trying to appear stronger than he felt. "King Solem will not stand for such terrible behavior as has taken place with *his* Lowlander population once he discovers it. I am sure that he will want to return them to their homes almost as much as I do once he discovers how they have been stolen."

"He is already prepared to hear this," said Manda sheepishly. "I kind of...warned him that something like this might happen day before yesterday."

Armando shook his head and smiled. *She is exactly like her mother,* he thought. Within an hour, Armando and Manda were bundled against a storm that now threatened. They were on their way to River Kingdom Castle again.

PREPARING FOR PROTECTION

Under orders from Thorndike, his militia men were getting themselves ready for a mission they felt they had been preparing for all their lives...going to battle to protect River Kingdom, of course. They had no idea who they would be fighting, but Thorndike's antsy behavior indicated that their first engagement would be coming soon.

The militia sharpened their blades and polished their armor which had now moved from storage in the manor to the sleeping areas of the barracks. The horses were brushed and groomed. They pranced in their stables and reared with excitement. Their grooms led them carefully into the pasture but they didn't turn them loose. They would have sprinted off in every direction. The excitement was contagious.

The twins were in a quandary. If everyone was getting ready for a battle, who were they going to protect? Their Consciences had told them through years of training that their job was to serve and protect. That was what militias were for, right? Now they were going to do it...but who?

Everything they had was polished. It always was. They kept things that way. Since their fellow militiamen were abuzz with just the anticipation of possible combat, the boys decided to talk to Flo. Maybe she knew who needed protection. Flo was on the inside. Maybe they could finally get a hint of who it was that they were going to fight.

Imagine their surprise when their father greeted them at the kitchen door. He had been on his way to find them and announce a decision he had just made.

Slapping his sons on the shoulders, he grinned but in a strange and secretive way. These were his sons, and they were going to be the first to know of plans he had formulated, plans that no one knew of but him. Very excited Auras flocked around him.

Flo was laboring on the next meal for all of them. She looked up in surprise. Early morning meetings were common for her with her brothers, but Thorndike was with them today. She noticed that her brothers seemed almost as surprised as she. *They were intercepted at my back door by Thorndike,* she reasoned. *They were on their way in to see me. What is Thorndike up to now?*

Flo turned back to the stew that would be tonight's dinner for the militia. She would pretend to pay no attention to what Thorndike was doing with her brothers, but she had her ways. She would find out what this was all about.

Thorndike was overly intense again. He slurred his words. His Auras were completely in charge. Drum had made him wait long enough, they said, encouraging the growing rivalry. *He* had paid homage to Drum's leadership more than he should have. Now it was *his* turn to lead.

The gathering up of the Lowlanders had bolstered both him and his Auras to the point that they now realized that he, Thorndike, and his enormous militia could now do anything they wanted to do on their own. He didn't need to wait for Drum any longer.

Majestically Thorndike seated himself at the head of the large conference room table, motioning his boys to be seated on either side. His Auras curled around him like a snake, relishing their control.

Thorndike smiled down on his sons as manic pride grew. This was the moment he had been preparing for all his life he now realized. He had created these sons for a purpose. He and *his sons* were in position now to accomplish the goal that had been stolen from his gtandfather one hundred years ago...the conquering of Urhonordo. The Auras filled the room with expectation.

"My time has come. Urhonordo should have been my legacy and yours, boys. Now, at last, it will be!" Thorndike began. "It is time that a white man rules the Kingdom of Urhonordo. It is time that The Great River is controlled by white men ruling kingdoms *on both sides*. It's time to put Lowlanders like the incompetent Prince Armando back in their place where they have always belonged, in The Lowlands."

"Yes, yes, yes," cheered the Auras that only Thorndike could hear.

The twins listened in silence. *Does he remember that we are half Lowlander?* they asked themselves. *And what has Urhonordo done wrong? Have they threatened someone that we haven't heard about? And why are we alone here and hearing all of this? Does this make any sense?* The boys were thinking questions but saying nothing.

Thorndike stood and smiled down on them with a look that they interpreted as pride. For the first time it felt that he was truly acknowledging that they were his sons. This was a heady experience, one they had always wished for but never dared to request. ..but why now?

Jackson felt that he was most in line with his father's thinking. Now he dared to cautiously speak up. "Do you mean that we are going to invade Urhonordo, Father?"

Thorndike sneered. "Urhonordo has no militia to speak of, son. I believe that we can simply walk onto their castle grounds, lock up their Royalty and assume leadership. From what I have seen of Prince Armando, we should not have any problem. He is just another Lowlander like the ones we escorted to The Wall."

The boys stared at each other, their thoughts in turmoil. For a moment, the way their father explained it, it all sounded so logical. They would simply be moving into a kingdom that should have belonged to their grandfather one hundred years ago and Prince Armando didn't deserve now. There would be no shots fired. It would just be a peaceful change of leadership.

"Are you crazy?" shouted the twins' Consciences. "Do you hear what your father is suggesting? He just said that he wants to take over a peaceful kingdom that has done nothing wrong and make himself King of it! We just thought that Lord Drum was the only one who thought he had 'King Potential'," the astonished Consciences added. "Your father has lost his grip on reality!"

But he is our father, the twins thought defensively. *Isn't he trying to use us now for something good...for protecting River Kingdom? Maybe we just don't understand...*

"Think!" pleaded the Consciences. "Your father wants to invade Urhonordo which is not invading anyone. Do you think that is a good, protective idea?"

The twins said nothing. They nodded their heads and looked obedient, a demeanor they had perfected over the years, but slowly the potential for possible violence began to sink in. Even a small kingdom would fight to keep someone from coming in and taking over.

Thorndike concluded his battle plan discussion with his sons with a very direct and complete announcement. "Now that you know the total plan, sons, I am swearing you to military secrecy. Only the three of us will know all of the plan so there will be no leaks to the enemy.

"And we won't use The Bridge. That would be too obvious and it will not hold men in heavy armor. Instead, we will sweep down the new Royal River Road and through the gates of The Wall. We will then skirt around Drum's mote and up into the southern part of Urhonordo for a surprise attack... just like my grandfather did.

"Urhonordo will never know what hit them. Then we will capture all the Lowlanders in Urhonordo, clean the place out, and march those Lowlanders down to Drumsworth to live with their relatives in the Lowlands.

"Now, you two tell the others to get ready for conquest, but don't tell them where it will be. This is a battle plan that only the three of us family members will share. Tomorrow morning, we will move out!" His eyes gleamed with anticipation.

The twins shuddered. As much as they had respected their father, their Consciences were right. They knew this idea was insane. This was not a battle for the protection of River Kingdom. It was an invasion of another, innocent kingdom. Their Consciences knew this but also knew the boys felt trapped by their father's manic obsession with Urhonordo. The Auras were obviously completely in charge.

With polite and obedient salutes, the twins left the conference room. They had their orders. "This may be a time to disobey your father," the boys' Consciences whispered.

PREPARING FOR LIFE

The twins had never disobeyed their father. That was a rule that had been drilled into them early. Disobedience was unthinkable.

By the time they reached the kitchen, however, they were in complete turmoil. Flo could see it. Quietly she motioned them to the back door. She had been listening.

Flo studied her brothers carefully. She was in turmoil, as well. She had waited too long to tell them the whole truth about their father. Now she had to tell them that awful truth…when she, in her sisterly way, had raised them to obey and follow orders. If she didn't stop them now, however, they would be going to war.

Flo sat down on the back steps of the manor and motioned her brothers to join her. She looked at the boys, one on each side of her, and began a story that was a confession as well. Her confession was all she had to prevent the horrors that their father had planned for them.

She began with the most astounding fact first. "Our father is a monster," she began clearly. "Long ago, he wanted to kill his wife, Lady Wilhelmina, Willa's mother. But my mother wouldn't let him. She never died. She is still alive."

To her shock, the boys did not seem that surprised. They simply looked at her and nodded. "I thought that might be who I was looking at the other day," said Jorge calmly. "I found a lady who was still kind of pretty and looked a bit like Willa. I thought that might be who it was, remember, Jackson? I told you she was scared to death of me. I think it was because I look like father looked when he was young. Why in the world was she stuck back there in that room?

Jackson nodded. "I bet you're right. I was going to visit where you said she was, but when I opened the room where you said she was, there was no one there. Did she die?"

Flo just sat between the boys in silence. They had found the lady but had said nothing. She began to cry. "Thank you for telling me that you found her," she began. "And no, she didn't die. But she has been a prisoner in that room until just recently. *I* have kept her prisoner there for twenty years. I have been a terrible person."

Jorge and Jackson looked at their sister with disbelief. Their sister was kind. Why would she do something like that?

Flo wiped her tears and continued. "Our father wanted her out of the way. She couldn't produce a son. He yelled at her, and beat her and she cried all the time. Our mother talked Father into banishing her rather than killing her. To keep her safe, our mother decided to hide her in a vacant part of the manor instead of banishing her. She was a bit unstable at the time. Mother was trying to keep her safe."

"And you were trying to do that, too?" asked Jorge gently, wiping his sister's tears. "That was not a bad thing, Flo. You weren't being mean. You took care of her."

"Yes," sobbed Flo. "I took care of her, but I never told Willa. Willa has had a living mother that she knows nothing about all this time. She has been told that her mother died. She has even been told that by me. I have been lying to my own half-sister all these years. I am a terrible person."

The twins looked at each other, then they looked at their sister. Their father had banished Willa's mother. *Because she couldn't produce sons?* they wondered. *Why would their father have done something that mean just to have sons in the family?*

"You are completely dense," whispered their Consciences. "You boys are your father's heritage. When he dies, you will become kings of Urhonordo...which is the role he thinks his father should have inherited back in the day and given to him. This fantasy has driven him insane."

Wanting sons to inherit a kingdom that you don't have is stupid, thought Jackson.

Banishing a good woman so that you can make sons with another is just plain cruel, thought Jorge.

It sounds like he was crazy then and worse now, the boys each thought.

"Exactly," said their Consciences.

Their Consciences now came on strong. "We have always counseled you to obey your father in the past. That is usually sound advice, but now we see that our thoughts about you and your father must change. We are now sad to conclude that your father is no longer (and never was) worthy of the sons he created.

"But your sister did the best she could do under a difficult challenge," the Consciences continued. "She now feels guilty which is a good thing. The question is, what are the three of you going to do about it? Are you two going to follow your evil father to an evil war?"

The siblings sat on the back steps of the manor in silence. They had all shared a lot of information, but the brothers were still unclear.

Why had Flo told them this now?

Where was Willa's mother now?

What were they going to tell their fellow militiamen?

What were they going to do tomorrow?

Flo had been thinking about this for many days now. "I have needed to tell you about Lady Thorndike for a long while," she began. "The reason that you didn't find her in her little hiding place, Jorge, is because I finally set her free.

"The sight of others and particularly you, looking like a young version of Father, jarred her into rational thinking that she had not shown for years. When I noticed this, I told her about Willa. She of course, wants to see her but amazingly, she decided to wait until she thought her thinking had completely cleared. She has been quietly moving around the manor, listening...watching...and occasionally speaking up...slowly trying to catch up with what is happening now.

"When she heard that you all had driven the Lowlanders back to their homeland, she was shocked. She decided that she had to come forward to warn the world about him, so that his evil would not continue.

"Now he wants to fulfill his insane destiny. That's why I had to say something now. Tomorrow may be too late."

Jorge smiled a sad smile. "Our father is really crazy, isn't he?"

 "Yes. And Lady Wilhelmina needs to be returned to her daughter so that they can finally unite, and her story can be told." Flo said with quiet determination. We need to do it tonight before she is discovered, and while we are all here to do it. Tomorrow we may 'go to war' but I, for one, will not fight this insane war of father." said Jackson gallantly.

"Neither will I," sighed Jorge, "so that may mean we will not be able to come home at all."

PREPARING FOR CHANGE

There was no moon at all, and the winter sky was black as three horses were quietly led from Thorndike Manor to the main road. The new Royal River Road was already ready for travel, between Thorndike Manor and The Castle, the brothers noted with uneasy pride. They had worked on it. They had pushed themselves and others to get it done before the snow fell and they had succeeded. They were proud of that, at least.

Lady Thorndike was sitting on one of the horses already. Flo was riding with her. They had all expected that she would ride with one of the boys, but they still frightened her with their 'Thorndike look'.

She had managed a small smile for them both but looked to Flo. Flo had been her only friend for years. Flo was now taking her to see her daughter...a daughter she had thought was dead...a dream come true. She trusted Flo.

The tiny older woman trembled with anticipation. Flo had fussed over her all evening after serving dinner and cleaning up. She had also found a lovely dress of Willa's that was still in the closet. It fit almost perfectly.

Wilhelmina Thorndike did not feel confident, but she did feel alive. It was finally happening. She was on her way to see her daughter.

Flo and the lady were nervous but as ready as they could be when they arrived at The Castle and guards met them at the front gate. With a small amount of panic, however, they all realized that they had not known what they would do at this point. What should they say to gain entrance? What could be the reason that they had all come to the Royal Entrance at this time on a cold winter night?

The bundle that was Lady Thorndike had ridden quietly from Thorndike Manor but now the bundle came to life. "Tell the guards to tell the Queen that *her mother* is here to visit her."

Jorge and Jackson looked at each other and at Flo. "Please, boys...do as I ask," came the small voice again from under the wraps. "I have been waiting twenty years for this."

Jackson jumped down and approached the guard at the gate with respect. "I know this sounds strange", he offered with a smile, "but Florence and us boys are here from Thorndike Manor. It would be jolly if you could tell the Queen that her sister, Florence, is here for a visit...with her mother."

The three young people waited tensely with their special cargo as the guard grinned. He recognized the markings of Thorndike on the boys' clothing and decided this must be a joke among the young people. As far as they knew, the Queen didn't have either sister or mother. The young guard opened the gate and then hustled to find Horace. Horace would escort them inside if the Queen said she would allow it.

The guard told Horace the joke, but Horace didn't think it was funny. He wondered about the decorum of young people these days...no appropriate protocol at all.

Horace passed the joke to Agnes with a snort. She didn't think it was funny either, so she shortened it. "There are four young people waiting for you at the front hall," she announced briskly at the sitting room door. "It is an inappropriate time of night for an unannounced visit, but one of them says she is your sister." Agnes rolled her eyes and waited for a response.

Willa was sitting by the fire. Both she and Solem were deep in history books. Manda's cryptic warning about the possibility of rebellion based on that history had made the musty history books suddenly more interesting.

Willa went to the door of the sitting room. "Who did you say was here, Agnes?"

"Horace says that he was told that the girl among the visitors says she is your sister."

Color drained from Willa's face. If Flo was here at this time of night, there must be something terrible happening. Without further thought, she ran down the stairs.

The visitors were shivering in the cold entryway. Willa immediately recognized Flo and the boys. Now the wrappings on the fourth person fell away.

Before her... in the middle of The Castle entryway... stood *her mother*. Her mother? Willa dropped to her knees on the stairway. She must be seeing things. This could not be happening. Her mind reeled into a faint.

Solem saw what Willa saw and could not believe who he was seeing. This was surely an older version of his wife. He reached down to Willa and pulled her gently to her feet. As illogical as it seemed, he knew at once that this frail woman that stood before them was, indeed, Willa's mother.

The entryway now spun into activity. "Please escort yourselves to the conference room immediately," he ordered. Protocol suspended, Solem picked up his quaking wife and carried her to her usual chair beside his. Horace read the situation and placed a chair close to Willa, nodding instructions to the two young men who were now on either side of the little lady they had brought with them.

Flo did not wait for an invitation. There was room in the chair next to Willa for two and she unceremoniously deposited herself there, carefully snuggling the older woman beside her as one would a child.

The room grew silent. Solem looked with concern at his wife and then looked sternly at the twins who obviously wore Thorndike insignias on their clothing. He waited for an explanation.

But from the little lady cradled next to Flo, a tiny hand emerged and reached out tentatively toward the Queen whose eyes remained closed. The hand touched Willa lightly. "Open your eyes, my beautiful daughter. I *am* your mother. At last, we see each other again. Please...just open your eyes and see me like I see you."

Willa did as she was told. Slowly she opened her eyes and looked into the eyes of the woman she had always thought to be dead. It *was* her mother.

Willa had a mother. Slowly her senses revived, and she sat more erect, taking the offered hand in hers." How can this be, Mother? How can you be here with me and alive?"

"It is a long terrible story," Flo offered. "Too long and terrible to tell on this wondrous night. "The important thing is that the two of you are finally together as you always should have been."

Tears began to stream down Flo's face. She wouldn't ask for forgiveness. She didn't deserve it. The twins stood back and looked on with mixed emotions. Flo was obviously relieved that her terrible secret had ended in happiness for two people she loved, and the boys were happy for their sisters and the long-lost Lady Wilhelmina...but now their father's looming plan reclaimed their focus.

We should be back at Thorndike now, getting ready to invade Urhonordo, thought Jackson. *That is where our father thinks we are.*

"You can't do that now," stated Cjackson bluntly.

Our father has no idea that we are here at The Castle right now. What would he say if he knew? Jorge asked himself.

"He expects that you will be readying your fellow militiamen for a war. You know now that war is wrong, right? countered Cjorge.

It all came together now. The momentous meeting they had had with their father this afternoon was not one of family pride as their father had thought to present it. It was rather the culmination of a life of selfish insanity that had ruined the life of the woman here in front of them.

"I am proud of your thinking," whispered Cjorge.

"Your sense of right and wrong has survived a major test," agreed Cjackson.

The boys knew their Consciences were right, but now, what? Their father was expecting them to lead a charge down the new Royal River Road... toward war.

The boys approached King Solem and bowed solemnly. Jackson spoke for them both. "We apologize for this late-night visit, Your Majesty, but Flo thought it was important enough to bring Lady Wilhelmina here tonight.

"Lord Thorndike was not informed that she was found or that she was leaving. He will not be pleased that we are here. We should leave immediately...and our sister, too, we think." Both boys bowed their heads and waited for a response.

Flo stepped forward. "My brothers are right. We came here tonight to help our sister and her mother come together, a meeting that should have happened years ago. But we should return to the manor tonight before we are missed."

Solem had so many questions that he could scarcely contain himself but, as he turned and saw the love that was now pouring from his wife's face, he decided that he could get the details behind this amazing night later. He nodded and dismissed the three young people with thanks for their endeavors and a request for a return visit when they were able.

The siblings mounted their horses and left the colonnade, riding out into the black winter night, worrying about war in their future but grateful for the joy they had just delivered.

THE CASTLE REACTS

As the sun came up the next morning, The Castle was buzzing with news of the new arrival. Queen Willa had a mother, a mother that she had not seen since she was a very little girl. Questions among the servants skittered happily from room to room making the morning's chores almost enjoyable.

"Where has she been all these years?"

"Did the Queen really have no idea that her mother was still alive?"

"Was she a captive somewhere?"

"Who would have done such a thing?"

"Did you see how lovely she is?"

"She looks just like Queen Willa."

Willa was in heaven. She had tucked her mother into her own bed that night, and now a beautiful, new room was being made ready for a very special occupant. All night she had maintained vigil over this miracle mother of hers. She still couldn't believe this was happening.

Mimi was enthralled with both the new arrival and the mystery that surrounded her. She sent to town for a seamstress immediately to come and do measurements for "the Queen-mother's new wardrobe".

In the meantime, Willa brought out a lovely gown that had been recently stitched for her and insisted that her mother wear it. "You will look splendid," she whispered lovingly. They turned out to be the same size... "almost replicas of each other," the servants clucked happily.

Willa and her mother sat across from each other in adoring silence. They were mother and daughter again for the first time but so much time had passed between them that they hardly knew where to begin. Love and longing poured from both faces but also the pain of separation and not understanding how this had all come about. At last Willa began the questions that were on the mind of a lonely little girl, now grown.

"Where have you been all these years, Mother?" Willa asked gently. "Didn't you never wonder where I was or what I was doing?"

Wilhelmina's face crumpled into tears "I don't know", she answered miserably. "All I remember now is that I was in a room by myself, all alone and miserable forever and ever. I know that my mind was scrambled and filled with the fear that I would be killed at any moment. I was told that you died, you know...that you had been trampled by a horse and drowned."

"Trampled by a horse? Who told you such a terrible lie?"

"Your father," Wilhelmina sobbed. "He told me that you had been trampled to death and, since I could give him no useful sons, I was of no use to him. I believe that is when I lost control. I don't remember much past that."

"And you believed him?" asked Willa. "Didn't you ask to see my body? Didn't you mourn for me?"

Wilhelmina covered her tear-stained face. She remembered that horrible day clearly. "Indeed, I mourned. I thought you were dead. Your father told me that he had taken you for a ride in the little carriage that I had made for you. It tipped over and threw you into The River. He said that The River was too swift and that you were washed away. The Auras of his evil surrounded me and wouldn't go away. I'm sure you know nothing of Auras, but I was surrounded by them for years and years."

Willa moved to her mother's side, her arms circling her mother's shaking shoulders. Her father had cruelly, *intentionally* caused her pain, all because she could not bear him a son.

Willa had always assumed that her father was merely thoughtless. Now she knew that he was so much worse... thoughtfully, *purposefully, insanely* cruel to both his wife and child. She, too, had been lied to and told that her mother was dead. How could a man live with such cruelty for so long?

Thorndike must have been addicted to the Auras all that time, Cwilla thought to herself.

"I'm afraid that I lost my mind that day", the fragile Wilhelmina continued, now quieting her sobs to stare into the past. "I remember running to The River and jumping in after you. I was sure that he was lying and that I would find you, but I don't recall much more. I believe my mind couldn't accept what he said and I... lost control.

"Once I thought that I saw you," she continued trans like... "but I thought you were an angel who had come to visit me and that we were both dead. Another day, I thought Thorny had come for me and was going to kill me like he killed you. Flo told me that that wasn't true. It was only one of her brothers...I do think that I began to recover my senses then, however.

"I guess Thorny still doesn't know that I am alive" the older woman stated matter of factly. "I have tried to let him know in secret lately. I'm not sure but I think he heard me. I stalked him while I tried to gain back my thinking. I even tried to make him understand his insane obsessions. I'm not sure that had any positive effect," she added with a small smile.

Willa smiled too, remembering the mysterious voice she had heard at a meeting.

"Flo has been helping me," she went on. "Poor Flo. She has been so good to me, but her secret has made her so angry with herself. And when I began to get my wits back, she completely unlocked the door to my room and became determined to fix what she and her mother had done. They *did* save me, you know. They saved my life."

"Minds are very strange things," Wilhelmina continued sadly. "I think I am well now and that my mind is back to being clear...but sometimes I'm not sure. I do know that I am thrilled to be aware and mindful at last if only for a moment. I now know joy...real joy. I am here with *you.* I bless Flo for bringing me to you at last."

 Horace knocked gently at the door and Solem followed him into the room. "I have biscuits and jam for you both" he announced. "You two young ladies need to eat as well as talk." He placed his tray of biscuits on a nearby table. Another visitor appeared behind him.

"I am here just to be nosey," offered Solem as he entered and grabbed a biscuit. "I want to hear *all* that you have to say, Lady Thorndike. Am I right in assuming that his lordship has no idea that you are here?"

Wilhelmina smiled at his question and picked up a biscuit. "You are right, Your Majesty. There is no doubt that the twins helped Flo sneak me out of the manor last night at great peril to themselves, especially considering the march today that they were preparing for."

Solem's face turned quickly in her direction. "What was that, Lady Thorndike? Did I just hear you say something about a march? That's very interesting. Was Lord Thorndike escorting more of our Lowlander citizens down to Drumsworth again?"

"Oh no, Your Highness. He did that *last* week. *This* week he is going to Urhonordo. At least that is the general gist of what he was telling the two boys. And please call me Mina as you used to when we were young. Do you remember?

Solem smiled at the memory, but his casual demeanor had already changed to concern. His biscuit was placed on the table, and he sat down, leaning in toward the older woman and speaking intently.

"Yes, 'Mina', I do remember. I remember when you visited here, and Linda and I visited your manor. Those were lovely days, my friend, but these days are very different now. Lord Thorndike is no longer a young squire holding parties as he was then. Most recently he and I have had our differences. And now you say that he will be on a march heading south?"

"Indeed," 'Mina' answered as she munched on her biscuit. Her eyes closed.

Her Conscience, long inactive, came to life. "Tell them what you heard yesterday, Mina. I think they need to hear it."

Really? Who said that? thought the woman, looking around.

"I did. I'm your Conscience. Remember me?" answered Cmina.

Oh, for heaven's sake. I had forgotten there was such a thing as a Conscience, Mina thought. *I have been surrounded by Auras for twenty years.* "You are right. Yes Solem. I was eavesdropping yesterday on a meeting that he was having with his two boys. I may have got it wrong, but I believe that they were getting ready to march all the way through Drumsworth and on up to Urhonordo."

"Urhonordo? Why in the world would they be doing that?" Solem now asked seriously.

"I believe the wretched man thinks it's time for him to take over Urhonordo. He always did want to do that," Mina remembered without blinking. "That's what I believe he told his boys yesterday, too. By the way, Flo told me that those boys were his sons. Is that true? That must have happened after I was dead. That's amazing. They *do* look like him. I guess he finally got the sons he wanted, after all. Amazing."

The frail image was beginning to become a functioning human again. Her lined face smiled a forlorn smile at Willa and asked sincerely." Can you forgive me for my weakness, my love? Can you ever forgive my withdrawal into the hiding place of insanity all those years? I will never forgive myself, but I will do everything I can to make it up to you." Mina's Conscience smiled.

Willa just folded her arms contentedly around the mother she had always wanted. "Now you are my Mother Mina, here with me where you belong." 'Mother Mina' would now be her name.

Solem was happy for them both, but he began pacing the floor of the sitting room impatiently. He had just learned that Thorndike and his militia were on the move... and conquering Urhonordo might somehow be the goal of this move. He studied his new family member for a moment. She was an older version of his wife...and he knew instinctively that what she had overheard was true, at least from her perspective. The question now was, what should he do about it?

A rumble from the downstairs entry hall was interfering with his concentration. Horace hurried down to identify the source of the rumble. Prince Armando and Manda were in the entryway and asking to see The King. Solem rushed down to greet them. Without a second thought, he knew that this visit was not a casual one.

"Armando... Manda... Come with me into the conference room. I have very worrisome things to tell you and it looks like you are in a hurry to tell me something as well."

"Did I hear someone say that Urhonordo is here?" shouted June from the top of the stairs. "Horace just told me we have company in the sitting room. I don't know who that is, but things are looking up in this boring castle...lots of company. Great! It sounds like old times!"

Solem smiled up at his son and shook his head as he followed Manda and her father into the conference room. *Being young and full of life is a blessing,* he thought to himself. *He will need all that youthful energy, soon, if I am not mistaken.*

June immediately realized that his good mood did not extend to the others in the conference room...even though the smile from Manda was warm. He would sit next to her.

Everyone went to their usual positions around the table and Solem began. "Urhonordo does not arrive at my castle unannounced without a reason," he said seriously. "I know this, and you are always welcome, of course. Please tell me the reason for your visit."

Armando noted the solemn tone of his host and began without preamble. "Urhonordo scouts have been to Drumsworth and reported three things to my wife: There is a wall being built on the border with The Lowlands that is designed to be used as a barrier which will keep all Lowlanders behind it. This barrier is being built by captive Lowlander *slave labor*. All the River Kingdom citizens that are of Lowlander descent that were living in the town above Thorndike Estates have been rounded up and stuffed behind that Border Wall. As King of River Kingdom, the Queen and I thought that you should know."

"It's just as you predicted, isn't it," June said to Manda. He had slid into the chair beside her and now nudged her like the buddies they had become. "Thorndike, like the rest of the nobles, hates Lowlanders and now looks at them as part of his road to popularity. The nobles will love him for that."

"I heard of this kidnapping of Lowlanders from town yesterday," Solem responded. "The town from which the Lowlander population was stolen came here en masse and complained to me. They are angry. Most of them were not nobles and they want their people back."

Armando smiled and nodded. "Of course, they do," he responded. "And I came here today, hoping that you and your militia would help me bring them back. Drum has no militia...only a few Drumsmen that he uses to do his dirty work. I would dearly love to accompany your Royal Militia down to the border to free Lowlander slaves and hardworking Lowlander citizens of River Kingdom."

"I would like to punch Thorndike in the nose," growled Manda.

Armando, Manda and June all grinned at the thought. This would be something they could all enjoy. Their Consciences chortled within them at the anticipation of this righteous activity.

Solem was the only one who was not smiling. He looked around him at his son and his friends. Why could their solution not be the simple activity that they all anticipated?

"Because taking care of this much evil is not going to be easy, Solem," counseled Csolem.

Solem nodded ruefully. "I, unfortunately, have some information to share with the two of you, as well. Last night this Castle got a visit from a ghost. I also need to tell you the information this ghost brought with her was important information from Thorndike Manor."

"Maybe my mother would like to tell you, herself," said Willa from the conference room doorway. "I tried to make her rest, but she would hear none of it. Please allow me to introduce you to my recently freed mother, Lady Thorndike."

Everyone at the table rose and stared. June's mouth dropped open. He was looking at two Willas, women so much alike that they almost appeared as mirror images. They all inhaled at once and bowed but couldn't seem to speak.

Willa's mother smiled and nodded at their astonishment but entered the room with the seriousness that the situation deserved. Willa had filled her in on the details of the slavery situation and she was now not only 'back' mentally but concerned.

Armando bowed respectfully. He had known that she was dead for years and was shocked that she was now very much alive. There were almost *two* Willas standing before him. Manda was fascinated with the resemblance. There was no mistaking who was who. When she opened her mouth to speak, there was still no doubt how very much alike they were.

"Hello to all of you. Thank you for allowing me to attend this meeting. My daughter has told me of the importance of what I overheard yesterday, and my Conscience informed me that it was my duty to make what I heard clear. I now understand that what I heard was a plan for war."

Solem went to her side and assisted her to be seated at the end of the table where Queen Luna usually sat. Her demeanor was confident and strong, but Willa took her place beside her mother for support.

Folding her hands in front of her, Mother Mina began: "I heard Thorny bring his sons into his conference room yesterday and speak to them with a very definite plan. It was a plan that I remember from my early days with him.

"He has always resented the fact that King Sole of one hundred years ago returned early and defended Urhonordo when his Thorndike ancestor was ready to conquer that kingdom. He had caged the Lowlanders to keep them quiet then and he is doing the same thing now.

"Thorny and the sons he always thought were essential are on the march today to circle behind Drumsworth Castle and complete the conquest that his grandfather couldn't do. He thinks he has all the Lowlanders imprisoned at Drumsworth and Urhonordo hasn't got a militia big enough to defeat him just like before.

"This has been his quest since he was a boy and now, he has two sons to help him take over and rule the whole Urhonordo Kingdom. This was always more than a plan. It is his life's mission."

"And the dream of the Auras on both sides of The River," whispered Csolem."

Everyone in the room strained to hear each word that the new voice in their midst spoke. She was new and yet she wasn't. She was Willa with an old memory and now a very strong *new* understanding of something she had heard. Although fresh out of years of confinement, she remembered the man who had silenced her then and what he had done then matched with what he was doing now.

"History is repeating itself, just as you and Mother predicted," whispered Manda.

"Your mother read about it years ago and remembered it. Now it's coming alive again," marveled Armando.

June looked squarely at Manda. "You thought this was going to happen, too, didn't you?" he asked in awe.

"I read a lot too, June," she said with a look.

But now the table was full of people who were realizing a terrible truth. Thorndike and his sons were on their way with his militia to conquer Urhonordo., right now.

THE DRUM BEAT

Drum spent the day in The Crown Room, secretly trying on the crown of his grandfather again and again. It was his 'touch stone'. It was his link to the glory that his family had once possessed and would again soon. Carefully his returned it to its vault behind his grandfather's portrait. "I'll be back for you soon," he promised with a smile.

Drum was waiting impatiently for a signal from Thorndike that his militia and the road were both ready for the main part of their Plan...The Invasion of River Kingdom Castle. The road must be finished. It was time to get serious. Drum and his Auras were getting restless.

From Drum's perspective, Thorndike was still overly proud of achieving the beautiful new road. *What good is a new road,* Drum grumbled to himself, *if the troops of the Thorndike Militia are not using it to get me and my crown back where we belong?*

Drum sent a couple of his Drumsmen north to tell Thorndike that *now* was the time to implement the most important part of The Plan, the attack on The Castle. He visualized the scene. His Auras had polished the view for him. The combined militias would sneak up quietly on the smooth, wide surface of the new road, complete with a huge number of men. The Royal Guardsmen would be asleep, not even suspecting they were being invaded until it was too late to retaliate. The sleepy Guardsmen would be so easy to conquer that Drum could be installed as King, that very day. It was a brilliant plan.

Was Thorndike ready for this? He had wasted some time independently gathering up all the northern Lowlanders and bringing them down to build The Wall. That had been unexpected, to say the least. The militiamen had explained that they understood this was just a part of the bigger plan' and was done to keep the nobles happy.

Drum had been a bit unnerved by the size of this operation which had taken place without *his* authorization, but he had decided that the idea made sense. Happy nobles were Drum's main asset. They were important.

And so far, the move had been working just fine. The militia men that were housed in Drumsworth Castle were content, sleeping in shifts and shouting at the

Lowlanders from town as they slaved away to increase the height and width of The Wall. Excellent. The Wall was a beauty and was now stretching a long way to the west.

Drum strolled out to the drawbridge to enjoy watching the slaves at work but suddenly he saw... his Drumsmen...back so soon?

Good, he thought to himself at first. *Thorndike must have signaled that it's time for me to go up there and supervise the invasion of The Castle immediately. I can't wait to see the look on that smug King Solem's face when he sees me at his gate... while his Royal Guard is still asleep.*

But the Drumsmen came to a clattering halt in front of Drum in a hurry. "You won't believe this," shouted Roman, the head Drumsman breathlessly. "Thorndike's militia is already on its way down *here!"*

"Sounds like old Thorny is ready to go! That's great!" Drum shouted back with a grin.

Then he had a second thought. "Why is that stupid Thorndike coming all the way down *here* first? That's a waste of time. I should be going up *there*. Can't he even figure *that* out without help? Thorndike Manor Castle is practically next door to The Castle. If we start at Thorndike Manor, The Castle can be mine within hours."

"Actually" said Roman, "From what I gather, they are coming down *here* to go through The Wall, *around* Drumsworth and up to Urhonordo. *Their* plan is to invade Urhonordo!"

Drum staggered backward. "What?" he yelled. "What do you mean by that? He wants to conquer Urhonordo first? He knows better than to go off on his own with *his own* plan when *I* have already told him that *my* plan is going to go first! You must have got it wrong!"

Roman shrugged. "According to his militia, he's planning on conquering Urhonordo first instead. The rest of the Drumsmen nodded.

Drum was beside himself with anger. *How could Thorndike have got this so wrong? He must be much more ignorant than I thought,* he thought to himself. *I let him have too much control up north. Now he thinks he knows better than me. I should have spotted it when he got so pushy at the road signing. Now he thinks he can call the shots. Well, he is definitely in way over his head with this maneuver.*

Drum began pacing angrily. "How dare he put *his* stupid desires in front of *my* claim on The Monarchy of River Kingdom. Where are his priorities?" he shouted.

Drum's Auras had been relaxing near Drumsworth Lake, waiting for some kind of action to start. Now, they came to attention. What they heard was not the way they had been planning with Drum. This was not it at all.

The Auras' plan called for an attack on an unaware King Solem. They hadn't heard anything *at all* about Urhonordo. Urhonordo was not even an important place. There must be another bunch of Auras in the works.

Rushing to Drum's side, Drum's Auras began immediate counseling. "Don't over-react, Drum. Thorndike may be stupid, but *he* is the man with the troops. *You* are not. You can't let him know that you need him...but you do! Get smart. You've always had 'the power'. Now use it. Turn this situation around and make use of it for yourself. Get Thorndike back in line!"

Drum and his Auras were nothing if not opportunistic. Suddenly a new, ingenious plan began to take shape. Drum didn't have a militia, but he *did* have the support of the nobles from this southern area...and there were a lot of them. They hated the King as much as Drum did. He had seen to that with his planning meetings.

Immediately Drum directed some of his Drumsmen to another mission. "Go now to every manor in the south and tell my nobles that *their* future is at stake. If they want to rid River Kingdom of its useless King *and,* if they want to keep the Lowlanders in place behind this wall, they must rally on the new Royal River Road *now!* Tell them to meet there and head north, immediately!"

With that, Drum rushed back into his manor. He stopped and looked around. What he was about to do was daring, creative...exciting! His Auras loved it. He stood there in his entryway. He was now in the center of his own 'castle' but River Kingdom *South* was not ever intended to be enough. He had built a mote and a wall for strategy and protection, but these things were not enough, either. Royalty in River Kingdom Castle was his destiny, and he would have it!

His Auras gathered around their champion. They knew that he would not let them down if they continued their support. And they were up for it! With determination they coiled themselves around him, filling him with his own inherited feeling of self-importance.

Slowly Drum began to feel their strength. It was exhilarating! Drum's years of planning for his rightful place in The Kingdom were coming down to this amazing day. He had not spent years cultivating the respect of disgruntled nobles for nothing. They all loved him now and he deserved it! They all hated King Solem and his wishy-washy tolerance of Lowlanders and Urhonordo and he, Drum, had built that! Drum knew that now, in spite of Thorndike, his Royal future was his!

The Call of the Auras

The Auras sensed that their approach was working. But there was one more piece of motivation available and they Intended to make use of it. "You know that today will be *your* day, don't you, Drum?" one of the Auras crooned in his ear. "Now... think of the moment when King Solem is *at your feet,* and *you* are on the throne. What will be missing at that moment? What do you already have in your possession to make it happen?

"Bring the crown with you, Drum. Keep it with you so that you can bring it out at the right moment. When you declare victory, you can simply reach down and put it on."

Drum gathered his most majestic demeanor together and did what he was instructed to do. Reverently he pulled back the portrait of his grandfather and opened the vault. The crown gleamed welcomingly. Carefully Drum slipped the treasure into a velvet bag and tied the cinch cord to his waistband. It bumped reassuringly against his leg. Drum was ready to go after his goal and his Auras were ready as well.

Several Drumsmen were already on the move to inform the southern noblemen of their need to gather, but the remaining men were still gathered on the lawn, their horses munching lazily.

Drum now emerged, ready for action. He had donned a particularly regal robe for the occasion and the Drumsmen immediately stood tall and came to attention. They knew his look. It was one of confidence, even in the face of Thorndike's unpredictable actions. With fascination, they awaited his direction.

"Listen carefully, men. Your orders today are unique. Round up the Thorndike militiamen that are upstairs in their quarters or down here supervising the slaves. Leave a few to maintain order with the Lowlanders but tell the rest to saddle up. They are going with us. We are all about to move out and drive north to victory!

"The southern noblemen will meet us on the road and follow our lead. Eventually we will meet up with the ridiculous Thorndike and his militia. Do not draw your weapons, however. They will not be looking to fight *us...nor their own men.*" Drum chuckled a demonic chuckle.

The Drumsmen scratched their heads but did as they were directed. Drum headed out on horseback, cautiously encouraging his impatient horse over the drawbridge, and checking down the line of The Wall.

"The Wall looks good," commented the Auras. "The noblemen will finally be happy that these annoying foreigners are corralled and out from under foot."

Drum grinned with satisfaction. If this plan today worked as he expected, he, as king, would not need to worry about Lowlanders and the nobles that he found annoying would be solidly behind his reign.

Rapidly, the off-duty militiamen tumbled out of their housing in Drumsworth Castle, pulling on their boots and tucking in their nightshirts as they came. Drum was impressed. *At least Thorndike can do one thing right*, he thought. *These men are well trained and disciplined. It will be a pleasure to command them all.*

At last, all were on horseback and ready for instruction...and Drum was more than ready to give it to them. "Listen up, men," he began. "Thorndike is already traveling on the beautiful, new, Royal River Road but on his way in *this* direction. Apparently, he got my orders mixed up. Things like that happen sometimes.

"As many of you know, you have all been training for an eventual battle. It is now time to tell you where the battle will be and with whom."

The militia began to rustle with excitement. They had always expected something like this. They were anxious and ready for it.

Drum could sense their excitement, but the Auras were cautious. "Remember, Drum...These men are loyal to Thorndike. He has always been their leader. Be careful what you say."

Drum was not deterred. "Your Commander, Thorndike, is a man of action," he began. "I am sure that he intended to meet us here to create a battle plan, but *I* am sure that the plan we made is already ready...if we choose to take advantage of it *now*.

"We have planned enough. Now we are ready for action. What do you say men? Are we ready?"

The Drumsmen and militia erupted together. They were ready! They were *born* ready! On the wings of this enthusiasm, Drum delivered his surprise announcement.

"Today Combined Militia of River Kingdom will march on The Castle, itself! Today, we *men-of-action* will show the ignorant King Solem how to rule River Kingdom... and we will take charge!"

All the battle-ready men on horseback glanced from side to side. Had they all heard the same thing? They were going to *show The King how should do things?* Really? They had been told that their mission as militiamen was always to fight a dangerous foe and *protect* The Kingdom. Was this an act of protection?

The Drumsmen were the first to react, however. They were with Drum every day. They knew him. He loved the support of his men, whether they understood him or not.

"Wahoo!"

Yeah, Drum!

It's about time!

And the chorus of enthusiasm spread quickly. Soon all the militiamen were yelling and throwing their hats in the air. "It's about time that we are used to protect The Kingdom like we were trained for!", shouted the militia. There was much back slapping and friendly, manly punching. Even the horses seemed to catch enthusiasm.

The militia didn't know The King, but that didn't matter. Their mission was to do what they were trained to do and now they were finally going to have their chance. "Wahoo!"

Drum was thrilled with the response. He had expected it. He had envisioned something like this all his life. Now it was finally real. Gallently and with a broad, confident smile, he crossed the drawbridge and headed north ahead of them... to destiny.

And was Balynn excited too? Sadly, no. Drum and his Auras had given little thought to his daughter, Balynn, throughout this entire build up to glory. Today, she was peacefully reading in her room, totally unaware of the whole thing.

THE MEETING OF THE MINDS

Destiny was on the move on the Royal River Road up north. Thorndike was in *his* glory and living out his lifelong dream. His militia stretched out in front of him as far as he could see. His sons were at his side. His plan was simple. When he arrived at Drumsworth, he would inform Drum of his Urhonordo plan. He and the militia might spend the night in the impressive 'castle' that Drum was so proud of...or not, depending on his mood.

Thorndike was in charge of this maneuver, and he relished his authority. He had allowed Drum too much sway over the planning process, he now informed himself. *He, Thorndike,* was the one with the militia. *He* always should have been the one to set up the timing of The Plan. But now he was taking over.

The only thing that concerned him at all was Drum's amazing control over the nobles. It was uncanny. Wherever Drum went, the nobles cheered for him and clapped, shouting his name.

The stupid nobles are brainless sheep, he growled to himself. *They are all infatuated with the myth that survives about his grandfather and the wonderful way he treated nobles when he was king, back in the day,* Thorndike acknowledged.

Once Urhonordo was captured and in control, Thorndike dreamed that he would hand off temporary supervision of Urhonordo to his sons who were now riding at his side. This was his dream come true.

After he and his militia conquered Urhonordo, he would give Drum what he and the nobles wanted...'King Drum' returning at last to the throne. With the Thorndike Family in charge of Urhonordo and Drum was King of River Kingdom, white supremacy would rule both sides of The Great River, at last.

Riding along beside their father, however, neither this mission nor the reason for them being on it were clear to Jackson and Jorge. The only thing they had focused on was the order to depart that had come early this morning. By habit, they had moved out with their fellow militia men, but their minds were still churning with what they had heard from Flo. This Man was a monster, and they were riding with him. Now they were asking themselves why.

Cjackson crawled up into Jackson's ear. "What do you think will happen in Urhonordo," the young Conscience asked. "Do you think people will be killed? I have never seen a real battle, so I don't know but if there is fighting, killing seems likely. I don't approve of killing at all. What do you think?" Jackson rode on in silence.

Jorge had equally mixed emotions. He already knew that *he* wasn't going to fight in Urhonordo. The only information that held his positive focus was the fact that they were heading for Drumsworth and that was where the beautiful Balynn would be.

Cjorge decided that he needed to focus Jorge on the real reason that they were traveling. Jorge had never been much for training in such pursuits as sword fighting or hand-to-hand combat and had only participated to please his father. "Have you given any thought to what will happen in Urhonordo?" Cjorge began.

I haven't thought that far ahead, Jorge thought casually. *Father said that they don't have a militia and we might just walk in and take over, so I am spending my thoughts right now on better things...like what Balynn might be wearing when I see her and what she might say to me. I know that she likes me.*

"So much for deep thinking" Cjorge sighed.

Thorndike smiled at his daydreaming sons with pride. They were tall and strong and handsome, just as he had always dreamed they would be. And they were good at fighting. He had made sure of that. He had trained them in the arts of combat, himself. Their leadership ability, however, was something that had never been tested. *I should have done more of that,* he thought with a frown. *If I expect them to lead in Urhonordo, I should begin to test that ability right now.*

With that thought in mind, Thorndike made a decision. He would use this trek to Drumsworth as the beginning of leadership training for the boys.

"Move up to the front of the line and take the lead, men," he commanded them as though they were not his sons at all. "Keep an eye out for problems up ahead, answer questions, keep the group tight and compact and ready for surprises. I'll stay back here and keep straglers in line."

The boys nodded and moved up through the lines of men on horseback, speaking with friends along the way. Most of the militia now knew that the boys were Thorndike's sons though this had never been formally announced. The boys were "pretty much okay" according to common opinion. They never tried to act special or get special favors.

The older regulars of the militia nodded to each other and knew why the boys were moving up to the front. *Their dad has put them there to teach them,* they thought. *That's a good idea. They will probably take command someday.*

The twins moved to the head of the militia without incident, but feeling for the first time that they had been put in a position which could be thought of as one of authority. It felt good.

They were now leaving the northern part of the kingdom and entering the Center portion. There should really be no major difference between these sections that were next to each other, but the boys noticed a definite change right away.

First, the road had become narrower, taking up less planting space. Jackson called back and ordered the riding formation changed from four abreast to three.

Next the boys noticed the raised curb on the side of the road that was inland. This would allow runoff from plant watering to stay with the plants and not sluff off onto the road. Additional boulders had been lined up next to The River to warn travelers not to get too close to the edge of The River. This was all very neatly done and very different from the looser, wider construction style that they had used in the northern section.

An old regular from back in the militia eased forward with a piece of information. "There's been Lowlanders buildin' on this-here road," he said quietly. "I traveled down in the Lowlands some in the old days and I remember them buildin' this way on their land. They're good at this stuff."

That's very interesting, thought Jorge to himself. *I wonder what Lowlanders were doing over here?* Now he saw something else that was very different up ahead. It was a bridge that was as wide as the road was wide and it was covering a very healthy little river that seemed to be flowing from the hills toward The River.

Now both the twins recognized the bridge they were looking at and who had built it. It greatly resembled the workmanship on the mote at Drumsworth. This was another thing that had been built by Lowlanders. They were sure of it.

As they led the way across the well-built bridge, the twins looked at each other with questions that neither could answer. How had bridges come to be built by Lowlanders this far away from The Lowlands? When had they been built? Were Lowlanders enslaved here that no one knew about before?

But they didn't have time to look for answers. Riders were approaching them fast from the south in clouds of dust. As the dust settled, it was not hard to see who the riders were. They were Drumsmen...and some of their own Thorndike Milit

The twins held up their hands to halt the men behind them and stared. They were stunned. Why were these men in front of them, blocking their way? They friendly and smiling, but what were they doing here?

Jackson recognized a few of his fellow militia comrades near the front of the approaching mass of manhood. This was as good a place to find out what was up as any.

"Hey, Jake," he called out. "What are you doing way up here? We thought you were watching the Lowlanders build The Wall."

"We were doing that until just an hour or so ago. Then Drum said to move out, and here we are. We don't know why exactly, but you should probably ask Drum. It was *his* idea. He'll be up here shortly. He's on his way but this part of the new road is skinnier here so it's taking him some time. Is Thorndike with you guys?"

Jorge began to backtrack through the militia on the new road that was so thin that his travel was dotted with 'excuse me' and 'pardon me' and 'watch where you're goin' kid' but at last he reached his father.

"I thought I put you boys up there to leads, " Thorndike growled. "Looks to me like you weren't up to the job. What are you doing back here and why are we stopped?"

Jorge cleared his throat and tried to sound authoritative. "It appears that the way is blocked, Sir," he said. "Our passage is blocked by another group traveling north on this road and there isn't room for us to pass by, Sir."

"If they're too stupid to get out of our way, *knock* them out of our way. We are on an important mission and no farmer is going to block us. Knock whoever it is into The River and let's move!"

"I don't think you want to do that, Sir," Jorge gulped. "Some of the people that are in the oncoming group are *our* militiamen, Sir. And I heard Jake say that Drum was with them."

"Drum?" Thorndike gasped. "Drum...is here on this road? ...heading north?"

Thorndike began to sweat again. What was happening? Why would Drum be coming this way now? They had made no special plans to meet...and some Thorndike Militia was with him? Why was Drum here with some of *his* men?

There was no explanation that made sense, but there was one thing that was sure. If Drum was up front waiting to talk to him, he needed to get there quickly.

It was one thing to assume leadership when Drum was *not* around. It was another to try to keep leadership when Drum *was* around. "Stand aside," he bellowed. "I am on my way to the front to meet with Drum. Stand aside!"

The militia jostled and squirmed this way and that trying to make way for a large, angry man and a large, nervous horse. The narrow road was not giving space away. It was well made.

At last Thorndike reached Jackson and the front of the militia. And there was Drum. He looked calm and unruffled, his hair well coifed and his robe immaculate.

And what was this? Behind him was a group of noblemen in carriages, each leaning out of their windows and clapping.

They're clapping again, groaned Thorndike to himself. *The damned noblemen are clapping for Drum again.*

"Hello there, Lord Drum," Thorndike called out in a tone that he hoped sounded jovial. "What brings you out on the road today?"

"Hello, yourself, Lord Thorndike." Drum responded in an equally jovial manner. "I was just on my way up to your abode to put together the final stages of what we hope to be our surprise victory over The Castle. Why on earth did you think that we needed to discuss our plans at Drumsworth? That would have been a waste of time...but we understand your eagerness to get started...don't we?"

Drum turned and waved to the noblemen behind him. They erupted into cheers..." Victory! Drum forever! Victory!" climbing out of their carriages and stabbing the air with their fists. Drum brought his horse up close to Thorndike. His smile was broad, but his eyes were keen and sharp.

"The lords from down south *insisted* upon accompanying us part way," he announced meaningfully. "And, of course, I have *my* Drumsmen and *your* militiamen here with me, ready for battle, as well. What say you? Is it not a perfect idea that we have met here with the same goal in mind? Shall we simply join our troops here on the new Royal River Road and charge northward?"

Thorndike looked behind Drum at those who were enthusiastically milling behind him. In addition to the Drumsmen, the nobles that loved Drum, if only for his family name, smiled in support. They could not be ignored. They were the base of their actions against King Solem. In addition, even his own militiamen were here in support of the battle Drum had already shown them was ahead.

"Northward, northward, northward to victory!" chanted the Drumsmen, militiamen and nobles together.

Drum smiled with satisfaction. "It sounds like I am not alone in this idea," he offered calmly. "What do you say, my friend? Are we ready to put *our* plan into action? Are we ready to conquer The King together?" His gaze returned to Thorndike, a gaze that Thorndike read as Drum intended. This gaze was a threat.

Lord Thorndike had been bested and he knew it. His Aura and Drum's Aura now joined in a common cause. (Auras always sided with winners when they had a chance.)

"Indeed," mumbled Thorndike painfully, his Urhonordo dream fading before his eyes. "Indeed, this was my plan all along," he lied. "Your advance to meet me today will save time from unnecessary planning and give our cause its victory that much sooner.................... Victory!" he shouted in conclusion.

"Victory!", the assembled shouted.

Drum nodded happily and turned to the nobles in true gratitude. They had accomplished what he had thought they would. He had expected that they would stand strong for him personally as he had trained them, and they had done so. This was his base and would continue to be once he was king. This was his power.

"Thank you for your support this morning," he said, trying to catch the eye and pointing to as many as he could see. "*You* are the ones we will be fighting for. Because of *you,* we will take back our kingdom and restore the leadership that puts white men first!" (Another hearty cheer from the nobles.)

"Now, I ask you to return to your homes and think positive thoughts on our behalf. Lord Thorndike and I will proceed to The Castle to take back what is rightfully ours...and yours...our Kingdom! This is *your* revolution! I am doing this for all of you! Thank you again!" The mighty cheers rose again.

The noblemen continued to cheer as they dispersed. The Drumsmen and Thorndike Militia streamed by them single file on the narrow road as the nobles returned home. They were inspired. Drum was their man.

The Thorndike Militia had been willing to take over Urhonordo first, just for Thorndike. It would have been an easy, bloodless win. But now it seemed that they were truly heading for an actual Revolution! Again, their training as 'protectors of River Kingdom' came into their thoughts. How did revolution and protection go together? Surely Thorndike would tell them soon.

PPREPARING FOR REVOLUTIONP-1

There was no happiness in The Castle after the woman they were now calling 'Mother Mina' made her observations clear. She was Queen Willa's mother. There was no reason to doubt the truth or importance of what she said.

A cloud of gloom settled over the conference room. Even Horace and his cheerful pots of tea could not chase it away. Urhonordo would soon be under siege by Thorndike.

Prince Armando looked at King Solem in misery. "Luna and I have discussed the need for a militia many times," he said. "We always concluded that we didn't need one. Urhonordo is so small compared to River Kingdom. The only thing there to conquer is our peaceful people."

"But that is what's special about Urhonordo," June said, his eyes settling on Manda. "Your people are kind and honest and peaceful. They aren't prejudiced. They aren't greedy. They're just good people."

"And they have good Consciences," whispered Cjune. "Don't forget that. That's what helps them be the way they are."

"What makes the Consciences in Urhonordo more successful than Thorndike's Conscience?" grumbled June.

"Practice," responded Cjune, "and no nearby Auras." But the question hung in the air.

"According to my wife," Armando answered, "Thorndike has harbored a resentment against the old king that prevented his grandfather's takeover of Urhonordo back in the day. Now he is going to take advantage of the strength of his militia to finish that takeover while nobody's watching. And I am afraid that my kingdom is ill-prepared to fight against an armed group that big."

King Solem turned and looked Armando in the eye. "River Kingdom is not ill-prepared," he stated." My Royal Guardsmen force is bigger than the militia of that blowhard."

Everyone stared at Solem. What was he saying? Was he making an offer? Silence settled over the room. People drank their tea. Each person there was coming at the answers to their own questions from their own points of view.

Suddenly there was a very strange little noise. Mother Mina jumped and grabbed her daughter's arm in fright. Something or someone was outside. Everyone looked from side to side...everyone but Horace and Willa. They had heard that sound before.

Horace bowed quickly and dashed for the colonnade. The sound they heard was pebbles on the window. Hector was outside. He might have news.

Willa provided the explanation. "Hector is a friend of mine and Manda's," she began.

Manda nodded and smiled. "Hector brings news from his part of the kingdom whenever he thinks it's important."

"Hector is a good man," Solem agreed. "If there ever was a time when we need news from that direction, this is it."

Horace dragged the reluctant Hector into the conference room. Now everyone but Mother Mina recognized him from The Gathering. Hector stood still with his hat in his hand and stared at the floor.

"Welcome, Hector, King Solem said warmly. "Thank you for coming. I hear that you may have some information for us that you think would be useful. Is that so?"

"Yessir, Your Majesty," Hector replied. "Yessir. I came as quick as I could after I seen' em. The whole Thorndike Militia is on the new road and headed down to Drumsworth with Thorndike himself bringin' up the rear."

Hector stopped and looked around. What he saw made him proud. He was there, in The Castle, and the royals were all paying attention to him. He was proud.

Willa smiled at Hector directly. "It was so good of you, Hector, to come and tell us what you saw. Did you hear anything? Did they say anything while you watched?"

Hector blushed and smiled back. "Nothing official, Miss Queen", he stuttered. "I did hear some of the militia grumble about Urhonordo. I think that was where they were headed for an invasion, but nobody seemed to know why." Hector shrugged his shoulders. "That's all I heard for sure, Miss Queen."

Willa grinned. "You were so good to listen to what you could, Hector. I know The King appreciates it, don't you dear?

"Don't mention it, Your Highnesses," Hector said, bowing to Armando and Solem. "After you two Kings fixed my water problem in the Central farms, it's the least I can do."

Solem and Armando smiled and nodded. Hector and Horace bowed their way gratefully out the conference room door.

"Well, that settles it," said Armando, rising from his chair. "*I* am Crown Prince Armando. Urhonordo is *my* kingdom and Queen Luna is *my* Queen. There is invasion headed toward my kingdom and *I* am sitting *here*. This is a situation that cannot stand. I must go now to Urhonordo and defend my kingdom the best I can."

Solem only thought for a moment and then he stood as well. "Wait, Armando," he said quietly. Let's think this through. We now know that Thorndike is, indeed, marching to conquer Urhonordo, but he has taken the long way around to get there by surprise, as they did in the old days, I have been reading. He will probably stop at Drumsworth to rest and then make his assault on Urhonordo.

"My Royal Guard is highly trained and can move quickly. If they begin now, they can catch up to them when they get to Drumsworth and make them change their minds. What do you think of that idea?"

Armando stared at his fellow monarch in amazement. *Solem is willing to send his own militia to save Urhonordo,* he thought. The Prince was stunned...and grateful...but proud. "I appreciate the offer, Your Highness. I am truly grateful for it. But I cannot sit here and depend on others to defend what is my responsibility to defend.

"I will send out a call. The men of Urhonordo will rally. We are few but we are strong willed. I must leave immediately. The men of Urhonordo must save themselves."

King Solem nodded. He understood Armando perfectly. He was a proud man.

Armando bowed to all and then looked specifically at Mother Mina with a bow. "I thank you so much madam, for your honesty and warning. I know now why your daughter is as she is...fine, honest, and brave. Your warning may have saved Urhonordo." With that, he bowed and left the room.

Manda was waiting to go with her father by the door. "I am coming with you," June announced quietly.

"No, June," Manda said staring into his eyes. "You must stay here. You are heir to the throne of River Kingdom as I am to the throne of Urhonordo. It isn't right for you to risk yourself for the fight of another kingdom. For the sake of *your* Kingdom, you must stay here." In a few minutes, she was gone.

King Solem knew the route would follow. They would first travel down the new road but would soon cross over The Bridge, the ancient structure that had always linked the two kingdoms for centuries. From that point on, they would be in Urhonordo and on a war footing, warning their Urhonordo citizens and pulling them together to fight as they could for their kingdom.

Solem shook his head. Their fate did not look good.

The Consciences in the conference room sprang into action.

"When do we get *our* Royal Guard ready?" asked Willa and June. "I don't think we have any time to lose."

Solem stared at Willa. "You are in favor of using *our* militia against *your* father?" he blurted. "I thought you would hate the idea."

"I do hate it, but my Conscience and I insist," answered Willa. "We can't just sit here and wait for disaster to invade Urhonordo. We *must* help."

"I am ready to pick up a sword for those people, myself," Mother Mina said as she headed toward the door. "Women can fight, once they get their heads on straight." With that, she lifted her skirts and marched toward the stairway.

Willa hurried after her. "Come, Solem. I must catch up with Mother, but *you* must raise the militia right now! We don't have any time to lose."

"I can't just go barging into Urhonordo," Solem said gruffly. "Urhonordo is not our kingdom. You heard him. Prince Armando wishes to defend his own land. Interfering in someone else's kingdom is against all protocol. He will be offended."

"Damn your protocol," June stated. "If he is dead, you won't care if he is offended, will you? We have to do something right now to help our friends or we may lose them to crazy Thorndike. What are you going to do, Father?"

Solem sat in his large chair at the end of the table. His head was in his hands, and he was deep in thought. June, however, was ready for action.

"Father," he said slowly. "How would it be if you helped Urhonordo by simply 'following Willa's father and his militia down to Drumsworth'. If the Thorndike Militia see the Royal Guardsmen with you, they will probably think twice about invading Urhonordo. Following Thorndike on River Kingdom land has nothing to do with 'going to Urhonordo uninvited'. There's no protocol problem that I can see."

Solem looked closely at his son. Never before had his son attempted to plan a military operation. This was new and something he had been waiting to see.

Solem raised an eyebrow. "Thorndike will deny what he's doing. He'll just say that I am interfering."

"And that is just exactly what we *will be doing!*" June replied, coming directly in front of his father to make a point. "But *you* are the King. *You* can tell him that you are just testing out his new road...which wouldn't be a lie...exactly... But you could also order him to turn around and stop what he is doing. *You* are his King!"

Solem's Conscience decided to take a stand. "Protocol is important, Solem, when you are trying to maintain order. But the order is already broken here. Citizens of *your* kingdom are getting ready to invade another kingdom for no reason except greed. In this case, Thorndike's planned treachery trumps your protocol. Do not hesitate one more moment. Move!"

Solem stood shoulder to shoulder with his son. "Tell Willa and her mother to dress to defend The Castle. You and I and The Royal Guard will travel down the Royal River Road...to 'inspect the thing!"

June did as ordered and went to prepare for conflict. Hopefully, he could protect Urhonordo...and Manda.... by stopping Thorndike's invasion before it started.

PREPARING FOR REVOLUTION-2

Manda and her father galloped into the courtyard of Urhonordo Castle and were greeted by cheers. The courtyard was teeming with activity. Men, women and children were working on the courtyard wall, arranging ladders and hauling stones and rocks to pile on its top. There were baskets of arrows and bows hanging from the ladders. Buckets of oil were balanced along the walls as well. Urhonordo was getting ready for an invasion.

Armando rushed into the castle and was greeted by cheers from the people around him. Manda followed close behind with eyes wide. They all knew! The people of Urhonordo knew about the coming invasion and they were already getting prepared!

Luna was not in her map room as usual. She was now firmly planted on the castle porch and giving orders like an army general.

Now she stood and rushed forward as Armando rushed to her as well. "Thank goodness you're here at last," she whispered into his chest. "I was afraid that you might not have received word. We just heard this from our runners this morning. Thorndike is heading this way with his whole militia, and they are planning to invade us."

"I know. I know," Armando said as he held her close. "Everyone at The Castle just found out today, too. We were there discussing the possibility of such a thing when we heard that it was already happening. How did you learn so quickly?"

"I sent out your wonderful scouts when you left and they came back with the news," Luna explained tearfully. "That's when I asked everyone to spread the news. We may be small, but we will put up a fight. I am so glad you and Manda are here."

Manda looked fondly at the couple as they hugged each other, but this was not a time for affection only. She had been thinking about this invasion as she rode here. Urhonordo was acting brave, but they would be outnumbered. They needed more manpower. She now had an idea.

"Mother," she began. "I see the preparation that our people have made in just a short time, but I also see that there are few fighting men among them. I know the size of the militia that will be approaching, and I am worried."

Armando smiled sadly. "I'm worried, too. I wish there were more like you here, but we will do the best we can."

Manda smiled. "I know where there are more like me, if you will let me go to get them, tonight," she said.

Armando looked at the young woman in front of him. *She's young and full of ideas*, he thought. *She thinks she can do anything. I only hope she will live through what happens tomorrow. This is such a sad and frightening day.* He smiled at her but said nothing.

"Thank you, my dears" answered Luna, seeing Armando's mood. "I know you always do wonderful things when you put your mind to it. If you think of a good solution to this problem, you have my blessings to do it. I'm sure that whatever you think of will help." She walked out into the courtyard with Armando.

"I am being dismissed," chuckled Manda to herself. 'They don't think that 'young people' have any idea what to do here, but I think I know something that will help...and I'm going to do it."

Quietly Manda slipped through the busy courtyard to the stables. She led a fresh horse quietly down a path and then hopped on, walking south into forest lands. She was on her way to Drumsworth.

The night was cold and crisp as Manda reached the flatlands and tied her horse to some branches. Little by little, she crawled toward the back side of Drumsworth Castle. As she neared the prison, the sounds that she knew she would hear became louder and more miserable.

There was the crying of children and the attempts to care for them by their mothers coming from the prisons. There were hopeless sobs weaving in and out of frustrated growling and the shaking of bars that held them. It was a whole town full of miserable Lowlander prisoners in dark, cold prisons.

It' worse than before, Manda thought. They've opened up all the old parts of the prison that are still standing and squashed the poor townspeople into them until they can get The Wall completed. How can one group of people be so cruel to another?"

"I hate to state the obvious, but Drum and Thorndike have no working Consciences at all," whispered Cmanda. "We must hurry. All I hear is pain and misery. We've got to get them all out. Do you think that the key is in the same place?" she whispered. She was now Manda's only co-conspirator.

"I have my old key. What do you want with it?" said a voice from out of the dark.

The voice came from the bushes behind them. Manda flattened herself to the ground. She had been discovered! Would she be joining the prisoners soon?

Slowly she rolled over and glanced up. It was Paulina, smiling down on her, with her fingers to her lips. "Shh," she advised. "Somehow, I knew that I had not seen the last of you. Why are you here tonight? Are you on a mission to free slaves again?"

"Yes, indeed. I heard that a whole town full of Lowlanders was kidnapped and driven down here," whispered Manda. "I came down here to turn them all loose."

Paulina shook her head. "Don't bother," she sighed. "They'll just get caught again in the morning. I come here every night to bring them what food I can, but I know what will happen if someone frees them. They'll be caught and beaten badly. That's already happened twice. Letting them out will just make things worse for them."

"Not if I take them to Urhonordo," whispered Manda

"Urhonordo? Why way up there?"

"First, it is a long way from *here*. Second, The Kingdom of Urhonordo is about to be invaded by the same people that put your friends here in prison. Now Urhonordo needs help to fight off the monsters before *they* get conquered, too. They'll be under siege by Thorndike's militia tomorrow and they will need help," whispered Manda.

Manda stated it plain. "Our plan is to free everyone here. We'll hide the children in Urhonordo Castle. Then the Lowlander adults can join Urhonordo's citizens and help Urhonordo fight off the invaders...if they will. They'll be fighting the same terrible people that kidnapped them."

"Lowlanders don't fight. They didn't even stand up *for themselves* when they got kidnapped," explained Paulina. "Now they are letting themselves be turned into slaves. What makes you think that they'll fight for Urhonordo?"

"Because they know they can...but they don't have to. They will no longer be slaves," said with conviction.

"Humph," snorted Paulina. "We Lowlanders know right from wrong, but in our culture, we usually try to avoid confrontation rather than fight. Sometimes we don't understand that 'standing up for what's right' is the right thing to do until it's too late. Then we're stuck. I'm the same way."

"Maybe you Lowlanders have not been the best fighters in the world" Manda acknowledged, "but tonight, Urhonordo really needs your help. I understand that the prisoners never really learned how to fight back in The Lowlands," Manda insisted "but maybe they'll figure it out now, after being put in prison and enslaved."

The Call of the Auras

Paulina stared at Manda. *Is this girl right?* she thought with a start. *I have always given in or run away before and so far that has kept me alive. Is it dangerous to change?*

"You have been doing what you just said the others do," grumbled Cpaulina. "It's time for all Lowlanders to figure it out. You know what is right here. You should never be slaves. You and the others must stand up for what you know is right. Don't give in anymore. It's time to stand up."

Paulina straightened her back. She had a question. "Who did you say that my Lowlanders would be fighting...if they fight in Urhonordo?"

"Thorndike and his militia."

Paulina squared her shoulders. "Then it's time for all of us Lowlanders to stand up for ourselves *and Urhonordo* now," she announced. "Follow me."

Together the stealthy pair approached the prisons, their fingers to their lips for quiet. The 'prisoners' understood immediately and even the children were silent. Quietly, without any noise at all, one by one the Lowlanders tiptoed out of the prison and slipped away into the underbrush where Manda gathered them up and led them north.

A completely silent game of follow-the-leader twisted and turned through thickets and forests and even a very damp marsh but, by daybreak they had almost arrived.

"Where are we? What should we do now?" they asked as Paulina brought them to a stop. "We are here to help Urhonordo stand up against Thorndike," Paulina answered. "We all need to do that, too...unless we want to go back to prison and be slaves." Adults nodded. They had figured it out.

Cpaulina smiled.

*****

Careful nose counting was under way among the Lowlanders to make sure that all were accounted for as the unsuspecting Queen Luna and Prince Armando stepped out of their castle to an astounding, *crowded,* battle preparation site.

Over in a far corner, Lowlander children were gathering under quilts brought to them by neighbors. Lowlander and Urhonordo mothers scampered from child to child, singing little songs and passing out biscuits.

But, most important and most surprising of all, the Lowlander men had climbed the ladders against Urhonordo's perimeter wall and perched themselves on the top as lookouts. A Lowlander army had taken position, and all were ready for combat. They were 'standing up'.

Armando closed his eyes and then opened them again in total disbelief. Where had all this additional manpower come from? He had reinforcements. How had this happened?

Then he noticed Manda. She was out among all the new recruits, talking, explaining, comforting, encouraging.

"Your daughter recruited the Lowlanders from Drum's prisons," explained Carmando. "They have decided they want to help."

Lowlanders are standing up for us...and for themselves. This is new. Armando thought to himself. *This is a good sign.*

PREPARING FOR REVOLUTION-3

At The Castle, King Solem walked among the Royal Guard confirming old relationships and answering questions. Now he stood before them and raised his hand. The mass of male humanity turned obediently in his direction and silenced themselves.

"We will be traveling south today on the new Royal River Road," he began. (A cheer went up) "It has just been completed, making it smoother and wider and easier to travel for troops on horseback such as yourselves. "(Another cheer)

"Unfortunately, this is not just a scenic tour. As your King, I must warn you that our march today may include violence."

The guard gasped. There had been no real violence in River Kingdom for years. This was a peaceful kingdom. Had there been an invasion? Why should they expect violence...and with whom?

"I hear, gentlemen, that the Thorndike Militia is headed for Drumsworth Lake to launch an attack on our friends, The Kingdom of Urhonordo, for no reason." The Guardsmen gasped.

"Thorndike is a citizen of River Kingdom. If what I heard today is true, this will mean that River Kingdom is invading Urhonordo. Should we allow such a thing?"

The Guardsmen looked at each other for only one astonished moment. Then they began to growl.

"No!" they shouted unanimously.

"Never!"

"Thorndike doesn't speak for us!"

"The Kingdom of Urhonordo is friendly!"

"I agree with you," said Salem with satisfaction. "Now, what about slavery? It has come to my attention that Drum has gathered up a large number of Lowlanders who were peaceful citizens of River Kingdom until a few days ago. His plan is to wall them off into The Lowlands so that they can only come north as slaves. No matter where *you* came from, how many of you are in favor of Drum's plan?"

"No!"

"Hell, no!"

"That's against the law!"

"Slave holders should be locked up!"

"A lot of us are of Lowlander blood!"

"That is what I hoped you would say," Solem said gratefully. "The reason I ask you is this: Our goal today is to head south and verify whether or not these two things are really taking place with some of River Kingdom's very own countrymen. You answered my question about Urhonordo. Will you also fight to end slavery, as well?"

The Guardsmen came alive.

"Let's go!"

"What are we waiting for?"

"How far down the road are they?"

"Saddle up!"

Within the hour, the complete River Kingdom Militia was ready to be on the move with King Solem at the lead. Solem left his men for only a moment, but he needed to say goodbye to Willa.

"It's time for us to go south," he announced bluntly. "I just wanted to say goodbye." He reached down and pulled her from her chair, closing her in his arms.

Instinctively, Willa knew that he was worried. He was on his way to a situation that had been brewing for months and it was likely to end in confrontation, even battle. And who would that confrontation be with? Her father.

Willa wound her arms around Solem and snuggled into his chest. "Wish him well and tell him you love him," whispered Cwilla. "That's all you can do."

Csolem heard Cwilla and agreed. "Tell her you love her, and you will be home soon...even if you're not sure about that," counseled Csolem.

"Come home to me, Solem," Willa whispered.

"I will. Stay inside The Castle, so that I know you're safe," Solem whispered back. And he was down the stairs and out onto the colonnade and his waiting troops.

TURN ABOUT

It took some time and some determination but the mass of humanity that was the combined forces of Drumsmen and Thorndike Militias were now heading in the same direction. Slowly but methodically, they lumbered back northward, but there was no doubt now that they were headed together for the same mission, the conquering of River Kingdom.

The satchel that held Drum's ancestral crown thumped persistently against his leg as he rode. *It is there to remind me of my destiny,* he mused contentedly to himself. *We are only a half day away.*

Thorndike, however, was not as sanguine. Though he had never really been in a battle (much less with his own troops), he was certain that victory would not be easy.

Drum had always regaled him with the prediction of this battle's simplicity. It would be unexpected, of course, because they would be arriving quietly on a road made for that purpose. He, Thorndike, would surround the unprepared Royal Guard and hold them at bay while Drum would simply walk into The Castle with his Drumsmen, inform The King that he was no longer King and take the throne.

The Auras were thrilled with The Plan. They were now the leadership of both Thorndike's and Drum's combined goals, and these goals were what they had been working toward for years. It was an Auras' dream come true. In only a few months, they had managed to corral the two forces of discontent into one, strong unit.

Making sure that Drum was enthroned as King might not be easy, but it was within reach. Drum had whipped up a love fest between himself and most of the nobles. Once they had their white supremacy in place, the northern and southern nobles would credit Drum for it and simply sit back and wait for the "good times" that everyone knew would follow. Drum was exactly like his grandfather. You could see it.

The nobles could hardly wait. As a strong man, Drum would write favorable rules without interference from Lowlanders r Urhonordorians. The nobles would be able to do what they wanted, whenever they wanted and use The Kingdom's money to do it. Slavery controlled by unquestioned white supremacy would be the crowning glory of their victory.

And remember that stupid habit King Solem had in which he always wanted to 'feed the hungry?' That would be gone, immediately.

The Auras had only one small question today. What was going on with the nobles from the Central farms? As they had passed by that area, they had not heard a thing, but their portion of the new road was finished beautifully. How had that happened? Oh, well...They had most of River Kingdom angry and ready for rebellion. The quiet middle of the kingdom wasn't that important.

The combined forces of Drum and Thorndike rode north in silence, absorbed in their own thoughts and giving relatively little thought to the easy invasion they believed to be ahead.

But the peaceful procession was now in for a surprise that they never could have imagined. Up in front of them, one of the twins noticed a cloud of dust, the kind of thing one sees when a herd of animals stampedes.

"Look there, Father," said Jorge. "The road might get wider up here in the north, but it doesn't get firmer. Look at that cloud of dust up ahead there. What do you think that is?"

Drum reigned up and squinted into the distance. "I don't know," he answered ominously, "but whatever it is, it's big. You don't suppose somebody let your sheep loose, do you?" he asked Thorndike.

"I don't think so, but some of the shepherds snuck out and joined the militia yesterday when they heard there was going to be action," Jackson snickered. "Do you want Jorge and I to go up ahead and check it out, Father?"

"Indeed," young man, said Drum irritably. "And if it's sheep, get them out of our way. Drive 'em into The River if you must. We are about to get down to some serious conflict soon, and the last thing we need is a bunch of stupid sheep in our way."

Thorndike stared straight ahead. *What this ignorant man doesn't know about farming is huge, I see,* he thought with disdain. *He has no idea how to do anything but tell other people what to do. I wonder if he has ever done any real work in this life. Without me and my militia, he could do nothing. I wonder if he is even aware of that?*

"Go take a look, boys." Thorndike ordered.

The boys streaked off, goading their horses into a fierce gallop. But they returned as quickly as they had left, skidding to a halt in front of both Drum and their father.

"It's The Royal Guard, Father!" they declared in unison. Their horses pranced and reared with excitement. "It's the whole Royal Castle Guard, dressed for battle and headed this way!"

Their voices gave away more than excitement. It was fear. The boys had come face to face with over one hundred men, marching in formation and heavily armed. Breathlessly they described the scene that they had just seen. Fortunately, they had reigned up in time to be able to see without being seen.

The Call of the Auras

"There were over a hundred of them!" chattered Jorge as he calmed his horse enough. "They were wearing helmets and armor. They carried lances and axes and swords were strapped to their sides. They were a huge force, Father!" His horse reared again.

Jackson took a turn at the description of what he had just witnessed. "They were massive, Father! They carried heavy shields and marched in unison, waving their axes in the air like they were twigs! And ahead of them all, The King rode high on a stallion that looked like it was ready to trample everything in its path!"

Cjorge had to say something to his young human. "Now you are getting a taste of what this wonderful 'action' that you have been looking forward to might really be. Those men you saw were ready for war...with you!" Jorje shuddered.

Drum shrunk back in his saddle. *What is this?* he thought fearfully. *This is not the way things are supposed to be. Where is my triumphant entry through the gates of The Castle? I should not be getting ready to lead a charge. This is definitely not my strong point!"*

Thorndike was just as alarmed at the prospect of their foe coming at them now on the road. Although they had fancied themselves marching toward battle, they were totally unprepared right now. Drum's position in the face of fear had always been anger. Now that habit served him well. "Halt!" he yelled to the men behind him. "Halt, you mindless idiots! Do you not see that there is something approaching us? Are you blind? Halt, I say!"

The militia came to a noisy, clanking halt, bumping into each other with horses and equipment up and down the line. "What is this all about?" they grumbled, unable to see what Drum and Thorndike now saw. Impatiently, horses and men jockeyed each other for position.

The fact that the road was wider here helped some. They had arrived at the northern part of the new road but the men groused to each other.

"What now?"

"We turned around once."

"Are we going to do it again?"

"Who are we going to fight now?"

"Does anybody know what's goin' on?"

Drum and Thorndike were both pierced by a sudden flash of reality. They had been casually aimed at a military triumph over an ill-prepared Castle. Now their foe was approaching <u>them</u> by surprise. And, to make matters worse, The King and his Royal Guard were once again thwarting Thorndike's dream of an easy victory. This was just what had happened before to his grandfather.

Thorndike's scrambled brain couldn't move. He was not ready for this. His nightmare of his grandfather's loss froze him in his tracks.

Drum, on the other hand, was his grandfather's heir. He was planning to be the monarch of wealth and power, not of battle. He halted only momentarily...for inspiration. His Auras were ready. "Your rise to power is not cancelled, because of this surprise up ahead...only postponed. You are close to Thorndike Manor. Go there immediately. Tomorrow is another day."

"Take our forces back to your manor, Thorndike," Drum ordered loudly. "This was enough militia formation practice for one day." Slapping his horse hard, he charged off across an empty field toward Thorndike Manor without looking back.

Gratefully, the twins understood immediately that Drum was making the smart call and followed him at a gallop. After only one or two additional seconds, Thorndike also understood and gave the order. The fearsome invasion force of the 'Thorndike combined militias' cut across the same field, ducked behind a stand of trees ...and vanished.

King Solem had passed the gateway to Thorndike Manor without incident a way back. He had thought that some men might have been left behind for protection of the manor, but he saw nothing. They were not traveling at a rapid pace. He didn't expect contact with Thorndike until they were much farther south. He did notice, however, a cloud of dust up ahead.

That could be the militia already, but I doubt it, he thought. *Although it would be nice to have caught up with them without going too far.* Quickly he sent the men next to him back to rest of the Guard to warn them. They must be ready for...he was not sure what. A battle was possible, but he hoped not. *Thorndike's men are probably good men but misguided,* he thought to himself. *If only we could stop a battle before one starts.*

"That's a generous thought but be careful. Trust should be earned," whispered Csolem.

I know. I know, Solem thought as he slowed the march to a crawl and sent scouts ahead to get an idea of numbers, readiness, leadership. To his surprise, the scouts were back in a flash.

"There's nothing there but dust and hoof marks on the new road," the scout reported breathlessly. "Somebody *was* there, but whoever it was, was warned and took off across a field and out of sight. We might be in for an ambush."

Solem halted the march. "No," he said thoughtfully. "There will be no ambush. I think I know what has happened here. We surprised Thorndike's militia, and they retreated to Thorndike Manor which is just behind those trees. For some reason, they changed direction and decided to come in our direction instead of to Urhonordo. But they weren't ready for a battle with us on the road." He shook his head sadly. "I don't think they're through with us, however. 'Bout face, men!"

Solem threaded his way slowly through his men to the back, explaining what he believed to have happened. "Turn back toward the Castle, men. The militia we were following fled back to Thorndike Manor. Take up defensive positions at The Castle. They were coming our way, but I believe they may have pulled back only for a better advantage."

 June solemnly agreed. He was not sure when Thorndike would strike, but he was as sure as his father that it would happen. He was glad he had not chosen to go to Urhonordo, but he was perplexed. What had happened to the Urhonordo invasion? Why was there any invasion at all?

THE URHONORDO INVASION

June was not the only one who was perplexed. The Urhonordo citizens and their newly added Lowlander forces waited all day for an invasion, but by nightfall, nothing had happened. Prince Armando sent out scouts again.

What was happening at Drumsworth? Had Thorndike just decided to wait there for a day? Had Urhonordo messengers simply misunderstood the information they got? It was all very strange.

Queen Luna circulated among the Lowlander women and children, inviting them into the castle to keep warm. The men would have to keep watch outside, cold, or not.

At last, Armando's scouts returned. "Thorndike never made it to Drumsworth," they said. "And there was no one there...not even the militia men and The Drumsmen. The place was deserted." But this was not the end of their surprise announcement. The inquisitive scouts had ridden north from Drumsworth, all the way to the Central portion of the kingdom. There they had met up with Hector and his fellow farmers, people that these scouts had worked with to fix the road they were now riding on.

The farmers were perplexed, as well. They had watched the Thorndike militia storm past them. Then they watched Drum and some more militia meet up with the first batch there on the road. According to the landholders, the whole group had turned around right there and gone back to the north. They told the scouts that it was the strangest thing they had ever seen.

Armando shook his head. There was no doubt in his mind what had happened. Drum had intercepted Thorndike and now they were together...and they were all armed for combat and headed north. He looked around. All the Lowlanders and his own Urhonordorians were still armed and waiting fearlessly for an invasion. He and they were ready for a battle that was not going to happen here...but it was going to happen. The two masses of fighting men that the farmers had seen were not going north together for no reason.

Manda was at his side as she had been all day. She thought she knew what was happening and there was obvious pain in her eyes. "They are headed for The Castle, aren't they, Father?" she sighed.

"Indeed," mused Armando. "I wondered about the invasion of Urhonordo as we thought last night. We knew it was happening, but that was Thorndike's dream. I believe that Drum is the man who has really been leading since before The Gathering. He has been courting the nobles toward what I now suspect has always been his main goal. Invading The Castle would surely come first for him."

"That now shows itself to be exactly the fact, doesn't it dear?" Queen Luna, complete in riding gear, was standing at his side. "Thorndike has had Urhonordo in his sights personally, particularly since he has appeared to lose some of his mental faculties. But Drum's vision has been going on longer and stronger. Drum intercepted Thorndike's drive toward us and now they're headed toward The Castle. All we can hope is that we will get there in time to keep Drum from reaching his goal."

"Armando and Manda stared at Luna in amazement. She was dressed and already prepared to do what both of them now knew they needed to do...go north to help Solem...and June."

Manda hugged her mother, but then drew back. "The Thorndike Militia is huge, Mother. Do you think we dare to ask the Lowlanders to go with us?" They are already prepared to fight here."

"That's exactly what you should do," contributed Paulina. Her new-found strength and self-determination had already taken root. "I'm sure that's where the kidnapped Lowlanders *want to* go right now," she added. "I know they will be willing to show the militia that they are now ready to stick up for themselves. They were sure ready to fight if Urhonordo got attacked, here, she added proudly as she looked around at the battle-ready force."

"I think they have learned, after what happened to them at Drumsworth," agreed Armando. "I think they have decided that they have a right to freedom and as much as anyone else and they are willing to use the power that goes with it. I will ask them what they think of the idea. Will *you* be willing to lead them?"

Surprisingly, Paulina shook her head. "No, Your Majesty. I am on my way down to Drumsworth," she announced staunchly. "My fellow-inside slaves are still in Drumsworth Castle. We didn't free them when we freed the others. And Balynn is down there, too, I believe. This is not where any of them will want to be if Drum comes back there."

The attitude displayed by this young lady was what Armando had been wanting to see for years.

Paulina didn't wait for permission from anyone. Hastily she and Manda ran from person to person, climbing up walls to the fighters, even scooting into groups of ladies preparing weapons, explaining the situation to Lowlanders of all shapes and sizes. She just asked them two questions. "Are you ready to go after Drum and tell him we won't be slaves anymore? Are you ready to join our Great Leader and go north now to fight for that?"

Within the hour, a motley crew of commoners and Royalty, young and old, Lowlander ex-slaves and even Urhonordorian noblemen had gathered what they needed for a trip they had never thought of taking. As the word spread, everyone decided that it was time to stand up for what they believed. Many were persuaded by an inspired Lowlander girl, a princess and queen in britches.

They all believed in a good King, honesty, basic goodness, and most of all, the end of prejudice. The fact that they had been so threatened by only two selfish nobles made them think...and made them finally decide to do something about it.

<u>ONE MORE GOOD THING</u>

The Thorndike Militiamen routed themselves back to the manor. This day had not gone as planned in any way. First, they had headed south, then made a turn and gone north, only to veer east, back to Thorndike Manor. Why had they done these things? They had no idea.

The confused militia put away their horses and returned to their barracks. The men pulled off their helmets and body armor and stretched out on what served as their bedding. It was only straw and blankets but, after today, it felt good. They must get their rest now. Thorndike had already announced that tomorrow was to be the long-awaited assault on their major foe. Surprisingly, however, they still didn't know who or why that would be.

Jorge and Jackson couldn't rest. They were more than just tired from today's cancelled invasions. Their minds were questioning everything. What would they have done in Urhonordo if they got there? What would they have done if they had met up with The Royal Guard on the road? Would they have fought and killed people that they didn't even know?

"Don't be absurd. You are good boys. You can't kill anything," their Consciences assured them.

But "good boys do what their father tells them to do." That's what their Consciences had always told them. Tonight, the definition of "good" required some extra thinking. And what about tomorrow? The twins knew who Drum was aimed at. They knew where the battle tomorrow would be. Were they ready to capture or even *kill The King of River Kingdom?*

Capture or kill? Why in the world would they do that? And to do that, they would have to fight The Royal Guard. Now they had not only seen The Royal Guard but seen them in their battle gear.! Fighting them did not seem like a good thing to do *at al!* And they had no reason to fight or kill anybody...no reason.

This was their problem tonight. There was no reason for any of this. The only thing that sounded reasonable was talking to Flo.

The boys found Flo where they knew she would be. She had not expected to have anyone to feed tonight but suddenly they were all back and hungry.

Flo's reaction was joy. If they were all here, they were not off fighting in Urhonordo. That was a good thing.

The boys lowered their exhausted bodies onto the kitchen chairs and waited. Flo was working over a steaming pot, adding what seemed to be pieces of meat.

Finally, she turned, the steam dripping from the hair around her pink face. "So? So, what happened? You all went galloping off to conquer the world and now you're back. What happened?"

"What happened today isn't important," Jackson grumbled. "Tomorrow is what's important. Tomorrow we're probably going to attack The Castle and King Solem. *That's* what's important."

"You will do no such thing!" snapped Flo. "Don't you joke with me, brother. This is not something to joke about. Close your mouth until you can make some sense."

"He's not joking, Flo," inserted Jorge quickly. "He's telling the truth. Drum is here at the manor tonight. Father and Drum are charting things out for the attack as we speak. Sometime tomorrow, I bet we'll be invading The Castle and fighting the Royal Guard to take over River Kingdom."

Flo stared at her brothers carefully. "You're serious? You aren't making a stupid joke? This is really true?"

The boys nodded. "It's true, Flo. We almost got into it with the Royal Guard on the road an hour ago. We practically met them face to face. If Drum hadn't taken off for here like a scared rabbit at the last minute, and if father and all the rest of us hadn't followed him, we might be in the middle of a battle right now."

"Or dead in the middle of the beautiful, new, Royal River Road," added Jorge sarcastically.

Flo stared at Jorge. "You would have been fighting your own King's Guardsmen? Why would you ever do something stupid like that?"

"Because now we can see that that's what Drum and Father have been planning for years, Flo," answered Jorge. "I could see it in their eyes all day. "Remember hearing about "The Plan"? Well, this is what The Plan has been all this time."

"Father and his militia were always to provide the muscle to get it done while Drum took over as King," Jackson added. "Father wanted Urhonordo first, though, probably for old time's sake, considering he has always thought he should have inherited Urhonordo. Drum just wants to be King of *all* of River Kingdom."

Flo turned around and went back to the hearth and her boiling pot. Then she picked up a basket of potatoes and onions and placed them on the table. "Chop these up," she ordered. The boys complied. They had done this many times. Flo was thinking.

Now she stood at the head of the table, her spoon in her hand like a scepter and issued a statement.

"The wealthy plot...

Force carries out the plot...

The thoughtless suffer and die."

"It is time for all of us to stop being 'the thoughtless', boys," she declared. "We have been thoughtless too long.

The boys were bright but now they were confused by happenings that were out of control. What did Flo mean? What was she trying to say?

"We three have thoughtlessly let everything around us happen, day after day, as though what we were doing wasn't wrong or if it was wrong, it would fix itself," Flo declared sadly. "We have lived with evil for so long that it seemed normal. Now we may have waited too long to avoid a catastrophe."

"Once you figure out what needs to be done, it's never too late to start fixing it," said Cflo, bluntly. "You did something good for Lady Thorndike and your sister the other night. Now you need to find out what your next 'good action' should be."

It took me years to get my Lady back to Willa, thought Flo.

"But you did it. You and your brothers did it because you knew it was right. "*Now* you understand," offered Cflo. "You *did* something to be proud of. Now it's time for you and the boys to do something to be proud of again."

But what about the invasion? My brothers have been taught to fight all their lives, Flo grumbled to herself. *That's all they know.*

"Indeed. This is true but some of what they learned might come in handy. They have learned to fight. Now they just need to figure out *who they should fight* and *what* they should fight for. They have Consciences. Ask them to think about right and wrong."

Flo went back to the stew pot and stirred some more. It helped her think.

Then she got an idea. "What do you boys know for sure?" she asked as she stirred.

"Not much," Jorge answered sadly. "We know that we are going to invade The Castle tomorrow for sure. We know that we don't want to fight. We know that the Royal Guard is huge...but not as big as our combined militias, though. Our combined militias will probably wipe out everyone in The Castle."

"We know that a lot of people will get killed tomorrow including us," added Jackson... "and we know there is no reason for any of it."

Flo kept stirring. "If you die tomorrow, Jackson, what do you think you have done in your life so far that you can be proud of?" she asked.

"Like what?" Jackson laughed. "I've never done anything yet."

"Yes, you did," Flo tossed back. "Yesterday you took Lady Thorndike back to her daughter...who happens to be your sister. You made them both very happy. That was something to be proud of, don't you think?"

Jorge spoke up. "Yeah...you're right. I did that, too. That was *one* good that we all did that we can be proud of...which isn't much...but so what?"

Cflo was puzzled. Where was Flo going with this?

Flo stopped stirring and faced the boys squarely. "The fact is that the three of us haven't done much in our lives to be proud of...until yesterday... and then we all did just this *one good thing.*

"Tomorrow there's going to be a battle and the one good thing that we three did yesterday will be wasted. What do you think about that?"

"I have no idea," answered Jorge. "Why will our 'one good deed' be wasted?"

"Because there is going to be a battle in The Castle and both the people we helped yesterday may be killed, that's why." Flo sobbed. "Lady Thorndike and our sister Willa may be trapped in the fighting and when it's over, they will probably be dead and so will the two of you." Flo slammed the big ladle to the floor.

But she wasn't finished. She looked at them fiercely. "I have another question for you, my brothers. What is *worst* thing that you two 'half-Lowlanders' have done in your life? What are you most ashamed of?"

Strangely, the boys came up with two answers immediately. "Being slave masters at Drumsworth and kidnapping the Lowlanders from town,' they answered in unison.

Flo smiled at them. "I thought so!" She grinned with pride. "And you haven't killed anybody, yet. So... tonight we are *all* going to stop being thoughtless. I have a plan for us to do at least *one more good thing* to be proud of before we die. And I have a plan...a plan for *good*.

"I have friends in town that I have worked with, and they know me...*white friends*. My idea is that we take our sister and her mother out of The Castle and into town...We take them out of tomorrow's battle *tonight...to keep them safe.*

"You boys will have to apologize to the town for your kidnapping you did before but we all must ask the town folk to hide the two of them until the fighting is over. That way the 'one good thing' that we did already won't be wasted. We will save our sister and her mother, and we will have done at least *one more good thing* before we die. What do you think?"

The boys stared at Flo. "You are crazy? You aren't serious?"

"Completely, serious. The guards at the gate of The Castle know us now. We can just tell them that we need to see our sister like before. They will let us in, and we will tell them our idea. They all know that there is going to be an invasion tomorrow, right? I think The King will like the idea. He really loves Willa, they say. He will be glad to see her hidden in a safe place."

At first the boys just stared at their sister. She went back to stirring the pot. After a few minutes, however, the crazy idea grew on them. The truth was that they were not ready to fight tomorrow, either for or against their father *or* The King. This idea was a good thing...a thoughtful thing...and if it didn't work, at least they would die trying.

They began to plan quickly...while they ate at least *some* of the stew. Their Consciences had not coaxed them. This was a crazy idea, but at least their humans were thinking with their hearts.

THE PLAN

Drum and Thorndike hunched together as usual among their maps and plans, but after today, their confidence was badly bruised. Today, they had both momentarily entertained notions of grandeur for themselves, only to have those notions dashed. Where was their expected grandeur now, and after today, what was their plan to regain it? Did they even have a plan?

Drum and Thorndike were understandably depressed, but their Auras would not allow it. They needed a joint conference. "Today was a fiasco," they grumbled to both men. "You both know that. Thorndike, you could have defeated the royal Guard right then and there on the road, if you had just forged ahead and done it," growled the Auras. "Why didn't you?"

Thorndike looked at the battle plans scattered around him. His plans were for an assault on The Castle. They were not for a troop-to-troop field battle with a prepared enemy on the road. That would have been messy and wasted some of his troops.

"Fighting on the road was not in my plans," he stated to Drum. "I have planned all along for a dignified charge on The Castle, itself."

"Me, too." Drum growled grudgingly. "I have always dreamed of swooping in on The Castle and conquering it. So, is that what you are going to do tomorrow, Thorny?"

"If we attack The Castle tomorrow, they will expect us," Thorndike shot back. "Now The Guard will be on alert."

"So what?" scoffed Drum. "Your combined militias have twice as many men as The Royal Guard. Just march our men in and take their men out."

"That would lose too many of my militia," Thorndike responded angrily. "I wouldn't have any leftover for Urhonordo. That won't happen." The Auras shook their heads. He was being too cautious, and the Urhonordo mania was still there. Thorndike needed focus.

"The white folk in the town watched your men move into town to do what they pleased," the Auras recalled confidently. "They looked terrified when you stripped them of *all* their Lowlander trash, and they did nothing, just because they saw the frightening size of your militia."

Drum heard the Auras and followed their lead. "That's right, Thorny. Size matters. So what if you lose a few, especially if they're from other militias?" Drum said casually. "You'll still have your own men left over to mop up and secure The Castle for my coronation when we win.

"This Castle Assault has always been The Plan, Thorndike and you know it. All you must do is do it." Drum was through with the discussion.

Thorndike's Auras now quietly draped themselves across Thorndike's shoulders like a cape of honor, crooning to him, alone. "*You* would have been King of Urhonordo by tonight *with no loss of men* if Drum hadn't got in your way," they whispered. "Now he is trying to push you around. *You* have the power that he needs. *He* has no power. Think of that now and take advantage of it.

"Because of history, The Plan has always been to make Drum King, but Urhonordo should be yours *also,* because of history. Tomorrow, you can give Drum what he wants first to keep him and the nobles happy but then they must move on to the glory of *your* goal... but do it *your way.*

"You don't need to lose *any* men", Thorndike's Auras declared. "You have a military mind. You know strategy... finesse. Use it. The size of your militia alone will win tomorrow for you. Spread your men out. Use your new road. When the King sees your force as huge and intimidating, he will know how many more men you have than he has, and he may decide to avoid the slaughter of his men and give up".

The Auras glowed in the fire light, thrilled with their own ideas. "You can crush The Guard by dividing them into two pieces. Conquer The Guard that is protecting the walls of The Castle as you know they will do. Once they are conquered, you can storm the front gate and remove the pitiful number of Guard that will still be with The King. It will be glorious! And Drum's nobles will hail *you* as a mighty warrior... as you go on your way to Urhonordo to claim *your* prize."

Thorndike sat back and smiled. The fire was beginning to crackle in the fireplace. Flo would be bringing in some warm stew soon. He turned toward Drum.

The Call of the Auras

Strangely, Drum was smiling from ear to ear. "I have just decided on the perfect climax for your successful invasion tomorrow," he announced, beginning to stroll back and forth in front of the flames.

"Picture this... as my prize for your winning the battle, I will offer the lovely *current* queen of River Kingdom a choice...between reigning with me or losing her head." (Drum's Auras chuckled.) "And I will allow The King to stay alive long enough to watch me *in my crown* as I accept her as my Queen. What do you think?"

"Are you insane, Drum? You're talking about my daughter here! You can't do that! My daughter is already the Queen of River Kingdom."

"Very convenient. All I will have to do is sit down beside her and declare myself her King," Drum added with a smirk. He stopped strolling and looked Thorndike in the eye.

"Tell the blowhard how it is," hissed Thorndike's Auras angrily.

Thorndike stared back at Drum with a look of manic power. "*I* put my daughter *on that throne* to be loyal to *me,* Drum. *You* can declare yourself King as you have always planned, but *my* daughter will remain on the throne as Queen...*my* flesh and blood... Queen of River Kingdom. I planned it that way from the start and there she will stay."

Drum's face twisted into a ghoulish snarl. He strode to the seated Thorndike and leaned over him like a bird over its prey. "*You put her there*, you say? Very clever...... but that was then and now is now."

Drum reached down and pulled up the crown that had been comforting him...and reminding him...and *goading* him all day. He thrust it forward in Thorndike's face.

"Do you see this?" Drum snarled. "Do you see it? *This* is my power, Thorndike! This crown is not your woe-be-gone, unwanted girl-child foisted into position to gain *you* power. This crown is *history* reclaiming its due! Stand back tomorrow and watch me take *my Castle and my* Queen!"

Drum stormed out of the map room and up the stairs. His Auras applauded as he went...but that left Thorndike alone in the room with *his* Auras.

<u>CONSCIENCE ALERT</u>

The Consciences in Urhonordo were pleased with the decisions made by their humans to try to make it to The Castle in time to help, but it was a long way. This was the kind of dedication to right and fairness that the Consciences stood for, but would they all be too late? Was there nothing that the Consciences could do to help The King?

The Consciences were concerned but creative. Quickly they called an Emergency Conscience Alert at Urhonordo. Carmando was the first to speak. "I feel that we may be too late to help the good people of The King fight off Thorndike's Combined Militia," he said sadly. "It will be massive. There is no way that King Solem will be able to conquer them all.

"Time is not on our side. We are too far away to be of any use. For that reason, I suggest we skip Drumsworth today and take Manda's cutoff across The River. At least that will save us a little time."

"If we do that, we cannot rescue Baylynn and her house slaves," countered Cmanda. "That wouldn't be right. If Drum takes over River Kingdom, they will be stuck in their terrible slave lives at Drumsworth forever. We *must* go through Drumsworth."

Cluna shook her head. "You Consciences have forgotten the power of ideas from one good Conscience," she announced. "How did our humans learn that they needed to go north to help The King?

"The idea was brought to them by a human messenger, right? Why can't we Consciences send a Conscience messenger up to the Conscience of a northern human the same way? Maybe *that* Conscience can help *that* human use *that* message to help The King. If nothing else, at least they will both know that we Consciences and our humans are on our way. "

Carmando and Cmanda were amazed. Why hadn't they thought of that?

Cmanda shook her head. This ability of Consciences to travel alone on occasion had slipped her mind...but Cluna was wise. Cmanda was immediately ready for the challenge that Cluna had delivered. If word needed to get up north to help the embattled Consciences and their humans, she was ready to go. "I will be on my way with a message as soon as we as a group are ready to send it, she declared.

Carmando was also reinvigorated. "You will be the perfect messenger," he told her, relief ringing in his voice. "My suggestion is that you go directly to Cjune. June will undoubtedly be in the center of the defense for The Castle. Maybe at least the knowledge that we are on our way will help him keep Thorndike and Drum at bay until we reinforcements get there."

The gathered members of the Conscience Alert breathed a sigh of relief. If they and their humans could not all be there right away as they wanted to be, at least northern Consciences would know that more good humans were on their way.

Without further delay and with the good wishes of all the Consciences swarming around her, Cmanda left for the northern part of The Kingdom. What was she going to say? Just what she had been told to say, Conscience to Conscience. She only hoped that she would be there in time.

THE OTHER PLAN

The fire was warm and welcoming in the sitting room of The Castle, but no one there appreciated the beauty of it. They were too worried. Willa and Solem both envisioned what was happening tomorrow. There would be an invasion by the Thorndike Militia, and many would be killed.

Willa was, of course, thrilled when Solem and The Royal Castle Guard had arrived home unharmed earlier. The Guard had gone straight to their quarters to take off their battle garb and rest. No one spoke about what had happened on the road. They were getting ready mentally for tomorrow.

Solem gave The Guard their instructions. They were to assemble their fortifications on the walls that surrounded The Castle and post guards throughout the night. They were to stay alert and ready. They were to take turns sleeping in shifts. Tomorrow would be a day of war.

There was no way to avoid the inevitable now, Solem thought. The Guard had all seen the dirt fly up behind an enormous number of combined militiamen as they galloped across the field toward Thorndike Manor earlier, but they were not relieved by what might have looked like cowardice. They knew Thorndike was a military man. This retreat was a strategic regrouping. All knew without discussing it, that tomorrow the battle would be on...but why? No one could understand why.

"It's all about power," Solem explained. "Drum has dreamed of power since he was born, He feels he deserves it. We feel that he doesn't, but he has Thorndike on his side to allow him to come after it."

"That's not all he has," commented one of the Guardsmen. "He has all the other militias signed up to help him, so I hear. That will make Thorndike's combined militia bigger than us."

"Indeed," agreed Solem. "We will just need to fight better."

"And smarter!" yelled another one of the Guardsmen.

"No problem!" shouted another. All present shouted in unity together.

Solem shouted his agreement with them and bid them goodnight. The only plans they could make were in place.

*****

Solem pulled off his boots and wiggled his toes in the warmth of the fire. Willa watched silently.

At last Solem looked up at Willa and smiled sadly. "To be truthful, I have no idea what will be happening tomorrow or why," he said. "Thorndike and Drum took off for Thorndike Manor as soon as they saw us. We were definitely a surprise to them, there on the road. I do believe they were heading in our direction to invade The Castle right then."

"You would have beaten them to a pulp," Willa announced staunchly. "They wouldn't have stood a chance."

"I'm afraid they would have thumped us good, and you would have a new King right now if we hadn't surprised them," Solem answered solemnly, "but why? Is it all about power only?"

Willa stared into the fire. The voices she had heard day after day as a lonely young girl wandering the halls of Thorndike Manor now drifted into her thoughts and demanded to be remembered.

"I know why, now," she said aloud. "It's because Drum wants to be King like his grandfather."

"Exactly," whispered Cwilla.

Willa suddenly knew what she hadn't really known until now. "Drum is the whole reason for tomorrow," she realized with a start. "In fact, I believe his dream of being king has been at the heart of the road, the prejudice, the cultivation of the nobles...everything." Solem listened in fascination.

"Think of it, Solem," she continued.

"He encouraged dissatisfaction with you among the nobles. Why? Because he wants them to support him as king.

"He has promoted feelings of prejudice among the nobles. Why? Because the nobles are prejudiced and so was his grandfather...'the King.'

He has been encouraging slavery. Why? Because his grandfather used slave labor... and his grandfather *was* King.

"He has even built a mote at Drumsworth. Why" To protect himself down there *and* up here both...as King."

"And, most of all, Drum pushed for improving the road so that the Thorndike's Militia could use it to move quickly to support everything *he* wants to do...not for safety and protection like they said, but for slavery transport and ease of militia control. Tomorrow they will both be on that road coming here to invade The Castle. Why? So that Drum can become King."

Solem looked at his wife. She was standing in front of the fireplace, waving her arms. She made perfect sense. Why hadn't he seen the way that these things fit together before?

Because Auras are smart and willing to engage in long term planning, thought Cwilla sadly. *And we Consciences were not suspicious of Auras enough.*

"I actually see everything you are saying, and I think you are right", Solem mused aloud, "but what about Thorndike? Your Father now feels like a bigger threat to me than Drum. He doesn't want to be King too, does he?"

Willa sighed. "If you count Urhonordo, yes. My father is a perfect partner for Drum. My father always wanted Urhonordo as his own kingdom. He thought the old King Sol cheated him out of it one hundred years ago. That's why he has always disliked *you*. You are that king's grandson, so, he and Drum became a team. My father is now Drum's muscle. They *both* decided to go after what they *both* wanted...and neither of them have a right to any of it."

"Well said," whispered Cwilla. "This was what I warned the other Consciences about way back when. I don't think Thorndike would ever have gone this far but the Auras encouraged his dreams of Urhonordo to become his obsession. This whole situation including tomorrow, is a production of The Auras."

Solem smiled. "You know, Willa. I am glad I married you. I think I should put you in charge of The Royal Guard from now on." He pulled her to him and there they sat together, but for only a minute.

Now Solem stood, his hands on her shoulders. "Now I must go to work, my love. I'm placing The Guard strategically at the walls around The Castle. We *do* have good, stout walls to begin with."

"That sounds like a good idea," Willa responded. "What else do you have in mind?"

Solem shook his head. "I don't know what else. I must admit, I never saw this coming. I saw no need to prepare...so I didn't."

He began to think out loud. "They will be arriving here by foot and on horseback. We will be behind the walls. They will be well armed...but so will we. *Their* main advantage will be their numbers. There are twice as many of them now as there are of us."

"We'll just have to fight harder and smarter than they do," Willa said staunchly. "We can't lose. They are wrong to do this, and we have 'right' on our side."

Willa suddenly heard Agnes outside the sitting room door, and she wasn't happy. "You can't go in there now, June," she was stating firmly. "Solem and Willa are having some alone time. Go away!"

"No!" came the response as the door abruptly opened. June stood in front of them both, solid and unmoving. "I have captured two Thorndike Militia men and tied them up down in the entry way," he stated. "What do you want me to do with them?"

Solem looked at June carefully. "Two Thorndike Militia men? Here? Where did you find them?"

"At the front gate," June answered..." with a woman servant. They were unarmed, but..."

Willa jumped. "It's my family, Solem. That sounds like my family again," xhe cried as she ran out the door.

Solem followed Willa down the stairs in his bare feet, leaving June and Agnes to trail behind them both in a quandary.

At the bottom of the stairs, a shivering trio looked up cautiously. Indeed, it *was* Willa's family.

Solem had no idea what was happening. These two very large men were now possibly part of Thorndike's enemy force. They were bound and standing in his entryway but ...family?

Willa reached the entryway first and pulled her sister to her for a hug. "What are you doing here, Flo? Are you in danger?"

The brothers looked embarrassed and bowed. She might be their sister, but they still didn't know her that well.

Solem took charge. He was King, after all. With a grim look on his face, he went straight to the brothers. "I know who you are, but why are you here again? I know your father is planning an invasion tomorrow. I repeat, why are you here?"

"We're here because we *are* Thorndike Militia, Your Majesty, but *we* don't want to invade tomorrow," declared Jackson in a clear, strong voice.

Jorge stepped forward. "Our Consciences said to choose who was right and wrong and what side we should fight for tomorrow and we have chosen your side, Your Majesty. *And,* the three of us would also like to do at least 'one more good thing' if we're going to die tomorrow...Our first good thing was yesterday."

"What are these crazy people talking about?" sputtered June. (He hadn't seen the arrival yesterday and he was looking at Thorndike Militia men right before an invasion for heavens' sake!) "What kind of nonsense do you three think we'll fall for? Shall I have them thrown in the dungeon, Father?"

"June, wait! Don't talk! You didn't see these people yesterday. You don't know them at all, but *I* do." Willa turned from her embrace of Flo to grab June's hand. "These are *my* brothers and my sister, June...honestly," she said softly. "I don't know why they are here again, now, and we all need to find that out. So please, June. Just wait."

June stared at Willa in disbelief. "These were *your* brothers and a sister? And they are the ones that brought your mother here? Why didn't anybody tell me?"

"Yes...They brought my mother back to me here last night," Willa whispered, rolling her eyes. "While you were upstairs sleeping off your last trip back from Urhonordo."

"I can see why you're confused," Flo contributed with a half-smile. "But *Willa* didn't even know that we were related to her until just recently. There are many things that many of us didn't know...and didn't want to think about...until now.

"But that's why we're here. My brothers and I have been thinking about what we know so far...and now we know that we're on the wrong side of...tomorrow. We also think that Willa and her mother are on the right side but they may be in great danger."

The twins shuffled their huge feet and glanced up at June. Was he believing them?

June scowled. There was too much here for him to swallow at once. " So, you think *your side* of tomorrow's invasion is going to lose and you want to be over here on the winning side? Is that it?"

"No, sir! That's not true at all," Jackson shot back indignantly. "We think that the Thorndike Combined Militias are now almost twice as big as The Royal Guard even with us missing. They're probably going to crush The Royal Castle Guard and try to kill everybody in The Castle...But you see...that's why we're here. We want to take Willa and her mother into town, away from the fighting, where they'll be safe."

Solem was snide. "What makes you think *you* can make them safer *there?* Didn't I hear that you two recently stole all the town's Lowlanders?"

"You're right about that," sighed Jorge. "We're sorry for that but Flo knows many of the town folk. She thinks that they will accept our apology and protect Willa and her mother, in spite of *our* bad behavior."

Flo turned and looked squarely at June. "*You* were the one that talked your father into sharing food with the hungry, aren't you, Your Majesty?" she asked.

June blinked. "Yes." *Why did this woman ask such a thing on a night like tonight?*

"Some of the people in town that you helped back then and even today are friends of mine," Flo explained. "They don't know about tomorrow, but they do know about *you*. Because of the food, they love you. I think that all you need to do is ask for their help and they will protect Willa and my lady. They might even help fight.

Willa was still holding onto June's hand. "Flo *is* my big sister, June. Before that, she was my best friend. She *is* telling you the truth. I just found out that I had these brothers too, so I don't know them as well...but my sister is speaking for all of them. *I* don't need protection but my mother..."

Solem squinted at brothers. "Now, after you kidnapped unsuspecting Lowlanders, you think the town will still help us?"

"Doing that awful kidnapping was what started us to rethinking about this whole militia thing that we have been part of all our lives," said Jorge.

"Thorndike is our father, but he was wrong about taking the Lowlanders then, and he is wrong now. He and Drum are both wrong about a lot of things. If they win in battle tomorrow, River Kingdom will turn out to be as wrong as they are. That's why we want to help. Flo thinks we can help...but we would need your permission, King Solem," announced Jackson with a formal bow.

"We need more than that," added Flo. "We need someone from The Castle to go *with us* so that the town will understand that we represent *your* wishes, Your Majesty. "The town people love you because you helped them with food. Now they will be glad to repay what you and Prince June did and are still doing, but *someone* from the royal household needs to go with us to explain about Willa and my lady."

June looked around. He thought that he understood what she was saying. Now he glanced at Willa, as well. She was staring at *him*.

"I think they are all telling the truth," whispered Cjune "You and The King did a nice thing back in the day and are still doing it. Sometimes people appreciate things like that. I think these three aew counting om that."

June looked at Solem. "These boys are going against their own father. Do you believe they are trustworthy?" he asked.

"I know their father. I tend to believe these boys," Solem said slowly.

"Please believe us, Your Majesty" coaxed Flo.   "You have given my sister your love. We see that. Now, tonight, please give us your trust. Tomorrow may be too late. We need to start into town tonight...door to door...in the town...when no one at Thorndike will see or suspect what we are doing.

"The Thorndike Militia is huge. I know. I feed them, and now they have all the northern militias with them. If they win tomorrow, all of you may die and River Kingdom will be lost forever."

 Csolem climbed onto Solem's shoulder. "I *feel* that we can trust them. Not only that, but I think that we *need* to trust them, Solem. Tomorrow will be harsh, but the town likes you. They may even provide increased manpower."

Solem didn't answer. He was trying to put it all together in his head...and make sense of it.

"I really think they are right Solem, Willa said. "Please let them save my mother. After all she has been through, please don't let my father harm her again."

Agnes was peering at everyone in the entry hall over the upstairs railing. "The townspeople don't know how to fight," she blurted without thinking. "Won't they b if they be in danger help The King?"

Solem looked up and shrugged. "I don't think so, Agnes. Tomorrow's fight has been built on prejudice. The townspeople who are still in the town are white. If Drum and Thorndike's militia see them, they'll just think they're among friends."

INVASION MORNING

The morning dawned bright and clear. It was beautiful, a perfect day for achieving all the goals that had motivated Drum and Thorndike for twenty years. Drum would be the new King, the white noblemen would rule over Lowlanders and Thorndike's militia would move on to Urhonordo. Whooya!

To Thorndike's surprise, Drum and his Drumsmen were already at the front gate waiting for them when he arrived. The militia made its way down the long entranceway. Thorndike was the last to arrive at the gate. He had been looking for his boys. They were nowhere to be found. Had they gone on ahead? Were they taking the lead?

Unwilling to show his concern to the others, Thorndike said nothing. The fact that Drum was already waiting at the gate was irritating to him. Who did Drum think he was? The force that was gathering near him was a Thorndike force, not Drum's. His jealousy mania was already beginning.

"What ho, Drum?" he called out sarcastically. "Do you think you want to take the lead this morning? Do you want to be the first to attack the Royal Guard?"

"Not hardly," Drum responded with a grin. "You know me. I know nothing of warring arts. I am a born appreciator of you men with spears and knives and swords...and I intend to appreciate...from behind."

Thorndike didn't smile. This morning was serious business. He and Drum had been plotting the elimination of this spineless king for most of his life. Drum would rule, Urhonordo would come next and he, Thorndike, had the well-trained militia that would make it all happen.

Thorndike's Auras had been with him through the night. The old invasion plans that he had tweaked and re-tweaked for years were lying on the floor, discarded in favor of the brilliance of the Auras' new and improved plan.

Since yesterday, a surprise attack was out of the question. Strategy and finesse would now be Thorndike's approach. The idea was amazingly complex for his undercurrent of mania to keep in focus: One half of his combined militia would fan out around the walls of The Castle like an enormous human net. Anyone trying to escape would be captured or killed. That half of the militia would encircle The Castle standing side by side, armed to the teeth with battle hammers and swords. This would be more than impressive, striking fear into an ignorant king with nothing but a small Royal Guard and no military leadership such as his.

At Thorndike's signal, that net (which would be made up of the militiamen provided by the neighboring nobles) would close, driving into the interior of The Castle for hand-to-hand combat. When The Guard on the walls was captured, Thorndike and his well-trained militia would then storm the front gate of The

Castle and round up the stunned Royals as they fled from inside. Victory! The plan was foolproof.

Thorndike and his militia began to move. The plan was perfect in Thorndike's tumbling mind as he led his troops out onto the new Royal River Road. The new road was perfect as well, just as he had dreamed it would be for this exact purpose...military conquest.

Thorndike was determined and now completely manic. He had his destination in mind...'Today, The Castle, tomorrow, Urhonordo', he repeated to himself over and over.

Drum began pushing him from behind. He and his Drumsmen were bored. The steady pace of the proceeding militia was not the glorious charge of *his* dreams... and Drum's beautiful crown was banging unmercifully against his leg.

AGAINST THE DOOM

Up ahead in The Castle, the mood was not positive. King Solem had deployed his men inside and atop The Castle walls in a protective stance. They were equipped with the standard tools of defense ...spiked barriers, swords, bows and arrows and barrels of hot oil. But vigilance was nerve wracking. The Guard knew they would be outnumbered. They would have to fight hard. They fidgeted at their posts.

Solem surveyed the colonnade and the front gate beyond from an upstairs window. Willa was at his side. They were nervous. Willa felt responsible for everything that was happening today. Solem was preparing for an invasion that was being led by *her father*. It seemed to be *her* fault.

"Horse pucky!" hissed Cwilla in her ear. "None of this is your fault. None! You have been a good wife but only an obedient daughter to a tyrant father. How could any of this be your fault?"

I just can't help but think that none of this would have happened if I wasn't here, Willa thought.

"You have too great an idea of your own importance," snarked Cwilla. "Drum and your father have been pushing for this battle since before you were born. They are Auras' pawns now."

If I had been born a boy, maybe none of this would have happened, Willa thought sadly.

"Indeed," snarked Cwilla, reading her thoughts. "And if you had been born a bird, maybe you could fly out to the incoming Thorndike Milita and tell them not to invade, right? Get real. You are way too focused on yourself. Stay focused on Solem. He needs you."

Willa continued contemplating her position. Maybe she was just an insignificant factor in a bad situation, but insignificance did not have to mean uselessness. She heard her Conscience speak. "If you were a bird, you could fly out to the incoming militia and tell them not to invade."

Solem returned to their bedroom to review the large maps that were spread out on his bed for him to study as usual. She was alone.

Quickly Willa slipped down the stairway and out the front door. She didn't attract attention as she sidled along the wall of The Castle and crossed the colonnade to the stables. She walked her horse quietly toward the front gate and pulled it open just wide enough for her and her horse. She pulled it shut behind her and climbed on.

The guard nearest the gate on the wall above her spied her at once. "Halt!" he shouted. "Who goes there?"

Willa looked up and smiled at the guard. "I am Queen Willa," she answered, putting her fingers to her lips. "Shh. Don't speak. I am on a secret mission for The King."

The guard breathed a sigh of relief. He knew the Queen. Maybe she was going to do something to stop this expected disaster! He was impressed and vowed to say nothing. He waved and mouthed a silent "good luck."

Willa's Conscience had been stunned into silence by Willa's sudden decision, but being stunned was now over. "What do you think you're doing, Willa?" Cwilla growled into Willa's ear. "Do you think that *you* are so smart that *you* can stop this invasion? I thought you had more common sense than that.

"Comon sense is of no use when you're dead." Willa answered aloud. She crossed to the few trees running along the new road's edge in front of The Castle. Here she would not be noticed by The Guard or anyone else for a while.

"What do you think you are going to *do*, now?" Cwilla said with disdain.

"I am just following your advice," Willa answered with a chuckle. "I am now 'a bird who is going to fly to the invaders and tell them not to invade'."

"Are you crazy?"

"Maybe, but *my* father has been planning this invasion from the start right under my nose. Now things may go badly for my Solem. I should at least *try* to turn this around."

"You *are* crazy," said Cwilla.

"Probably, but I can't just stand by and not try to do *something* to help."

"You sound like *me*," mumbled Cwilla. "But I understand. Okay. I'm with you.

THE TOWN

In the town to the northeast, a few townspeople began gathering in the square. June, Flo and the twins had been busy all night. They were even handing out a signed request from The King. The King needed their help, it said. Young boys began running from house to shop to house spreading the news.

"The King needs our help."

"The Prince is here. He's the one that started the food sharing."

"Yep, him and his friends need help from all of us today...right now."

"Uh huh, he says that this is an emergency."

"Good King Solem and our pretty, new Queen may be attacked!"

"What?"

The information spread and a larger crowd gathered.

Flo was there with her brothers. Their role in the past removal of Lowlanders was a problem to begin with, but Flo explained that the twins were truly sorry for what they and the militia had done, and the boys apologized. June even stood up for them.

The people discussed it among themselves with their Consciences. "These two were not 'the mean ones', they remembered. "Forgiveness is a good thing... besides, they did say they were sorry, and they *are* Flo's brothers. Let's forgive them."

And someone else was here with them, they noticed...a lovely older woman that resembled the new Queen Willa. The ladies of the town looked her over carefully and decided that she *was* who Flo said she was...She was Queen Willa's mother, and *she* trusted the twins...now *they* would also trust the twins, for sure.

The crowd continued to grow but, now that they were there, the people were not sure what they were supposed to do.

The fact was, June was not too sure, himself. He needed more information. Carefully he pulled the twins aside. "We need some recon, boys," he explained. "We need to find out what's going on near The Castle and with all the militias right now so we can see how we can help The King. Can you do that?"

The twins took off immediately. The crowd of townspeople continued to grow. Now some were arriving with their weapons...shovels, hoes, bread pans. "When the time comes, we will stand up for our King, Your Highness," they explained. June nodded. This was working just as Flo said it would.

"We brought The Queen's mother here to keep her safe from a Thorndike invasion that will be happening at The Castle," Flo informed the gathering townspeople. "Your help with that is *my* main concern today. June and my sons may be asking for more, though."

The townsfolk nodded. They remembered Thorndike as leading the kidnapping of their Lowlander friends. "He *was* one of the mean ones", they muttered.

June pulled the townspeople together "I don't know if we will need you today, but our scouts (the twins) are now on their way to check on the situation at The Castle."

Flo spoke up. "I have one question for all of you right *now*. Are you here in town *willing* to help your King? I have been a slave, so I no longer approve of telling people to do something unless they want to do it."

"Well said," her Conscience stated proudly.

"I remember you from when we used to work sometimes in Thorndike's kitchen before," a woman at Flo's side said shyly. "You were nice back then. The King helped us with food when we needed it. He also said that he was going to help us get our Lowlanders back. That's why I'm all for you *and* The King and I think the rest of us are, too. We'll help you and The King, both, Flo!" Flo smiled.

The crowd of men with shovels raised them. "Here, here!" they shouted in agreement.

One of the twins was back soon. "Part of the combined militia is deployed for invasion all around The Castle," Jorge announced breathlessly. "But they are not invading right away. It looks like they're waiting for a signal. My guess is that they are local militias that were chosen to do the dirty work...assaulting The Castle Guard that is posted on the wall of The Castle. I feel bad for them. They aren't the regular militia. They are men that were contributed by the northern nobles. They don't look like they know what they're doing."

One of the townsmen stepped forward. "You're right. I used to be in one of those militias. They're good hearted, but they're never trained. The nobles just have them for show."

"Yeah. They're nice guys. We see them in town all the time."

"Maybe we should sneak up on them before they get into trouble and see if we can work something out," one of the shovel carriers snickered." The town waited.

Suddenly, June appeared lost in thought. He had heard what the townsman had said about sneaking up on the militia around The Castle, but now he heard another voice. It was his Conscience. Cjune seemed to have an important message for him. Apparently, Manda's Conscience had just arrived.

"I just got an important message from down south," Cjune said in a very excited tone. "The messanger brings word that reinforcements are on their way!" CJune continued to listen to Manda's Conscience with fascination...and hope.

"Manda's Conscience also brings word from Lowlanders and Urhonordorians who want to be here to help. They are on their way. They just want to encourage you and The King to stay strong until they get here," Cjune announced.

"Are you kidding me?" June whispered back. "Mand'a Conscience is talking to you?"

"Would I kid you about something this serious?" Cjune asked, insulted. "I'm delivering this message from Cmanda who came all this way to tell *me* to tell *you* to stay tough with what you have around you until Prince Armando and reinforcements can get here to back you up. They all know that the town's militia will be joining the Royal Guard to make it bigger."

June was amazed. This was a message from *Manda's* Conscience to *his* Conscience. He smiled. *She loves me and is thinking about me,* he thought gratefully.

"Keep your mind on her message," growled Cjune.

"Right...Right...Tell Manda's Conscience, "Thank you...thank you very much.

"I just received a new piece of information," June announced, turning to the townspeople. "Let me ask you people something. Did I just hear you correctly? Did you say that some of you know some of the militia men that are going to fight against The King's Castle right now?"

The townsmen gathered around and listened. Several nodded.

"Do you think that those town militia men *want* to fight against The King?"

The townspeople scratched their heads. This was a good question. Many of these men were their neighbors. They liked The King. They were just doing what Thorndike had told them to do. *Maybe they won't fight if they have a reason not to fight,* they thought. It was worth thinking about.

June explained this thought to one townsman, then another, then another.

This might be worth a try, they gradually concluded.

Jackson arrived now and delivered the discovery *he* had made on *his* recon. "Guess what I just saw," he exclaimed. "I just saw someone come out of The Castle and ride straight to some *more* militia that are waiting with Thorndike at the front of The Castle! And guess who it was. It was my sister, Willa!"

"The Queen?" everyone gasped.

"Willa?", June croaked.

"Indeed," responded Jackson. "I'll bet she's on her way over to our father to try to talk some sense into him. She's going to try to fix this situation before somebody gets killed. That's pretty noble...and brave. I'm proud that she's my sister."

"How would you like to make her proud of *you*?" June now asked Jackson. "What would you say to sneaking over to the town militia men that are now surrounding the sides of The Castle right? Maybe *we* can talk them out of the invasion that they have been ordered to do."

"Do you think there's a chance that would work?" Jackson asked, as he climbed off his horse.

"Yes," answered one townsman...and then another...and then another.

With a sigh of relief and her own share of pride, the visiting Cmanda realized that what she had said to Cjune, Cjune had said to June himself. The Conscience-to-Conscience message had had a wonderful, energizing effect. Queen Luna's suggestion had paid off. It was Consciences helping Consciences. That was the way it was supposed to be, right?

Cmanda returned to Drumsworth and the Conscience Alert with the good news, just in time to see Balynn and the Lowlander house servants bundle themselves into a small carriage and head north, following the reinforcements for June and The Castle that she had promised June's Conscience. Time was still a factor, but now all these good people and their Consciences were headed in the right direction.

The Call of the Auras

*****

Solem came out of his bedroom scratching his head. The maps he had been studying were still draped across his bed, but the problem-solving he usually gained from such study had not materialized. In place of military inspiration, Solem had come to a sobering conclusion: A personal offering to step down as king was the only way he could think of to avert bloodshed and civil war.

"Admirable but not effective," commented Csolem. "Thorndike will probably just kill you and Drum will take your place as King. Keep thinking."

Solem shook his head. "Perhaps I should demand to meet with Drum. His desire to be King is at the heart of today's invasion and this would satisfy the nobles. Perhaps Drum and I could work out some sort of compromise."

"You know you're dreaming, don't you?" Csolem asked bluntly. "You know that Drum isn't interested in any compromise. He wants *you dead*. *His* mania has always told him that 'The Crown" was his birthright. Leadership is not even important to him. He only wants to be King."

Solem sighed. His Conscience was probably right. What else would explain the mindless drive of twenty years that brought Drum to this horrible day?

But what about Thorndike? "Willa is his daughter. Perhaps he will be willing to at least call off a bloody invasion in exchange for...peaceful transition."

"Now you're talking about surrender," snorted Csolem. "I am not in favor!"

Solem needed to talk things over these ideas with Willa and she always had good ideas.

But where was she? She had been there with him looking out the window only a moment ago...or had it been longer? He had been studying the maps...Solem now ran from room to room...the sitting room, the conference room, down the hall to Mimi's room.

"I haven't talked to Willa all day, Father," Mimi offered when asked. "I saw her out on the colonnade about an hour ago, but I haven't seen her since. I'm just hiding, hoping everything will blow over."

Did Willa actually go outside on a day like today? Solem wondered.

Solem was beside himself. Where was Willa? Had she been kidnapped? Had someone come and gone with her while he was staring at maps in his bedroom? How could he be so stupid as to lose his own beloved wife?

463

MEDIATION

Willa was not lost or stolen but was on her way to achieve peace... or rescue Solem... or fix this terrible situation that her father was causing...or *something*... What she saw up ahead was an apparently halted militia. *It's like they're waiting for me,* she thought to herself. *Did they know I was coming?*

"Hardly," sighed Cwilla patiently. "Their position on the road is not about you. They haven't even seen you yet. But be careful, Willa. They are waiting there in one place for a reason. We just don't know what the reason is yet. And *you* have brought yourself out here now to try to do something useful. Have you thought over what you're going to say?"

"I have thought about this my whole life," Willa answered. "I am on my way to speak my mind to a man who has always ignored me and pretended that I didn't exist...only because I was born a girl. Now I intend to *stand up right in front of him as a girl and say what I want to say!* That way he cannot ignore me."

"So much for 'standing where you want to stand'...but what are you going to *say* when you're there?"

"I am going to tell him that he will have to kill me before he attacks my husband!" she declared.

"Emotional, not thoughtful," sighed Cwilla. "Think again."

Two hardy militia men had now spotted her and reigned up beside her. "Who are you, woman?" they asked roughly.

"I am your Queen," Willa answered with dignity. "I am here to speak with your leader, Lord Thorndike. "Please take me to him immediately."

Surprisingly, the two militia men jumped off their horses and bowed. "We have never met a Queen before," the first one said. "This is exciting!"

Then the militia men re-mounted, smiled respectfully at their Queen and nodded. "Of course, Your Majesty. Does this request have something to do with the reason we are at The Castle today? It would be our honor to take you to our Commander to talk about it." Willa nodded in return and the three were off toward Thorndike and the waiting Thorndike Militia.

"How interesting," commented Cwilla. "These militia are quite courteous. They act like they don't even know that you are 'the enemy', Willa."

Drum now sidled his horse up to Thorndike's as she approached." Do you see what I see, Thorny?" he asked. "Isn't that *your* daughter that your men are bringing this way? Isn't that Queen Willa, herself?"

"Shut up, Drum!" Thorndike hissed. "I see her." He didn't need Drum's aggravating input. He had heard more than enough of it through the years. This was *his* show now, and he was in charge.

But Drum's question to Thorndike had filtered back among his troops.

"The Queen is here! The Queen is here!" they shouted happily to each other. Militiamen on foot bowed. Militiamen on horseback dismounted and took a knee as she drew near. Queen Willa nodded and waved.

"What?" Drum couldn't believe it." Your Militia is bowing before the Queen? Don't they know this is who they are fighting today? Are they idiots? Get up, fools! This is the enemy!" Drum hollered above their heads.

Willa smiled inside. *These are men who we expect to be fighting today but they don't seem to know it,* she thought. *They are as ignorant about this whole thing as my brothers were.* Could it be that *my father taught them all **how** to fight but failed to explain **who or why?***

"That's amazingly common with all fighting forces," sighed Cwilla. "Many men have gone to battle through the years and fought because they were told to without knowing exactly who or why."

" Each of *our* militiamen has a Conscience, of course, "commented the Conscience of a nearby militiaman. "We Consciences have all worked hard to keep our boys informed about right and wrong for years."

"*Our* men know how to fight and ride and follow orders", another Conscience said proudly, "but they don't like unnecessary violence. Thorndike has been so busy with Drum that he never took the time to tell them that they were going to fight The King today. They have no idea that they are going to be asked to do that."

"That is wonderful!" hooted Cwilla. "This is a Conscience-led militia! Amazing! They're good on following orders and protecting those that need it, right?"

"Definitely," the militia Consciences agreed proudly.

Consciences are a wonderful thing, Cwilla thought, with pride.

Willa's militia escorts placed her in front of her father. "Hello, Father," Willa began. "I see you have come to visit The Castle again. Your militia is very friendly. What do you plan to do?"

Thorndike glared at Willa in disgust. What was his stupid daughter doing now? She had never been the asset he thought she would be in her queenly position. Now she was unbelievably in the way.

He had worked toward today since before her birth. He was ready to kill to acquire Drum's stupid goal for him and move on to Urhonordo. Didn't this girl know that he had never wanted her...that he could do away with her right now and not miss her? A resurgence of his Aura-induced mania began to well up inside him. *Maybe I will just do that,* he thought fiercely.

"She *is* beautiful, though, isn't she?" whispered a quiet voice. "She looks exactly the way her mother looked when you married her. Remember? You couldn't really kill her beautiful mother then, either could you? You just banished her until she died. I lost my fight with the Auras then, but I am still trying. You couldn't possibly *kill* this girl, could you?" Thorndike's long forgotten Conscience had resurfaced.

"Shut up!" Thorndike yelled at his Conscience. "I didn't need you then and I don't need you now."

Drum had moved to Thorndike's side. "You don't have to be rude and yell at me, just because your daughter is standing in the way of *my* victory, *my* castle, and *my* crowning," sneered Drum. "By the way, when are you going to stop stalling around? What are you waiting for? It's time to charge The Castle. I want some action."

"I am waiting for you to stop nagging about what *you* want," hissed Thorndike, glaring at Drum. "I am in charge here, and my new plan for today is a matter of finesse and timing."

Drum rolled his eyes. "It just looks like your daughter is in the way of your 'new plan'," he snarked insultingly. "*And* it also looks like your militia would rather 'finesse' with her than fight her."

Thorndike's impatient Auras now were listening and ordering his next move. "Keep the Queen. You will know why later."

Now Thorndike gave an order. "Move this so-called Queen out of my way but don't let her leave! She may be useful." he bellowed."

The two militia men who had now decided that they were 'The Queen's Escorts' carefully guided her and her horse to the left side of the militia's ranks.

Drum snickered. "Not a bad idea, Thorndike. Keep her safe for *me* for later."

But the Auras of both men were losing patience with their star players. "Put our plan in motion now, Thorndike, and for Pete's sake tell your militia who they are here to fight. How did you manage to leave out that detail? They don't act like they have a clue."

"They will fight who I tell them to fight!" Thorndike muttered, glowering at his invisible partners. He did sheepishly decide, however, that it was time to yell out some important information. "We are here to eliminate a worthless king today, men...a man with no spine...a Lowlander lover! Get ready to ride right through the gates of yonder castle and *kill* everybody there! Sound the battle call!"

THE FINAL CHARGE

A trumpeter in the militia ranks blew a mighty blast of the battle horn. The cavalry militia mounted their horses again and faced The Castle, but looked at each other in disbelief, "Kill everybody in The Castle?

Really?"

"Why are we doing that?"

"Is he joking?"

The militia's Consciences were upset with themselves. "We forgot to warn you ahead of time, that you might be told to do some more bad things like when you were ordered to kidnap the Lowlanders the other day," the militia's Consciences informed their humans sadly.

"We should have warned you that Thorndike was acting crazy."

"Initiating a battle is what militias are ordered to do sometimes."

"Then you have to ask yourselves if the battle is for the right reasons...if it's for a good cause."

"But we didn't think today would get this far without Thorndike giving you a good reason, at least."

"And we haven't heard any kind of reason yet."

"Oh dear...Maybe there isn't one."

"We had no idea that a real assault was the plan today, even though Thorndike talked about it for years."

"We thought you were really supposed to protect The Castle,"

"Do you think Thorndike's serious today?" a few of the militia asked.

The militia began to grumble. "Protection of Thorndike and The Castle is what we were always trained to do," they whispered among themselves.

"This doesn't look right. We might be doing something really bad."

"Who knew Thorndike would 'pick a fight' with The Castle for no reason."

" And actually pick a fight with The King, himself? Are you kidding me?"

"Does he really expect us to hurt our own King who hasn't done anything wrong?

They shook their heads and frowned in confusion. "This doesn't seem right at all."

Drum misunderstood the grim looks on the militia faces. "That's more like it," he hooted. "You men look angry now! *Now* you look ready to take this Castle by storm! That's when *I* will show everyone who the real king is!"

Suddenly they all heard something new. But what was it? The echoes of a roar were rising from the left, back along the side of The Castle.

A triumphant smile broke out on Thorndike's face.

"The town militias that are surrounding The Castle have heard my battle horn!" Thorndike shouted as he relished his new battle plan coming together. "That was their victory yell! They must have captured the King's Castle Guard! *My Auras' new plan was genius!*" he gloated.

Thorndike was proud of the local militia's accomplishment. Those men were all the "combined members" of his militia that the other nobles had given him. He had not been sure of the quality of their training, but apparently, they had done well today.

"Hear that, Thorndike? You're already winning," bragged his Auras.

Although my daughter has tried to distract me, Thorndike growled to himself, *I have overcome her interference, my own slow-witted militia, and the pushy, ignorant Drum. I am now ready for the big event.*

His eyes glazed over as his increasingly tortured mind overflowed with his dream vision. *My men and I will charge The King and The Guard that he has left, at last. This is all I need to finally prove my inherited skill, skill that was thwarted by this stupid king's grandfather so long ago...Conquering this King today will make my ancestors proud...and Urhonordo ready for my taking, at last!*

Cthorndike pulled himself up to Thorndike's shoulder again. This ignored Conscience was still trying. "We have communicated very little over the years because you refuse to listen. But I am still here for you now to inform you of something important that you should know before it's too late."

"You can change. Is the vindication attached to an old story really worth this current, evil invasion? Today is only Drum's stupid dream" whispered Cthorndike, "not yours. The assault that you and the Auras have planned today will kill some of your men. The Auras don't really care about you. They are only in this world for the sake of pure, evil mayhem. Tell them they have gone too far. Tell them to go away. It's not too late."

The Auras could not hear Cthorndike, but they felt Thorndike's sudden hesitancy. Quickly they draped themselves heavily over his head like a hood. "Your plan is already in progress, Thorndike. Your men are now capturing the sides of The Castle", the Auras crooned engagingly." You can take what's left of this stupid monarch's preparations for battle, easily. Don't hesitate one more minute! Get on with some real Thorndike warfare! You know how you have been yearning for it!"

The Auras had blocked out Cthorndike's interference, and rejuvenated Thorndike's mania. The Castle seemed to beckon him. The goal glowed before him, and his eyes twitched. "All *you* have to do now is lead your own, trained troops *to and through* the front gates of The Castle, kill The King and take over!" the Auras growled seductively.

Willa was still with the two young militiamen who were now only respectful escorts. But the three remained at the side of the militia as it now lined up for whatever was going to happen next. Their Consciences had told them to protect her. One of them now alertly spotted something approaching from the far left.

In the distance, it appeared to be the **other** half of the militiamen. The escort squinted more closely. It *was* militiamen, to be sure, but it was also...what? Townspeople?

Thorndike saw them, too. *Townspeople! Wonderful!* thought Thorndike to himself. *The local militia that the Auras and I sent to conquer The Royal Castle Guard has also captured some townspeople. They must have got in the way. No problem... That will give me more prisoners. I wonder where they are holding The Guard that they captured.*

Quickly one of Willa's escorts obeyed his training to warn Thorndike of what he saw...

"Commander," he called out. "Our militia is approaching us from far left!"

"So what?" Thorndike shot back, looking annoyed. "They've got prisoners."

But...they haven't really got prisoners, the escort said quietly to himself as they neared. *They look sort of friendly.* But he really didn't understand this whole operation, anyway. He decided to remain silent.

Thorndike looked at the oncoming again, momentarily. His mind saw what it wanted to see. His militia was coming with prisoners. That was what he expected and now, this was what he saw.

"Is this part of the new battle scheme that you were working on last night?" Drum asked, glancing at those approaching. He was looking down his nose at these townspeople. "I don't get it. *Nobles* are our support base. What is our militia doing with a bunch of insignificant townspeople? Have we lowered our standards now to this target group?"

Willa started to cheer… inside.

"Who are those people?" asked Willa's escort. "Do you know them?"

"Yes, indeed," Willa assured him. "The twins at the front are my brothers and the woman in the center is my sister. The man in charge of that whole beautiful parade is Prince June."

"That's nice," smiled the escort. Now he thought he understood. "Looks like a family reunion and friends from town are coming to join us. Will there be a party?"

The Auras, who had been excited with their evil plan, had better eyesight than Thorndike. They were furious…and frightened. The approaching townspeople were not prisoners. *Nobody had been captured!*

This was not acceptable for an Aura inspired plan. Had the first half of their magnificent plan failed? This could not be true. Aura egos were too large to accept this as an answer. They were in denial. Whatever the reason was for what they were seeing, their response was ready…more violence *here* was the order of the day! "Move out, Thorndike! Let's get this invasion moving! What are you waiting for!"

Thorndike's mania accelerated. "With my successful militia advancing on foot from the left, it is time to use my own especially trained military to do what you, my Auras have instructed," he proclaimed. He stood up proudly in his saddle. "Onward to The Castle, men!"

He and his nervous steed stepped proudly out ahead. "Onward to crush anyone in our path, men! The Castle will be ours today! Onward to glory!"

The militia mobilized into action. They were being ordered to do something that was at least familiar! They had practiced this many times out behind the manor. They were all going to march toward The Castle! They wouldn't really hurt anybody, but Thorndike had always told them that militias follow orders, and this march must be a demonstration for The King. The ground rumbled as Thorndike's militia paraded grandly toward The Castle.

The earth in front of The Castle was firm and dry, just the way 'Commander Thorndike' knew it would be. As his horse pranced ahead, his addled mind acknowledged his own accomplishment. *This is what our new road was always for! This invasion is going to be easy now. And this is the way it will always be for me and my militias... wide, firm roads for my fighting units. We will swarm through the gates and straight into The Castle! That's what me and Drum fought for and here it is! We are here!*

Again, he glanced at the approaching townspeople and their 'captor militia'. *And I don't even need those men coming to help me now,* his mind told him grandly. *They have done what I told them to do. They have conquered my enemies inside The Castle and townspeople besides. Now I have only to drive forward to conquer The King! The front gate of The Castle is coming more clearly into view. My beautiful road is carrying me right to the gate. This is the way I always knew it would be.*

But now, the Auras panicked. The townspeople advancing from their left were clearly *not* prisoners. Couldn't Thorndike see that? That militia was not coming to help him. They were not even on his side. The Auras' lead man here was 'crazy Thorndike' and he was alone with *his* militia only... and he seemed to have completely lost his grip on reality.

"Focus, Thorndike! King Solem is ahead," they shouted. "He can still be beaten, but don't be a fool. *You* must strike the first deadly blow, but you must use these men only!"

You're right! The men I have here with me can handle this operation easily, Thorndike agreed with a fiendish grin. *The Royal Castle Guard has been depleted. I have twice as many men as I need for full control!*

But, as he looked more closely at the front gate, he was shocked at what stood before him. There was the *complete* Royal Castle Guard spilling out of the gate. Over one hundred men, standing at attention, full battle gear and shields at the ready. King Solem stood boldly in front of them, sword in hand.

"Halt!" commanded King Solem. "You are close enough, Thorndike. And I see Drum is at your rear as I thought he would be. What is your intention? Why are you here?"

Thorndike halted with a haughty smirk planted on his face. The militia came to a messy, clanking halt behind him.

Drum heard his name and stepped forward as though taking command. He was too close to his goal to be patient with Thorndike and the man had been slow all day. He was closing in on his grand and glorious desire. *He* would do the talking.

"Why do you think we're here, you lame excuse for a king?" he shouted for all to hear. "We are going to take you *down, remove* you and put someone on the throne who knows how to rule, for a change. That someone would be *me!*" He sneered at Solem mockingly. He had been wanting to do that for weeks.

Drum's Auras sighed in relief. At least Drum was clear headed.

The Call of the Auras

The group of approaching townspeople with their militia in tow heard Drum. June had explained what they might see at this point. They now became more of an army and the town militia beside them had no objection.

"Go home, Drum! King Solem rules!" they shouted.

Drum frowned but did not understand the meaning of the noise beside him. He was not deterred. *These common people who are gathered around know nothing,* he told himself with a shrug. *Thorndike must simply get on with his mighty assault and defeat these Royal Guardsmen in front of him.*

Drum stepped to the right to let the assault begin. *It's all up to Thorndike now,* he thought with confidence.

Solem surveyed the situation silently as Thorndike approached on horseback, but he noted that the man was slightly crooked in his saddle and his voice seemed slurred. Solem sensed Thorndike's slipping state of mind, so he stepped forward, Csolem at his side.

"Times have changed, Thorndike," he said calmly. "The power of your greed and jealousy has disappeared. Your corroded thinking is failing you. As you see, the militia you sent to the walls of my castle has been neutralized and there is no one left but you and the men behind you. Without you, Drum has no power here. And I have enough power to defeat both of you. Stand down now and go home. We will just call this 'Drum's Miscalculation.' Violence won't be necessary here."

Drum had not taken full note of their current situation. But now he did. Now he realized that what The King had said was true. But years of mindless politics to gain one goal...years of plotting with Auras... all had finally produced the same mania in Drum that had now completely conquered his partner. Drum's palms began aweating with desire, and the crown was clanking against his thigh.

"Don't listen to this has-been monarch, Drum," the Auras counseled him smugly. "King Solem doesn't want violence because he knows he will *lose* against Thorndike's forces. Look at them! Men on horses...strong young men in battle gear, just waiting for the order to charge and conquer the stupid man in front of you! How dare he and his troops stand there on foot only. Behind you is an armed militia on horseback!

"Sound the trumpet for charge, Thorndike!" yelled Drum. "There are not that many Royal Guard in front of you. When you and your men blast through them, The King will fall, and I will take his place!" he shouted. "Can't you see that, Thorny? He's bluffing!"

Drum turned and rode promptly back along the rows of militia. "Get ready men," he shouted confidently as he went. "Your Commander is getting ready for the charge that will show this king his last few moments on the throne! Keep your swords at the ready!"

Solem didn't flinch. "Your men have no positive reason to be this close to my castle, Thorndike. You and Lord Drum are 'trespassing without positive intent.' According to the rules of protocol, I must ask you to leave immediately before I use force to make you do so."

"Protocol? Really?" snarked the Auras. "At the beginning of a battle, this stupid king wants to talk of protocol?" The Auras laughed and jumped full force to Thorndike's shoulder. "How out of touch can this king be? If this speech is his royal idea of defense, maybe we can still help Thorndike pull this invasion out of the fire, whether he's crazy or not," they whispered to each other.

"Show him the card that you saved to play in just this kind of situation, Thorndike! Watch this love-sick weakling of a monarch crumble," they hissed in Thorndike's ear.

 Thorndike grinned and took a deep breath. He knew *just* what to say, now. He leaned menacingly toward Solem. "If you want us to leave, Your Highness," he said glaring at Solem, "we will. But *I will not leave alone! I* will take this useless wench that you call your wife with me!" He pointed with pride to Willa who was still on horseback at the side of his militia. He motioned his men to bring her forward in front of Solem.

Puffing out his chest with new bravado, Thorndike maneuvered his horse next to hers and circled around her. "Pretty isn't she, Your Highness?" he shouted at Solem with a leer. "I'm sure that she has won your heart already as *I* planned when *I* sent her to you. As you can see, she is back here again with me and my men now, thinking she can use her beauty to influence *this battle.*

"Well, she has done that, but not the way she figured. I know a good hostage when I see one. Give up your castle immediately, Solem, or I will take her with me when I go and you will never see her again. I will banish her useless beauty to my manor *forever as I did her mother.* There she will reside in seclusion until she dies, just the way her mother did before her!

"If you want her, leave your castle to my man, Drum! Willa is *my* daughter! She was mine before she was yours! Now I own her again!" he snarled belligerently.

Suddenly there was rustling sound among the townspeople to the side of the militia.

"Stand back, Aura-infected tyrant!" came a small but strong voice from their ranks. "She was mine before she was yours! I gave birth to her before you stole her from me and neither of us will ever be owned by you again!"

Mother Mina stepped bravely out from the crowd of townspeople. With a determined gate, she walked up to Thorndike's horse. There she stood before him, staunchly in his way. The militiamen behind Thorndike smiled at the little woman and backed away respectfully to give her space.

"Mother!" cried Willa joyfully. She jumped from her horse and ran to hug her mother fiercely. The militia smiled again. It was the Queen and her mother. The townspeople cheered.

The surrounding Auras shot into the air and stared at the ghostly apparition that now glided toward Thorndike, smiling. He seemed to slide sideways on his horse, giving his militiamen the strong impression that he was becoming even more unsteady.

Their 'Commander' wiped his brow as nearly incoherent words tumbled from his mouth. "Who are you?" he slurred toward the vision before him. His eyelashes fluttered uncontrollably. A mist seemed to gather in his mind. *She's here. She's here from the dead again, just as she was that evening on the colonnade of The Castle,* Thorndike thought fearfully. "Why are you here, ghost? You are dead. You have been dead for years," he whined.

"Don't listen to that ghost, Thorndike," the Auras hissed. "We remember her. She's not real."

Mother Mina *was* very real, however...and strong. Although they had stolen her youth and her rational thought twenty years ago, they had not thought to steal one other thing. She still had her Conscience.

HISTORY'S REVENGE

Mother Mina was now determined to confront the man who had ruined her life to save the people she loved. Willa understood and stepped back as her mother stepped forward alone.

"I am with you, Mina," whispered her Conscience. "The Auras that captured your mind for so long are no longer in control. Your own goodness has fully returned. This is the moment that you came back for. You will now conquer the Auras that once controlled your life and now want to control River Kingdom. Stay strong and be brave."

 Mother Mina smiled up at Thorndike charmingly. "No, Thorny. I'm not a ghost and I am not dead. I am back here today to claim my daughter, the daughter that you and your Auras took from me twenty years ago. My Conscience is with me."

Willa watched her mother with pride. She was strong and healthy and confronting her father face to face.

"I am not here to hurt you. I am only here to tell you that we have lives now, Thorny," Mina continued..." full, rich lives to be lived in the present, not in the history books of long ago or the dreams of undeserved futures, but *now*. Just stop this useless invasion peacefully."

Thorndike's twisted mind was in trouble. Kindness, sweetness and forgiveness... he couldn't understand it. And neither could his Auras.

"Focus, Thorndike," ordered the Auras. "You helped us dispatch this ghost once years ago when she wouldn't follow our plan. Remember? And we can do it again."

"Yes Auras, you did," admitted Mina, "but my Conscience was young then and it wasn't as strong as it is now."

"You are a withered ghost from my past, nothing more" Thorndike slurred." You could not produce a son for me...only a useless girl. The Auras and I got rid of you for good, back then and I got my boys without you!"

"I understand, Thorny. And where are your boys now?"

The twins stepped forward silently from the edges of the townsfolk. "Ah yes...Here they are."

Thorndike growled and clenched his fists. "Get over here, you laggards. I fed you all these years. Get over here now and act like real men." The boys looked directly at their father, perhaps for the first time, and backed away, joining Mina and Willa.

"Your children's Consciences are alive and working with them now. They have left you alone with your Auras, haven't they?" Mina whispered.

Thorndike glared at his children before him. Who were these people? They couldn't be who this ghost said they were. His mind's grip on reality was fading fast.

"Where are you, Auras?" Thorndike screamed. "We have come this far and now you show me these useless strangers? They are no good here. This is a failure of yours, Auras! Get these people out of my way so that we can conquer this King the way we have always planned. Do your job!"

But Thorndike's own Conscience was still with him. "This is not a failure of the Auras. This is your own failure, Thorndike. You have ignored me too long. You have listened to them too long. Now your Auras have completely ruined your mind."

"Forget about the ghosts and riffraff, Thorndike," ordered the Auras in frustration. "What do they know? We Auras got you this far, but if you don't get rid of them and lead this militia forward to win this battle, we're leaving. We Auras don't waste valuable time with losers."

Mother Mina had been tortured by Auras for twenty years, but during that time, she had developed an awareness that no one else had. In her painful exile, she had learned *to see Auras and hear them…and know them.*

"I hear your Auras, Thorny. We are not ghosts and riffraff," answered Mother Mina. "We are the family you didn't care for through the years. Now you have nothing left but the Auras and this invasion of Drum's that is no longer winnable."

She walked slowly toward Thorndike. "And picture this. Your Auras just said that they will be leaving you if you don't win today, Thorny. They have ruined you, but now you are being dumped. They used to love you in your manic state but now they are just tired of you. They don't like anyone but winners and, if you look around you, you are not winning.

"Unfortunately, I *do* understand what is happening here. They pushed me to insanity when you told me that Willa died, but they tired of my insanity and left me to rot, crazy and alone. They are about to do the same to you here, pushing you to a win that you cannot achieve.

"Listen to me, Auras. I see you," Mina continued. "I know your evil ways. Maybe I should encourage you to stay and continue to torture Thorndike as you did me, but my Conscience wouldn't approve. Be gone!

"Now I am uncovering what you have been up to for years. I will continue to do so wherever a love of evil like Thorndike's and Drum's might lure you back. You and yours have been the source of extended evil in the valley of The Great River for years. Now you have lost this failed attempt at invasion which means that your plans of prejudice, greed and power are through. I wish I could be present when you admit to your Aura Leader that you and all your Aura buddies have lost this invasion to Consciences."

The Auras backed away. "She can see us!" they shrieked. Their fearful words echoed among other Auras lurking nearby. "And she has given King Solem and the whole River Kingdom a secret weapon, she says! It's called Consciences."

"What is a Conscience?" said the nearby Auras.

"Show me one."

"I don't see a thing!"

" I don't either, but it must be strong."

"Right! This invasion is a failure, for sure!"

Thorndike's Auras searched frantically and then backed off in fear.

"Everyone has a Conscience," said Mina "but no one can see them. "Now that I have uncovered you, Auras and your treachery, the good people who live along the banks of The Great River will know all about you and their Consciences will cancel out your special evil. Your plans will run into more roadblocks quicker than has happened here. My Conscience has told me that passing out this information is my mission, and I have done so. She smiled and looked around her.

Thorndike slid to the ground. The power his Auras had given him was no longer with him. He collapsed on the new road in a heap.

Drum returned to the front of the militia just in time to stare at Thorndike's crumpled form in horror. This was his muscle, the manpower he needed, now lying on the ground immobile, his mind no longer responding.

"I told you were pushing him too far a while back" growled one of Thorndike's Auras.

"Tell that to Aura Leadership," said another. The Auras did not tolerate failure well.

"Thorndike just turned crazy. Drum still thinks that everyone loves him, but we Auras know that's a lie. We Auras lost this 'sure thing' invasion after months of work on our parts to a crazy man, an egomaniac and something called Consciences. The other Auras and Leader are going to be furious, but there is no real reason to stick around."

"Agreed. And we need to get out of here quick and find out about what this 'Conscience thing' is. If there is such a thing, we need to know how to fight it."

"Don't worry about it. Obviously, the old ghost was just babbling. There's no such thing as a Conscience. It isn't real." The Auras disappeared.

*****

Drum, however, had not given up. "Get up, Thorndike," Drum ordered. "I see your ex-wife here, but she isn't anything but an old hag they have brought here to scare you. I just got you militia all lined up and ready. What are you doing on the ground?"

Drum dismounted and strode confidently toward the townspeople. *There are people here to impress*, he thought, *even if they are only townspeople. The masses always believe anything I say. Watch me bring these townspeople to my side like I did with the nobles! I don't need Thorndike or Auras.*

Cdrum sighed. "Will this man never learn?" he grumbled to himself.

Almost as though on remote control, Drum smiled at the townspeople and began marching back and forth, clapping his hands as he had always done in front of adoring crowds in the past.

The Consciences of the townspeople whispered to each other and their humans in disgust. "Look at this heartless man," they snickered. "His friend is lying on the ground, and he doesn't even care. He's just out for his own glory." The townspeople listened to their Consciences and nodded in agreement. They understood.

"Boo, Drum. Boo," the crowd shouted. "Go home!"

Hearing the booing crowd of townspeople, the Thorndike Militia, still standing in formation, sighed in relief and relaxed. Drum's famous approach with crowds had failed him at last. The townspeople were actually laughing at him. Obviously, Drum wouldn't be leading an invasion, the militia told each other.

In shock, Drum stood and stared. The crowd stared back silently now. No one was moving.

But suddenly those present heard another sound. It started quietly but it seemed to be growing, approaching them steadily. It was not an unfamiliar sound... travelinging feet, people talking, an occasional shout. Who was this coming up on them now? Were they friend or foe? Did they have to worry about someone new?

THE PROMISED INVASION

A huge crowd now came into view from the south. As they came closer, the approaching horde became more visible. "It's our Lowlanders!" someone shouted joyfully. "Our Lowlanders are back!" The townspeople began to cheer. It was true. Their smiling brown faces were clearly visible. The cheering became louder as the townspeople ran to meet them.

"Those townspeople don't sound like they want to fight, Drum", commented Mother Mina. "They sound like they just want their Lowlander friends back,"

The leaderless Thorndike Militia stood still and waited. They had no orders. There was no one left to order them. They looked from side to side. The stolen Lowlanders were happily reuniting with the townspeople but still more people continued to approach. Who were all these people? What should they do?

"We should just wait and see what happens, suggested a few of their Consciences. "Maybe somebody will still have some 'protecting' for us to do." Some militiamen rolled their eyes, but they didn't leave.

At last they recognized someone. Manda and her parents now led a strange but formidable army of Urhonordorians and farmers toward The Castle. They passed the right side of the militia, arriving at the gate of The Castle. "All of us came to assist you in battle, Solem," announced Armando as he approached, "but it appears that our assistance may not be needed."

Drum rushed to Armando, fists jabbing the air. "What do you think you're doing here, foreigner?" sneered Drum. "Thorndike and I were about to charge the gate and eliminate The King, once and for all and that's what *I* am going to do right now. ... without him! The townspeople here will back me!" Again, Drum smiled at the townspeople and clapped his hands to rally them.

But there was no response. The townspeople were busy talking to their returned Lowlander friends. They didn't even notice.

Cdrum felt sorry for his human. After all that had gone wrong today and all the humiliation he had suffered from an invasion that was obviously failing, he still didn't understand that he had lost. Maybe it was his upbringing, his family heritage of never thinking of others. Maybe that was what had created this block, this barrier to looking at his own failings and blinding him to reality.

Drum took a stance at the head of the Thorndike Militia as though they would suddenly spring to life and obey his commands, but they didn't move. "You have lost, Drum," Cdrum told his human softly. "The Auras are gone. There will be no invasion. It's time for you to finally understand."

Drum looked from left to right. He appeared to be completely bewildered.

June quickly found his way past the immobilized Thorndike Militia to Manda's side. "Thank you so much for sending your Conscience to talk to mine, Manda. That was a master stroke. It came at a perfect time, too. But it looks like you brought most of two kingdoms here. Why?" he asked breathlessly.

The Call of the Auras

Manda responded with a hug. "My family and I spoke to our Consciences and decided that you and Willa and King Solem might need help with Drum and Thorndike. The freed Lowlanders heard us and decided that they wanted to help, too. They wanted to come home anyway and their Consciences said that helping The King was 'on the way so they could do both things," she laughed.

And the procession was still arriving. Following behind the Lowlanders was a group of farmers led by Hector. As Hector explained to the King later, he and his Central farming friends (including some nobles) had inquired as to the Lowlander's mission when they passed the Central farms, on their way north. When they heard that they were going to help The King fight off a possible insurrection, they decided to come along. They had made friends with the Lowlanders during their water search and their Consciences felt they needed to stand up with them for the King.

The still growing procession now covered the entire new road leading up to The Castle. There were representatives from both Kingdoms, brown and white, wealthy and poor. "Look here, Drum," said Mother Mina softly. "A beautiful road has been built and everyone is here to enjoy it. Perhaps it's time for you to learn something new. Thorny can no longer help you. The Auras' and their evil powers have disappeared. You are surrounded by too many good people with Consciences who don't want a new king.

"Consciences have won this battle against the Auras that were *your* fighting force... the fighting force for evil. Now you are alone...except for your faithful Conscience. I'm sure that yours is still with you as mine is still with me."

Drum was still standing alone in front of the Thorndike Militia. "Do you need a ride home, Father?" asked a voice from a tiny carriage that had threaded its way through the immense gathering that now filled the new road to The Castle.

Balynn, Paulina and the house servants from Drumsworth had caught up with everyone. Paulina had insisted that they follow the returning Lowlanders and Urhonordorians. Balynn didn't want to be left behind, either. She suspected that her father would be in trouble when Paulina had explained what everyone was going there to do.. Paulina and the other Drumsworth servants had joined the march of humanity that was ready to stand up for The King. Her Conscience had won out over lingering Auras who seemed to have faded in fear from the apparent goodness passing by.

Sadly, Baylynn and Paulina found Drum, standing alone and clapping quietly by himself. Gently they assisted him into their carriage and the door closed behind him with no objections from anyone.

Solem put up a hand and waited. The crowd immediately recognized the sign. Silence reigned. "The invasion is over," he declared. "Does anyone object?"

There appeared to be no objection. The relieved, good-hearted townspeople, the militias and even the kidnapped Lowlanders were in a forgiving mood. The Auras had taken much of the anger with them.

AFTER THE AURAS- A POSTSCRIPT

Solem looked over the heads of Thorndike's militia as Willa and her mother joined him in front of his Royal Castle Guard. He couldn't believe what he was seeing. There were hundreds of smiling brown and white faces looking back at him. All these people had come to help him defend The Castle…defend his reign. He had never felt closer to his people than he did at this moment.

"I guess most of the people think you are a King worth fighting for, dear," said Willa as she squeezed his hand.

The day that had started with so much foreboding and fear now seemed to bloom with promise. And indeed, everyone within Solem's eyesight now deserved that promise.

Townspeople had become warriors to protect The Castle. After talking to the twins, the local militias realized that they had never wanted to assault the walls of The Castle anyway

Hundreds of Lowlanders had marched north together, to help King Solem and now they had returned to their homes.

All of River Kingdo and Urhonordo had risen to the challenge. Their Consciences had spoken up in all of them and now were now proud. They had come together to fight for Solem and were reclaiming peace on both sides of The Great River.

Prince Armando and Queen Luna congratulated Solem on his victory, realizing that this victory was also a victory for them. They travelled home to Urhonordo now without fear of history trying to repeat itself. Hector and his fellow landholders joined in the happy discussions but soon traveled back to the newly irrigated Central portion of River Kingdom, complete with many thanks to and from King Solem.

Quickly the townspeople shook hands with the militia and returned to their town with thanks and hugs from June, Flo and her brothers for their bravery. The local militiamen wandered off as well, with tales to tell their nobles of the strength of The King and his supporters.

As they gradually discovered that no one was in the mood for an invasion or even protection any longer, the Thorndike Militia and The Royal Castle Guard shook hands with each other and compared notes on how confusing the day had been. Gradually, all militia personnel dispersed to their own housing, some to Thorndike Manor, some to The Castle.

There was still the unfortunate issue of Drum and Thorndike. Would they be severely punished? Thrown into the dungeon? Banished?

As no one bothered to watch, the twins and Flo quietly stepped forward to care for their disabled father. His eyes still held the vacant look that one gets when a severe mania calms but leaves little behind. Although they were embarrassed by his behavior, the siblings rallied and returned him gently to Thorndike Manor ... but not before the boys had contacted Balynn and Mimi and promised to visit them soon.

Balynn and Paulina had a more difficult time with Drum who was still grousing about Thorndike "letting him down on the invasion." He had lost more than the invasion, however. He didn't realize it until he arrived at Drumsworth, but the crown of his grandfather was missing.

Where and when the crown had fallen away, he had no idea, but it was gone. And that loss had an odd but welcome effect. Drum appeared to have lost his drive to become King. With Balynn's help, he settled back to claim that he never really expected to become a king, anyway. His Conscience still had issues with him.

Of course, the nobles of River Kingdom still harbored their prejudices, but, amazingly, as the Lowlanders and their Consciences continued to strengthen their self-determination, the Consciences of the nobles began to counsel their nobles toward more respect. It would be a long process, but both white and brown Consciences were waking up to their long-held errors in understanding each other. Without Thorndike and Drum constantly pushing prejudice, open-mindedness among the nobles seemed to take root.

With the Auras gone, Thorndike and Drum retired into the pages of history but maintained their value there as examples of the dangers of Auras. Their exploits were valuable lessons in Queen Luna's classrooms regarding the dangers of Auras.

King Solem and Willa decided that Luna's idea about education was wise. At last report, Auras were less of a problem among the educated of River Kingdom and Armando introduced the 'education concept' to the Lowlanders. June and Manda married and became traveling teachers and goodwill ambassadors together although everyone was sure that one or both would be a King or Queen someday.

The Call of The Auras was mostly silenced in River Kingdom on that fateful invasion day. Hopefully a large influx of Auras would never return, but most Consciences knew that they needed to remain watchful. Small batches of Auras would always tempt individuals, even whole groups, to dishonesty and greed, prejudice, and meanness. Sometimes, groups of Auras were noted, and Conscience Alerts would be held as a warning. With that in mind, we should all probably try to listen to our Consciences as much as possible and listen for The Call of The Auras.